SPACE SCIENCE, EXPLORATION AND POLICIES

DARK ENERGY: THEORIES, DEVELOPMENTS AND IMPLICATIONS

SPACE SCIENCE, EXPLORATION AND POLICIES

DARK ENERGY: THEORIES, DEVELOPMENTS AND IMPLICATIONS

KARL LEFEBVRE

AND

RAOUL GARCIA

EDITORS

Nova Science Publishers, Inc.

New York

LIBRARY OF CONGRESS CATALOGING-IN-PUBLICATION DATA

Dark energy : theories, developments, and implications / editors, Karl
Lefebvre and Raoul Garcia.
    p. cm.
  Includes index.
  ISBN 978-1-61668-271-2 (hardcover)
  1. Dark energy (Astronomy) I. Lefebvre, Karl. II. Garcia, Raoul.
  QB791.3.D3667 2010
   523.1'8--dc22
                            2010007598

Published by Nova Science Publishers, Inc. ✦ *New York*

CONTENTS

PREFACE

In physical cosmology, astronomy and celestial mechanics, dark energy is a hypothetical form of energy that permeates all of space and tends to increase the rate of expansion of the universe. Dark energy is the most popular way to explain recent observations and experiments that the universe appears to be expanding at an accelerating rate. In the standard model of cosmology, dark energy currently accounts for 74% of the total mass-energy of the universe. This book presents and discusses the nature and feasibility of laboratory registration using SQUID-magnetostrictor systems; a review on dark energy objects; the dark energy scale in superconductors; cosmic acceleration; a review of the quantum Yang-Mills condensate (YMC) dark energy model and others.

Physical principles underlying the concept of the dark energy are considered. Along with the history of the problem, the modern astrophysical observational data testifying to the existence of this phenomenon in the Universe are briefly discussed. Certain hypotheses clarifying the nature of the dark energy are analyzed from the standpoint of its internal microscopic structure. It should be noted that presently, in the great extent, the dark-energy concept is of phenomenological character. Thereby, it is distinguished with the well-known astrophysical concept of the dark matter whose nature was discussed many times in the framework of the microscopic approach (in physics of fundamental particles, quantum field theory, super-symmetry concept, etc., see review [1]).

In accordance with current concepts, the internal negative pressure intrinsic to the dark energy plays the role of indirect source of weak anti-gravitational repulsion of galaxies. In Chapter 1, arguments for possible interrelation between this internal pressure and the physical pressure of a massive medium are considered. Estimates of this ultra-weak contribution of the internal pressure to the total equilibrium pressure of the medium are made.

For the laboratory registration of possible dark-energy pressure fluctuations, the authors propose to use small-scale density variations occurring as a result of the complicated polycyclic Earth motion in the extraterrestrial space. The related measurements can be based on a supersensitive system containing SQUID magnetostrictors. Initially, this system was developed as a part of the gravitational-antenna device to detect gravitational waves [2]. According to the authors' estimates, the system response was quite adequate to the expected effect. Methods for increasing the signal amplification by the improvement of magnetostrictive properties of the sensitive elements have been proposed. To this end, features of the collective spin-spin interaction were used.

The hypothesis was also exploited that the nature of the dark energy is associated with the vacuum state of the quantized electromagnetic field. Within the context of this hypothesis, the possibility was considered to indirectly study dark-energy density variations on the basis of the precise pressure measurements in the case of the Casimir effect. This experiment can be also performed with the use of the SQUID-magnetostrictor system.

Since it was discovered that the galaxies are moving away from each other accelerately, researches have considered the hypothesis that the universe is predominately constituted by an exotic fluid, called dark energy, in order to preserve the general relativity theory. In this context, it is natural to ask about the connection between dark energy and the structures formation, as stars, black holes, naked singularities and including the possibility of the formation of new structures as the gravastars of dark energy. In Chapter 2 the authors present a review of the main results on this last troublemaker subject.

Chapter 3 revisits the cosmological constant problem using the viewpoint that the observed value of dark energy density in the universe actually represents a rather natural value arising as the geometric mean of two vacuum energy densities, one being extremely large and the other one being extremely small. The corresponding mean energy scale is the Planck Einstein scale $l_{PE} = \sqrt{l_p l_E} = \hbar G / c^3 \Lambda^{1/4} \sim 0.037mm$, a natural scale both for dark energy and the physics of superconductors. The authors deal with the statistics of quantum fluctuations underlying dark energy in superconductors and consider a scale transformation from the Planck scale to the Planck-Einstein scale which leaves the quantum physics invariant. The authors' approach unifies various experimentally confirmed or conjectured effects in superconductors into a common framework: Cutoff of vacuum fluctuation spectra, formation of Tao balls, anomalous gravitomagnetic fields, non-classical inertia, and time uncertainties in radioactive superconductors. The authors propose several new experiments which may further elucidate the role of the Planck- Einstein scale in superconductors.

The cosmic acceleration is one of the most significant cosmological discoveries over the last century. Following the more accurate data a more dramatic result appears: the recent analysis of the observation data (especially from SNe Ia) indicate that the time varying dark energy gives a better fit than a cosmological constant, and in particular, the equation of state parameter w (defined as the ratio of pressure to energy density) crosses -1 at some low redshift region. This crossing behavior is a serious challenge to fundamental physics. In Chapter 4, the authors review a number of approaches which try to explain this remarkable crossing behavior. First they show the key observations which imply the crossing behavior. And then the authors concentrate on the theoretical progresses on the dark energy models which can realize the crossing -1 phenomenon. The authors discuss three kinds of dark energy models: 1. two-field models (quintom-like), 2. interacting models (dark energy interacts with dark matter), and 3. the models in frame of modified gravity theory (concentrating on brane world).

In Chapter 5 the authors review the quantum Yang-Mills condensate (YMC) dark energy models. As the effective Yang-Mills Lagrangian is completely determined by the quantum field theory, there is no adjustable parameter in the model except the energy scale. In this model, the equation-of-state (EOS) of the YMC dark energy, $w_y > -1$ and $w_y < -1$, can both be naturally realized. By studying the evolution of various components in the model, the authors find that, in the early stage of the universe, dark energy tracked the evolution of the radiation, i.e. $w_y \rightarrow 1/3$. However, in the late stage, w_y naturally runs to the critical state with $w_y = -1$, and the universe transits from matter-dominated into dark energy dominated stage only at recently $z \sim 0.3$. These characters are independent of the choice of the initial condition, and the cosmic coincidence problem is avoided in the models. The authors also find that, if the possible interaction between YMC and dust matter is considered, the late time attractor solution may exist. In this case, the EOS of YMC must evolve from $w_y > 0$ into $w_y < -1$, which is slightly suggested by the observations. At the same time, the total EOS in the attractor solution is $w_{tot} = -1$, the universe being the de Sitter expansion in the late stage, and the cosmic big rip is naturally avoided. These features are all independent of the interacting forms.

In Chapter 6 the authors test models of cosmology for the best estimates of important parameters, that is, for matter, dark energy (DE), spacetime(ST) and cold dark matter(CDM) within variants and special cases of the Friedman-Robertson-Walker(FRW) approximation. The authors use the largest available collection of supernovae Ia data(SNe Ia), 307 SNe Ia along with 69 data pairs reported from gamma ray bursts(GRB) and the authors' present situation for a total of 377 data pairs extending back perhaps 10 billion light years. Modeling with this large set allows better definition of the limits of DE, CDM and ordinary matter than previously. While the ancient GRB data are quite noisy they do allow better estimates of the fitted curve asymptotes towards singularity than without. The results from models employing the commonly used luminosity distance moduli(log) *versus* redshift data slightly prefer the DE model with H_0 of 70.2 ± 0.6 km/s/Mpc at a normalized matter density, Ω_m, of 0.34 with significant negative spacetime (ST) curvature. The best fit calculation for the model without DE presents a much lower matter density (0.10 ± 0.01) and a very slow Universe expansion rate (41.4 ± 3.2km/s/Mpc). When tests were made using the actual distances (Mpc) *versus* frequency decline υ_0 / υ_e data, which is the preferred method not relying too heavily on the great errors from far distance emissions(some standard deviations are many thousand Mpc), the flat DE model presents the worst fit of all models with H_0 of 75.4 ± 0.7 km/s/Mpc at Ω_m of 0.12 ± 0.02. Most interestingly, the two best fits present Ω_Λ of 1.03 and 0.68 with significant ST curvature. These solutions suggests a Universe consisting primarily of DE with abundant ST. The authors suggest astronomers cease using the luminosity distance moduli from distant emissions for modeling cosmology and present results from models based on distance estimates *vs.* redshift or frequency decline. At least an order magnitude more and better astronomical data are required to unequivocally choose the best FRW model to refine for further work in cosmology. The authors' results suggesting a Universe with excessive DE should exhibit open or negative curvature might foster some interesting theoretical models in the near future. FRW models with the largest collection of SNe and GRB data tell us the Universe is not flat.

In Chapter 7 the authors study quantum mechanically our expanding universe which is made up of gravitationally interacting particles such as particles of luminous matter, dark matter and dark energy as a self-gravitating system using a well-known many-particle Hamiltonian, but only recently shown as representing a soluble sector of quantum gravity. Describing dark energy by a repulsive harmonic potential among the points in the flat 3-space and incorporating Mach's principle to relativize the problem, the authors derive a quantum mechanical relation connecting, temperature of the cosmic microwave background radiation, age, and cosmological constant of the universe. When the cosmological constant is zero, the authors get back Gamow's relation with a much better coefficient. Otherwise, the authors' theory predicts a value of the cosmological constant 2.0×10^{-56} cm^{-2} when the present values of cosmic microwave background temperature of 2.728 K and age of the universe 14 billion years are taken as input. It is interesting to note that in this flat universe, the authors' method dynamically determines the value of the cosmological constant reasonably well compared to General Theory of Relativity where the cosmological is a free parameter.

In the framework of noncompact Kaluza-Klein theory, the authors investigate a (4 + 1)-dimensional universe consisting of a (4+1) dimensional Robertson-Walker type metric coupled to a (4 + 1) dimensional energy-momentum tensor. The matter part consists of an energy density together with a pressure subject to *4D* part of the (4 + 1) dimensional energy-momentum tensor. The dark part consists of just a dark pressure $\bar{p}$, corresponding to the extra-dimension endowed by a scalar field, with no element of dark energy. It is shown in Chapter 8, that for a flat universe, coupled with the non-vacuum states of the scalar field, the reduced Einstein field are free of *4D* pressure and are just affected by an effective pressure produced by the *4D* energy density and dark pressure. It is then proposed that the expansion of the universe may be controlled by the equation of state in higher dimension rather than four dimensions. This may reveal inflationary behavior at early universe and subsequent deceleration, and account for the current acceleration at the beginning or in the middle of matter dominant era.

In Chapter 9 the authors survey the application of specific tools to distinguish amongst the wide variety of dark energy models that are nowadays under investigation. The first class of tools is more mathematical in character: the application of the theory of dynamical systems to select the better behaved models, with appropriate attractors in the past and future. The second class of tools is rather physical: the use of astrophysical observations to crack the degeneracy of classes of dark energy models. In this last case the observations related with structure formation are emphasized both in the linear and non-linear regimes. The authors exemplify several studies based on the authors' research, such as quintom and quinstant dark energy ones. Quintom dark energy paradigm is a hybrid construction of quintessence and phantom fields, which does not suffer from finetuning problems associated to phantom field and additionally it preserves the scaling behavior of quintessence. Quintom dark energy is motivated on theoretical grounds as an explanation for the crossing of the phantom divide, i.e. the smooth crossing of the dark energy state equation parameter below the value -1. On the other hand, quinstant dark energy is considered to be formed by quintessence and a negative cosmological constant, the inclusion of this later component allows for a viable mechanism to halt acceleration. The authors comment that the quinstant dark energy scenario gives good predictions for structure formation in the linear regime, but fails to do that in the non-linear one, for redshifts larger than one. The authors comment that there might still be some degree of arbitrariness in the selection of the best dark energy models.

In Chapter 10 the authors review some of the possible models that are able to describe the current Universe which point out the future singularities that could appear. They show that the study of the dark energy accretion onto black- and worm-holes phenomena in these models could lead to unexpected consequences, allowing even the avoidance of the considered singularities. The authors also review the debate about the approach used to study the accretion phenomenon which has appeared in literature to demonstrate the advantages and drawbacks of the different points of view. The authors finally suggest new lines of research to resolve the shortcomings of the different accretion methods. The authors then discuss future directions for new possible observations that could help choose the most accurate model.

The exixtence of dark energy is a serious problem in modern cosmology. For the origin of the dark energy, many models including a cosmological constant have been proposed. Although these models can explain the present acceleration of the Universe, some of the models would not be able to explain the observed large-scale structure of the universe. Therefore, in order to constrain the models of the dark energy, we should consider the structure formation in the universe. From primordial density fluctuation, the large-scale structure is formed via its own self-gravitational instability. Even though numerical simulations are necessary to follow the full history of the structure formation, in order to understand the physics behind the structure formation, analytic approaches play important roles. In Chapter 11, the authors summarize various analytic approaches to the evolution of the density fluctuation in Newtonian cosmology and show they can be helpful to distinguish models when applied to the quasi-nonlinear region. The authors also mention several applications of the analytic approaches including the initial condition problems for cosmological N-body simulations, higher-order Lagrangian perturbation theory.

Type Ia supernovae, acting as standard candles, play a leading rôle in the exploration of the Universe evolution. Initiated by similar stellar explosions whose physics is known in detail, they provide simultaneous measurements of the (luminosity) distance versus the redshift. Observations of this type of supernovae at high redshifts, being sensitive to the Hubble expansion rate, provide the most direct evidence for the accelerating expansion of the Universe and they are consistent with cosmological models proposing a dark energy component dominating the Universe energy budget. These findings have been corroborated by several independent sources, such as measurements of the cosmic microwave background, gravitational lensing, and the large scale structure of the Cosmos.

Chapter 12 focuses on the importance of supernova observations for exploring the nature of dark energy. It briefly outlines the procedure followed in order to extract information relevant to cosmology from measurements of supernova luminosity and spectra and addresses the statistical and systematic errors involved. A complete review is given on supernova observational evidence starting from the first observations presented in 1995 by the Supernova Cosmology Project and the High-z Supernova Search Team and reaching the recent developments by the Hubble Space Telescope, the Supernovae Legacy Survey and the ESSENCE project.

The cosmological implications of supernova observations in conjunction with evidence collected from other astrophysical probes are discussed. A survey of the theoretical approaches devised to address the dark energy problem and their relation to observational questions is given.

The prospects for future improved measurements by facilities such as the Supernova Acceleration Probe are also discussed, together with the possibility of exploiting observations of other types of supernovae to construct a Hubble diagram and determine the cosmological parameters.

In: Dark Energy: Theories, Developments and Implications   ISBN: 978-1-61668-271-2
Editors: K.Lefebvre and R. Garcia, pp. 1-19   © 2010 Nova Science Publishers, Inc.

Chapter 1

DARK ENERGY: THE NATURE AND FEASIBILITY OF LABORATORY REGISTRATION USING SQUID-MAGNETOSTRICTOR SYSTEM

G.N. Izmailov, L.N. Zherikhina, V.A. Ryabov and A.M. Tskhovrebov

Lebedev Physical Institute, Russian Academy of Sciences,
Moscow, Russia

Abstract

Physical principles underlying the concept of the dark energy are considered. Along with the history of the problem, the modern astrophysical observational data testifying to the existence of this phenomenon in the Universe are briefly discussed. Certain hypotheses clarifying the nature of the dark energy are analyzed from the standpoint of its internal microscopic structure. It should be noted that presently, in the great extent, the dark-energy concept is of phenomenological character. Thereby, it is distinguished with the well-known astrophysical concept of the dark matter whose nature was discussed many times in the framework of the microscopic approach (in physics of fundamental particles, quantum field theory, super-symmetry concept, etc., see review [1]).

In accordance with current concepts, the internal negative pressure intrinsic to the dark energy plays the role of indirect source of weak anti-gravitational repulsion of galaxies. In this paper, arguments for possible interrelation between this internal pressure and the physical pressure of a massive medium are considered. Estimates of this ultra-weak contribution of the internal pressure to the total equilibrium pressure of the medium are made.

For the laboratory registration of possible dark-energy pressure fluctuations, we propose to use small-scale density variations occurring as a result of the complicated polycyclic Earth motion in the extraterrestrial space. The related measurements can be based on a supersensitive system containing SQUID magnetostrictors. Initially, this system was developed as a part of the gravitational-antenna device to detect gravitational waves [2]. According to our estimates, the system response was quite adequate to the expected effect. Methods for increasing the signal amplification by the improvement of magnetostrictive properties of the sensitive elements have been proposed. To this end, features of the collective spin-spin interaction were used.

The hypothesis was also exploited that the nature of the dark energy is associated with the vacuum state of the quantized electromagnetic field. Within the context of this hypothesis, the possibility was considered to indirectly study dark-energy density variations on the basis of the precise pressure measurements in the case of the Casimir effect. This experiment can be also performed with the use of the SQUID-magnetostrictor system.

1. Introduction: Physical Prerequisites to the Appearance of Dark Energy Concept

Introduction to the physics of "Dark energy" concept on the border between the last and the present centuries [1,3-5] made it possible to get over a number of serious contradictions that had existed in interpretation of cosmological observations.

The first such cosmological paradox, which had in fact been discussed before the groups of Smidt, Riess [6] and Perlmutter [7] took their measurements, is that according to the data obtained by the mid-1990s the mean matter mass density ρ_M constituting the inhomogeneous structure of the Universe does not exceed 30% of its critical density

$$\rho_C = \frac{3H^2}{8\pi G} \approx 10^{-29} g / sm^3,$$ where $G=6{,}67 \times 10^{-11}$ m^3/kg/s^2 is the universal gravitational

constant and $H \approx 70$ km/s/MPc $\approx 2{,}3 \times 10^{-18}$s^{-1} is the Hubble constant. Such "shortage" of matter formally implied that the Universe must have a notable space curvature

$$R_K = \sqrt{\frac{3(\rho_C - \rho_M)}{8\pi G}},$$ which did not correspond to the observational data, and that without a

special fitting of parameters it would be difficult to reconcile this hypothesis with the Universe inflation models [8-9] (R_K is the space curvature radius).

The absence of considerable curvature was indicative of the fact that approximately 70% of the mass and energy density in the present-day Universe are provided by "something" which is not subjected to perturbations by the gravitational fields of produced inhomogeneous structures of matter and which preserves its nonclustered form in the course of cosmological evolution. To fulfill such "non-adhesion" condition it is necessary that the effective pressure of this "something", called Dark Energy, should be negative, $P = -\rho_{DE} \approx -(\rho_C - \rho_M)$.

Another cosmological paradox, which can only be understood with application of a weakly interacting substance, called Dark Energy, is demonstrated by the results of measurements of anisotropy and polarization of relic radiation. BOOMERANG (Balloon Observations of Millimetric Extragalactic Radiation and Geophysics) and MAXIMA (Millimeter-wave Anisotropy Experiment Imaging Array) balloon experiments performed in 1999-2000 [10-15] and WMAP (Wilkinson Microwave Anisotropy Probe) cosmic experiment of 2003-2006 [16-18], in which the position of lower (quadrupole and octopole) peaks in the angular spectrum of relic-radiation anisotropy (the peaks most sensitive to scales of space curvature) was determined [19,20], showed that our three-dimensional space is Euclidean to a very high accuracy $\rho_0/\rho_C = 1{,}015 \pm 0{,}020$. Complex processing of the relic background anisotropy measurements taken by the WMAP cosmic apparatus was performed on the basis of different theoretical models. In the course of processing, fine fitting of the estimated values of dark matter density ρ_{DM}, baryon density ρ_B, radiation density ρ_R, the Hubble constant, the Universe age, etc. was accomplished. The data obtained, $\rho_{DM}/\rho_C = 0{,}23 \pm 0{,}07$ $\rho_B/\rho_C =$

0,022±0,001 and $\rho_R/\rho_C = 7\times10^{-5}$, agree with the Euclidean character of space only in the presence of fourth substance – Dark Energy which constitutes approximately 3/4 of the total Universe mass.

The most striking cosmological paradox which was understood on the basis of the Dark Energy notion and played a particularly important role in the final formation of these concepts, consists in deviation from the Hubble law for remote objects with a high registered level of the red shift towards their accelerated recession. On the basis of Slipher's and Hubble's measurements of the Doppler shift $\frac{\Delta\lambda}{\lambda} = \frac{V}{c}\sqrt{1-V^2/c^2}$ of the radiation spectra of galaxies located at a distance of 20 Mpc from the Earth, Hubble established in the late 1920s the empirical relation between the object recession velocity and the distance to it. According to these astronomical observations it was precisely remoteness of the object that was fixed, to which there corresponds the spectrum shift towards the longwave "red" region and the velocity of motion determined from shift value turned out to be proportional to the distance V ~ R. The proportionality coefficient H_0, referred to as the Hubble constant, made up $H\approx62,3\pm6,3$ km/s/MPc after the scale improvement made by A. Sandage.

Such empirical dependence was immediately interpreted within the Friedman's Universe evolution model. This relativistic cosmological model rests on the Friedman-Robertson-Walker metric $ds^2 = c^2dt^2 - \alpha^2(t)\left(\frac{dr^2}{1+kr^2}+r^2(d\vartheta^2+\sin^2\theta\,d\varphi^2)\right)$ which is substituted into the Einstein equation $R_{ij} - \frac{1}{2}Rg_{ij} = \frac{8\pi G}{c^4}T_{ij} - \Lambda g_{ij}$ applied to the Universe as a whole.

The right-hand side of the equation involves the Λ–term corresponding to the empty-space background curvature and the role of gravitational impact source is played by the energy-momentum tensor written in the simplest hydrodynamic approximation $T_{ij}=Pg_{ij}+(P+\rho c^2)u_iu_j$ where all the spatial components of four-dimensional velocities u_i are set equal to zero. However, McCree and Milly showed later that the Universe evolution together with the Hubble law can be interpreted in a simpler manner in the framework of ordinary Newtonian mechanics [21]. Calculating the integral corresponding to the second Newton's law applied to a gravitating sphere of mass M and variable radius $R=R(t)$, we obtain the equation $\frac{1}{2}\left(\frac{dR}{dt}\right)^2 - \frac{GM}{R} = C = const$ describing energy conservation in this system. Expressing the sphere mass in terms of density $M = \frac{4\pi}{3}\rho R^3$ and substituting there the Hubble law $V = \frac{dR}{dt} = HR$, we arrive at $\frac{1}{2}R^2\left(H^2 - \frac{8\pi}{3}G\rho\right)=C$. In the case $\rho < \rho_C = \frac{3H^2}{8\pi G}$ it turns out that C > 0 and the motion will be infinite (because the constant C plays the role of total energy of a gravitating system), and the Universe so modeled will appear to be infinitely expanding, open and obeying the Lobachevsky geometry axioms. Otherwise, when the mean matter density exceeds the critical value $\rho>\rho_c$ the system's motion is restricted by gravity by virtue of C < 0. Sooner or later such a Universe will begin to contract infinitely and, being closed, it obeys the Riemann geometry axioms and has a finite volume associated with the space curvature value.

The galactic radiation spectra for a high-level (of the order of 70 %) red shift were measured quite recently (1998-2000) by the groups of B. Smith, A. Riess [6] and S. Perlmutter [7]. At high recession velocities they revealed nonlinear deviations from the Hubble law. The reason for more than half a century delay with carrying out astronomical studies corresponding to high $\Delta\lambda/\lambda$ values obviously comes down to the fact that according to the Hubble law $V \sim R$ appropriate objects for such studies must lie at huge (about 2000-5000 MPc) distances from the Earth. To register such a weak light signal with intensity decreasing in inverse proportion to the squared distance to the object, the "Hubble" cosmic telescope and an extremely powerful 10-meter ground-based "Kek" telescope were used. The silicon line displacement in Ia type supernova flare spectrum was measured. Observation of such flares showed that the recession velocity appeared to exceed the value corresponding to the Hubble law. This result pointed to the presence of additional acceleration in the recession dynamics of remote galaxies.

In Newtonian mechanics, an accelerated recession of material objects in a free space is provided by the repulsive force exceeding gravitational attraction. A formal introduction of repulsive force into the Newtonian law of gravitation requires admission of a possibility of existence of matter with negative density. The trial body of mass m located at a distance R from the massive center M will as usual be attracted to it with free fall acceleration $\dfrac{d^2R}{dt^2} = -\dfrac{GM}{R^2}$. If the massive center is surrounded with a sphere of radius R, filled with material substance of negative density $\rho_{(-)}$, then an "antigravitational" summand increasing proportionally to the distance from the center to the trial body $\dfrac{d^2R}{dt^2} = -\dfrac{G(M - 4/3\pi R^3\rho_{(-)})}{R^2} = -\dfrac{GM}{R^2} + \dfrac{4\pi G\rho_{(-)}R}{3}$ will be added to the free fall acceleration.

The role of gravitational source in the Einstein equation is played by the energy-momentum tensor T_{ij}. The simplest hydrodynamic approximation $T_{ij}=Pg_{ij}+(P+\rho c^2)u_iu_j$ shows that gravity is created not only by mass density ρc^2, but also by pressure P. But since the Einstein equation is essentially a generalization of the Newtonian law of gravitation, it follows that in this law; too, the pressure should be taken into account in calculation of the effective matter density creating gravitation together with the usual mass density. The expression for the energy-momentum tensor involves, along with the energy density $\rho_E=\rho c^2$, a threefold pressure value because the sum of diagonal tensor components is $-P+P+P+P+(P+\rho c^2)=3P+\rho c^2=3P+\rho_E$. Accordingly, the effective matter density which should be used in the Newtonian law of gravitation is $(3P+\rho_E)/c^2$. In equilibrium conditions, the absolute pressure value is equal to the energy density $|P| = \dfrac{|F|}{S} = \dfrac{1}{S}\left|\dfrac{\partial E}{\partial x}\right| = \dfrac{\partial E}{\partial V} = \rho_E$, and

therefore the negative pressure in the Newtonian law of gravitation will correspond to the negative effective matter density $(3P+\rho_E)/c^2=(-3|P|+|P|)/c^2=-2|P|/c^2$. Thus, to explain formally the antigravitational forces that provide accelerated recession of remote galaxies, one should assume the entire surrounding space to be filled with a weakly interacting substance with negative internal pressure, which is "Dark Energy". As the negative density $\rho_{(-)}$ one can substitute $-2|P|/c^2=-2\rho_E/c^2=-2\rho_{DE}$ into the Newtonian law of gravitation, and then the law can be written as $\dfrac{d^2R}{dt^2} = -\dfrac{GM}{R^2} + \dfrac{8\pi G\rho_{DE}R}{3}$.

Since the antigravitational component of gravitation increases proportionally to the distance and the mean "Dark Energy" density is negligibly small ($<\rho_{DE}> \approx 7 \times 10^{-30}$ g/cm^3) according to the estimates obtained from the red shift and relic radiation anisotropy measurements, it follows that the repulsive force will manifest itself notably in the interaction of massive objects only if these objects are spaced by a great distance

$$R_0 = \sqrt[3]{\frac{3M}{8\pi < \rho_{DE} >}} \approx 2,2 \times 10^8 \times \sqrt[3]{M}$$ (of dimension [R]=m, [M]=kg). For R=R$_0$ the

gravitational attraction is balanced by the antigravitational repulsion. For example, the three 600-kg satellites spaced by a distance of five million kilometers, which are planned by the LISA project to become the basis of a giant interferometric detector of low-frequency gravitational waves [22] will, under the impact of "Dark Energy", repulse each other with a force exceeding by an order of magnitude the usual gravitational attraction.

The meaning of introduction of "Dark Energy" compensating gravitational attraction of bodies at large distances turns out to be "akin" to the Λ-term introduction to the relativistic gravitational equation, Einstein undertook to compensate the mean space curvature generated by the presence of matter. Formally, the "Dark Energy" density can be "replaced" by the corresponding Λ-term which is introduced to the Einstein equation after recalculation by the formula $\Lambda = 8\pi G \rho_{DE} c^2$. Essentially, it is precisely in this way that "Dark Energy" becomes accessible to consideration in the language of relativistic theory.

There are no unified thoroughly developed notions of the physical structure of "Dark Energy" as such. The nature of this everywhere penetrating weakly interacting substance is associated with such popular attributes of modern physics as Dark Matter, supersymmetric partners of "ordinary" elementary particles and other WIMPs [1], quantum-field vacuum [23], Casimir effect [24-27], etc.

A new conception of so cold chameleon particles popular recently is a characteristic example of hypothesis of the Dark Energy "microscopic structure" [27-35]. This popularity is exited not only by the extravagance of Hamiltonian used in the field theory basing of possible existence of such particles but is related with the appearance of new observations indirectly proving their existence [34]. A chameleon scalar field ϕ coupled to matter and photons has an action of the form

$$S = \int dt\, d^3x \{L(\phi, \vec{A}_v, \psi^i)\} = \int dt\, d^3x \{L_M(\phi, \psi^i) + L_{EM}(\phi, \vec{A}_v) + L_{Ch}(\phi)\},$$ where

$L_M(\phi, \psi^i) = L_M(\phi = 0,\ \psi^i \exp(\phi/M_M))$ and $L_{EM}(\phi, \vec{A}_v) = -(1/4)\exp(2\phi/M_{EM})F^{\mu\nu}F_{\mu\nu}$ - matter and electromagnetic contributions into Lagrangian density, correspondingly, $L_{Ch}(\phi) = -(1/2)\nabla_\eta \nabla^\eta \phi - V(\phi)$ - chameleon Lagrangian density, M_M and M_{EM} – constants of chameleon coupling to matter and radiation [35-36]. Matter and electromagnetism contributions to Lagrangian renormalized under chameleon action could be combined with the chameleon potential $V(\phi)$. In this case the effective potential will be written as the following $V_{eff}(\phi, x) = \rho_M(x)\exp(2\phi/M_M) + \rho_{EM}(x)\exp(2\phi/M_{EM}) + V_{eff}(\phi)$. The

effective potential permits one to get expression chameleon field mass $m_{eff} = \sqrt{\dfrac{d^2 V_{eff}(\phi)}{d\phi^2}}$,

that should increase monotonically with matter and radiation densities upgrowth So the chameleon field corresponds to the particles which mass depends on there "nearest environment. A high density of matter and radiation in the vicinity of the Earth, Solar system and etc provides a huge mass for these hypothesis particles in such "physically accessible" places. Energetically advantageous lowering of the effective potential $V_{eff} = V_{eff}(\phi, x)$ is achieved due to reduction of matter and electromagnetism densities in those regions where chameleons are present. Hence matter and radiation are to be forced out of these arrears. In means that homogeneous distribution of chameleons over the space filled up with matter and electromagnetic field will cause the effect of the "inside" pushing apart. Owing to the tremendous mass of chameleons on the Earth the radius of action of "additional" forces is cut down to the dimensions much smaller then characteristic inter corpuscular distances of matter and radiation and hence the repulsion dos not become apparent. However in the astrospace where the matter density is vanishing low chameleons become very light, in means that "additional" long-range repulsive forces appear, whose radius of action exceeds characteristic inter corpuscular distances of matter and radiation. This additional repulsion of chameleon origination acting on the matter particles and electromagnetic field quanta corresponds to the existence of negative pressure characterizing the Dark Energy.

The dependence of chameleon mass on the presence the matter in the "near-field region" permits to carry out a laboratory experiment (GammeV) [37] in which such particles are to be accumulated in the vacuum chamber that serves as a trap. The mass of captured chameleons should enhance near the chamber walls thus originating a potential barrier preventing their going out. It is proposed that these particles come into being when the beam of high-energy photons gets into the region of great (5-30 T) static magnetic field. In the total Lagrangian the chameleon originating is described by the term of chameleon to matter coupling $-(1/4)\exp(2\phi/M_{EM})F^{\mu\nu}F_{\mu\nu}$ that leads to photon-chameleon oscillations in the presence of large magnetic field. In this experiment it is supposed to register photons that arise as the result of decay of chameleons captured in the trap.

At the same time, "Dark Energy" is often associated with the concepts which have been known in physics for over approximately a hundred years, namely, the Einstein Λ -term characterizing the "natural" empty-space curvature or material ether whose laboratory search had been performed by D. Mendeleev [38].

Different generalizations do not deny the usual definition of energy as maximum work that can be done by a system when it leaves the state with energy to be determined, or as minimum work required for the system to be brought into this state. Such an approach implicates an obligatory comparison of all energy forms with its most fundamental form – mechanical energy. That is, to determine the energy of an arbitrary "non-mechanical" field, it is necessary to "force" it (at least hypothetically) to do the work against any ordinary mechanical conservative force (correspondingly, to determine the density of "non-mechanical" energy, some work against the "ordinary" mechanical pressure should be done). It is mechanical energy, its density and fluxes, as well as fluxes of its gradient written in the form of energy-momentum tensor that form the right-hand side of the inhomogeneous Einstein equation $R_{ij} - \frac{1}{2}Rg_{ij} = \frac{8\pi G}{c^4}T_{ij}$ because this equation itself is in fact a relativistic covariant generalization of the Newtonian law of gravitation $\vec{F} = G\frac{Mm}{r^2}\frac{\vec{r}}{|r|}$ relating in the original form only purely mechanical characteristics of a considered system. The nonlinear effects obeying the "rules" of relativism and becoming topical under conditions of strong attraction ($mc^2 \approx E_{grav} \approx Gm^2/r$) are added to Newtonian mechanics by the Einstein equation. The energy characteristics of a non-mechanical (e.g., electromagnetic) field can of course come out as a gravitational field source both in the right-hand side of the inhomogeneous Einstein equation and in the numerator of the formula for the Newtonian law of gravitation. However, the characteristics of this field must, at least by way of a hypothetical experiment, be made comparable with the mechanical force or the pressure against which some work should be done (in the case of electromagnetic field, ponderomotive force should be applied)in the conditions of such an experiment. In the case of neutrino field it impossible to do work against pressure corresponding to the flux of momentum of these particles because of an extremely small neutrino-matter interaction cross section (($\sigma_{ve} \approx 9 \times 10^{-44} (E_v/10\ MeV\ [cm^2])$). But comparison of intensity of this field with density of "ordinary" mechanical energy is however possible if we mean that the neutrino energy is in fact part of the kinetic energy which the corresponding lepton failed to receive during the interaction act. The lepton playing the role of mediator in such a comparison is not such a weakly interacting particle as neutrino. If, as in the case of Dark Energy, the mediator is absent in the field-matter interaction, the energy density of such a field must correspond to the physically observed pressure.

Introduction to physics of the of "Dark Energy" concept to explain some cosmological paradoxes implies that this ever-penetrating weakly interacting physical substance uniformly pierces the entire space of visible Universe. However, such uniformity of "Dark Energy" distribution is confirmed by observations of objects on scales higher than 4MPc [39]. Smaller-scale deviations of the "Dark Energy" density from the mean value ρ_{DE} can have practically no effect on the recession dynamics of galaxies because the antigravitational

repulsion force $F_{DE} = m\dfrac{d^2r}{dt^2} = \dfrac{8\pi}{3}G\rho_{DE}r$ increases proportionally to the distance r

between the objects (thus, this force subsides on small scales, and the action of "Dark Energy" can be ignored).

One of the hypotheses suggesting a nonstatic character and spatial inhomogeneity of Dark Energy density is a popular "quintessence" hypothesis [40-49] relating Dark Energy to a new superweak and superlight field. Another hypothesis admitting spatial inhomogeneity interprets Dark Energy as "phantom energy" [50-52]. If small-scale fluctuations ρ_{DE} do actually exist, then as a result of a complicated polycyclic Earth motion in space the "Dark Energy" density in the near-Earth space, together with the pressure, will change. The "Dark Energy" mean density was estimated as $< \rho_{DE} >= 1$ keV/cm^3, which corresponds to the pressure $|P| = <\rho_{DE}> \approx 1{,}6 \times 10^{-10}$ Pa.

Another approach to the question of physical pressure associated with the existence of "Dark Energy" is possible. One of the most popular hypotheses concerning the nature of "Dark Energy", in fact, identifies it with the vacuum of a quantized electromagnetic field. An indirect evidence of such identification is reflected even in the notation ρ_V which is used in most papers to designate the "Dark Energy" density, the subscript V standing for the vacuum state. In quantum field theory, this state corresponds to the zero eigenvalue of the product of creation and annihilation operators of elementary wave excitations $\hat{a}_p^+ \hat{a}_p |vac\rangle = o$ $p = 2\pi\hbar/\lambda$ $-\infty < \lambda < \infty$. At the same time, according to quantum electrodynamics, virtual photons with all possible values of momenta and wavelengths permanently appear and vanish in a free unbounded space in the vacuum state. In the language of secondary quantization this corresponds to a unit eigenvalue of another combination of creation/annihilation operators $\hat{a}_p \hat{a}_p^+ |vac\rangle = |vac\rangle$ $p = 2\pi\hbar/\lambda$ $-\infty < \lambda < \infty$. However, if the space is bounded by two parallel mirror surfaces, then virtual photons with wavelengths multiple of the gap d between the mirrors become resonant $\lambda_r = d/(n\pm1/2)$ while all the other virtual excitations "that do not fall into resonance" are suppressed. The space is unbounded from the outside, and the mirrors are under "light" pressure of virtual photons with wavelengths from the entire spectrum, whereas it is only resonance photons that "exert pressure" from the inside. As a result, the difference pressure $\Delta P_V = \dfrac{\pi^2 \hbar c}{240 d^4} \approx \dfrac{1,2 \times 10^{-27}}{d^4_{[M]}}{}_{[Pa]}$

occurs whose zero value constitutes the content of the Casimir effect [23-27]. For of a 50 mkm gap this pressure is about $\Delta P_V |\approx 2 \times 10^{-10}$ Pa. In the presence of small-scale "Dark Energy" fluctuations due to a complicated polycyclic Earth motion in space the properties of vacuum in the near-Earth space can change periodically, which can manifest itself as periodic pressure variations during a long-term recording of the Casimir effect.

Thus, a highly sensitive pressure meter is needed for observation of small-scale "Dark Energy" fluctuations in recording of its pressure variations $\delta\rho_{DE} = \delta P$ and indirectly in recording of the Casimir effect parameters δP_V. As has been shown above, the mean pressure is estimated in both cases as 10^{-10} Pa, and therefore the needed sensor should be capable of taking measurements at a level of 10^{-11} Pa. We propose to fix pressure variations at this level using the SQUID/magnetostrictor system designed for gravitational wave detection [2].

2. SQUID/Magnetostrictor System as a Sensor of Superweak Pressure Variations

It will be shown below how a system for measuring pressure P at a level of $10^{-11}\, \Pi a/\sqrt{Hz}$ or relative extension $\Delta L/L$ at a level of $5 \times 10^{-23}/\sqrt{Hz}$ can be designed using the modern superconducting quantum interferometer (SQUID) [53,54]) with resolution $\langle \delta\Phi \rangle_{/\sqrt{Hz}} = 10^{-6} \Phi_0 / \sqrt{Hz}$ Hz connected with a magnetostrictor functioning as a pressure- or extension- converter into magnetic signal. The quantum interferometer resolution is given here in millionth fractions of flux quanta, which is $\Phi_0 = 2\pi\hbar/2e = 2{,}07 \times 10^{-15}$Wb , the resolution of $10^{-6}\Phi_0$ in the unit frequency band being thought of as high but far from record for a modern SQUID. In the 1990s, low-noise two-stage DC (direct current) SQUIDs were worked out for servicing Weber type gravitational antennas with extension registered by the principle of mechanical dislodgement of a magnetic flux frozen in a reversibly deformable superconducting circuit. In such a two-stage cryoelectronic circuit the second DC SQUID plays the role of a super-low-noise amplifier of electric signals coming from the first DC SQUID [55,56], the attained magnetic flux resolution being $\langle \delta\Phi \rangle_{/\sqrt{\Gamma u}} = 2 \div 5 \times 10^{-7} \Phi_0 / \sqrt{Hz}$ [57-60]. In some cases, the resolution of such devices was higher than that formally admissible by the Heisenberg uncertainty relations $\dfrac{(\langle \delta\Phi \rangle/\sqrt{\Delta f})^2}{2L} \geq \hbar$ (L is inductance of SQUID's operating ring with Josephson tunnel junctions) [61,62], and no special quantum compression methods [63] were used to "overpass" the quantum limit. In alternating-current SQUIDs (RF SQUIDs), the condition $LI_J < \Phi_0/2\pi$ (L is inductance of RF SQUID's operating ring involving Josephson junction with critical tunnel current I_J) makes it possible to pass over to the anhysteretic regime in which no direct energy dissipation channel exists. The anhysteretic RF SQUID [64,65] will have a nonzero noise temperature only in the measure of connection with the "external dissipative electronics", and its resolution can reach the level $\langle\delta\Phi\rangle_{/\sqrt{\Gamma\mu}} = 10^{-9}\,\Phi_0\,/\sqrt{Hz}$.

The operation of the pressure- (or extension)-magnetic signal converter, which is intended to be used together with SQUID, is based on the inverse magnetostriction effect - the Villari effect (discovered by E. Villari in 1865). This effect shows up as a change in magnetization due to mechanical strain (i.e., pressure) induced under magnetostrictive sample deformation. In turn, the "direct" magnetostriction effect, i.e., "ordinary" magnetostriction (discovered by J. Joule in 1842) shows itself as a change in sizes of the magnetostrictive sample observed during its magnetization [66]. The idea of using the inverse magnetostriction effect in tensometric problems can hardly be thought of as original. The description of such systems (without the use of SQUIDs) can be found not only in special reviews [67], but also in general physics courses[68]. We shall estimate the limiting feasibility of this classical method in the case of recording the magnetic response by a superconducting quantum interferometer (i.e., under condition $\Delta\Phi=\delta\Phi$ when the magnetic signal is equal to SQUID's resolution).

The simplest characteristic of the Villari effect is the magnitude of magnetostrictive sensitivity $\Lambda(B_{ext}) = \dfrac{\partial B}{\partial P}\bigg|_{B\approx B_{ext}}$ which shows the quantitative relation between magnetic induction variations and elastic strain which induces this change in a particular material [69]. In the general case, the magnetostrictive sensitivity depends on the field of external bias B_{ext} and normally has the value of 2×10^{-9} T/Pa (and higher [70]). The change of the induction flux B in the cross section S of the magnetostrictive cylinder will be $\delta\Phi = \Delta\Phi = S\,\Delta B = S\,(\Lambda\,\Delta P)$, which allows the limiting SQUID/magnetostrictor system sensitivity during pressure recording to be expressed in terms of the interferometer resolution, $\delta P = \delta\Phi/(S\Lambda)$. For $\langle\delta\Phi\rangle_{/\sqrt{\Gamma\mu}} = 10^{-6}\,\Phi_0\,/\sqrt{\Gamma\mu} = 2,07\times10^{-21}\,Wb\,/\sqrt{Hz}$, $S = 10^{-1}m^2$, and , $\Lambda(B_{ext}) = 2\times10^{-9}$ T/Pa the system's limiting sensitivity in pressure recording conditions will be $\langle\delta P\rangle_{/\sqrt{\Gamma\mu}} = 10^{-11}\,Pa\,/\sqrt{Hz}$. For a magnetostrictor with typical Young modulus $E=200$ GPa, this pressure, which the system still can in principle fix, will, according to the Hooke law $P = E(\Delta L/L)$, correspond to the maximum registered extension $\langle\delta L/L\rangle_{/\sqrt{\Gamma\mu}} = 5\times10^{-23}\,/\sqrt{Hz}$.

Thus, the preliminary estimates, made so far without explicit allowance for magnetostrictor noise, show that the above-said recording system is capable of fixing pressure at a level sufficient for observation of small-scale "Dark Energy" fluctuations (with about an order-of-magnitude reserve in sensitivity). These estimates also demonstrate applicability of the SQUID/magnetostrictor system for detecting gravitational waves with intensity of nearly 0.5 erg/(cm^2·s), which at a frequency of 1 kHz corresponds to the metric tensor perturbation amplitude $|\delta g_{ij}| \approx 5 \times 10^{-23}$, the initially discussed system having been worked out precisely to detect gravitational waves (Figure 1) [2].

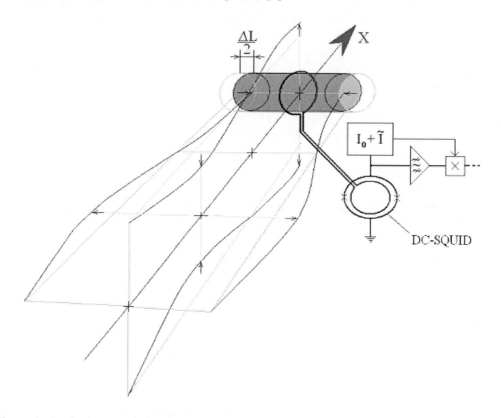

Figure 1. Qualitative correlation between the geometry of gravitational wave propagation and the magnetostriction antenna, the resultant change in the magnetic flux of which is registered by SQUID. A plane-polarized gravitational wave is shown schematically in the foreground, and in the background is the registering SQUID/magnetostrictor system consisting of a magnetostriction cylinder connected with the DC SQUID by flux transducer loops. Arrows show the regions of the trial body contraction and expansion caused by the action of the gravitational wave. The wavelength-to-antenna (cylinder) size ratio is deliberately distorted (λ is "actually" about five orders of magnitude greater than L), and for obviousness, the effect of change in the trial body geometrical sizes in the gravitational wave field is also enlarged by 20 orders of magnitude.

Next, we shall take into account the magnetostrictor contribution $\delta\Phi_{MS}$ into the resultant magnetic flux fluctuations $\delta\Phi_{MS}+\delta\Phi_{SQUID}$ restricting the system sensitivity in recording the pressure ΔP or the relative extension $\Delta L/L$. In order that the pressure and extension could actually be registered according to the above-mentioned estimates, it is also necessary that the magnetic flux fluctuations $\delta\Phi_{MS}$ generated by the magnetostrictor rod "itself" should be small compared to the SQUID's Nyquist noise $\delta\Phi_{SQUID}$ whose value was chosen earlier at the level $\langle\delta\Phi\rangle_{/\sqrt{\Gamma u}} = 10^{-6}\Phi_0 / \sqrt{Hz}$.

As the empirical estimate from above of the ferromagnetic-core noise density we can present the data from the earlier works on magnetic encephalography where the biomagnetic field was registered without a SQUID [53,71]. In these experiments, the core-coil total noise was $3 \times 10^{-17} B\tilde{\sigma}/\sqrt{Hz}$ at T=300 K ("room" temperature). According to the author [71], this value was determined by the Nyquist noise of coil resistance and, thus, the ferrite core noise appeared to be considerably lower.

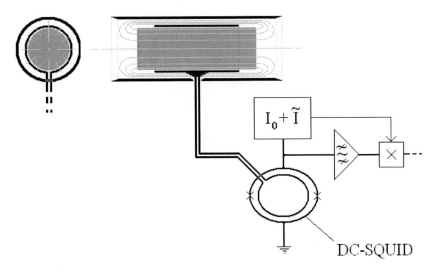

Figure 2. Schematic drawing of the superconducting magnetic flux concentrator allowing a considerable increase in the magnetic-response transfer constant of the magnetostriction cylinder into the working loop of the DC SQUID. The side view of the thickener is given in the left corner. The magnetostrictor is shown in grey, it is surrounded by a tube-shaped loop of the superconducting flux transducer (side view: inner. The external tube (side view: external black ring) with a hole for the twin-lead superconducting wire link of the transducer is the superconducting thickener. The right-hand side of the figure includes the DC SQUID and electronic server elements [53]. The daggers in the operating ring of the DC SQUID stand for Josephson tunnel junctions which are the basic elements of the superconducting quantum interferometer.

A theoretical $\delta\Phi_{MS}$ value can be obtained using analogy with the simplest way to derive the well-known Nyquist formula $\delta U = \sqrt{4\pi k T R \Delta f}$ which makes it possible to calculate the amplitude δU of noise generated in the frequency band Δf by resistance R at a temperature T. This formula follows, in fact, from the condition that some power W, which corresponds to the equilibrium value of energy kT falling at one degree of freedom,

$$kT = W\tau = \frac{1}{R}\left(\frac{\delta U}{\sqrt{2}}\right)^2 \frac{1}{2\pi\Delta f}, \text{ permanently dissipates in a unit band. An analogous condition}$$

for a paramagnet in the absence of external bias, when $<\delta\Phi>=0$, is written as

$$kT = W\tau = \left(E_{MAG}/\tau_\parallel\right)\frac{1}{2\pi\Delta f} = \left(\frac{L}{\mu\mu_{0)}S}\left(\frac{\delta\Phi}{\sqrt{2}}\right)^2 / \tau_\parallel\right)\frac{1}{2\pi\Delta f}, \qquad \text{which} \qquad \text{implies}$$

$$\delta\Phi_{PM} = \sqrt{4\pi k T\left(\mu\mu_0\frac{S}{L}\tau_\parallel\right)\Delta f}, \text{ where } \tau_\parallel \text{ is the time of longitudinal spin-system relaxation}$$

and $\mu\mu_0$ is magnetic susceptibility. For a ferromagnet with external bias, the averaged flux is nonzero $<\delta\Phi>=B_{ext}S$, and then $kT=\left(\dfrac{LB_{ext}\delta\Phi}{\mu\mu_{0)}}\Big/\tau_{\parallel}\right)\dfrac{1}{2\pi\Delta f}$, from which we have

$\delta\Phi_{FM}=4\pi kT\dfrac{\mu\mu_0\tau_{\parallel}}{LB_{ext}}\Delta f$. The longitudinal relaxation time can be estimated by the spin

resonance width $\tau_{\parallel}>\cup\approx\dfrac{1}{\Delta f_{SR}}$. For ferromagnetic resonance in yttrium-iron garnet

($f_{SR}(B=0,11$ T$)=3,3\times10^9$ Hz, $\Delta f_{SR}=5\times10^5$Hz), the longitudinal relaxation time is nearly 10^{-4} s. Using these parameters for evaluation of proper noise of a 0.1-m magnetostrictor core in a unit frequency band at helium temperatures ($T=4.2$ K) in a 1-T field we find that their

amplitude is restricted from above by the value $\delta\Phi_{MS}\approx\delta\Phi_{FM}=4\pi kT\dfrac{\mu\mu_0\tau_{\parallel}}{LB_{ext}}\Delta f<10^{-26}Wb$.

Thus, the magnetic flux fluctuations $\delta\Phi_{MS}$ generated by the magnetostriction rod "itself" turn out to be much smaller than the SQUID's Nyquist noise $\delta\Phi_{SQUID}=10^{-21}Wb\big/\sqrt{Hz}$ (by about

five orders of magnitude in the corresponding frequency band).

The limiting sensitivity was estimated above without allowance for the flux transducer coefficient. In SQUID, the superconducting flux transducer acts as a coupling and consistency element through which $\Delta\Phi$ are sent from the macroscopic region (where the magnetic flux is measured according to the experimental conditions) to the interferometer input circuit which, by the condition excluding multivaluedness of the operating characteristic $LI_J<\cup\approx\Phi_0$ must have a microscopic size (L is the SQUID input-circuit inductance and I_J is the critical tunnel current of Josephson junction). The magnetic flux transducer is a pair of loops closed in a united superconducting circuit by means of a low-inductance twin-lead line (Figure 1 – the circuit shown by the bold black line). Its first loop is connected with the microscopic region where magnetic measurements are taken, and the second - with the SQUID's working circuit where the Josephson junctions switch on (one junction in the case of RF SQUID and a pair in the case of DC). Since the resultant magnetic-field flux piercing the closed superconducting circuit remains unchanged, we have $\Delta\Phi_1=-\Delta\Phi_2$, which allows the flux changes to be transferred among spaced loops. The maximized flux transfer constant ($\Delta\Phi_{SQUID}/\Delta\Phi)_{max}$ appears to be proportional to $\sqrt{L/L_{tr}}$ and typically ranges between 0.005 and 0.05 [72]. During long-term pressure measurements, such sensitivity lowering can readily be compensated (without taking "special measures") by an increase in the signal storage time to about 10 000 s. However, to avoid the "1/F-noise" effect, instead of increasing the integration constant one can raise the flux transducer coefficient L_{tr} by decreasing effectively its coupling loop inductance. To this end, the coupling loop of the transducer should be screened by an additional external superconducting ring (Figure 2). The transducer coefficient can alternatively be raised though an artificial increase in the SQUID input-circuit inductance. To avoid the restriction $LI_J<\cup\approx\Phi_0$, it was suggested [73] that the separate Josephson tunnel junctions of the quantum interferometer should be replaced by chains of n junctions connected in series and having nearly equal I_J. For an interferometer with such chains, the restriction $LI_J<\cup\approx\Phi_0$ is replaced by a less rigid one $LI_J<\cup\approx n\Phi_0$.

3. Physical Bases of Magnetostriction and Criteria of the Optimum Trial-body Choice in the SQUID/Magnetostrictor System

In the light of what has been said above, of certain interest is the search for ways of improving the magnetostrictive properties of substances used in the presented SQUID/magnetostrictor system. According to the simplest phenomenological scheme for the description of magnetostriction effects, the conventional free energy decomposition [74] $F(T,M)=F_0+a(T-T_K)M^2+bM^4$ should be supplemented with the "mixed" summand proportional to the product of the magnetic moment M of the system by the relative extension $\Delta L/L=\varsigma$. Then $F(T,M, \varsigma)=F_0+a(T-T_K)M^2+bM^4+c_1 M\varsigma$ [75]. Such mixed summand corresponds to the case of both direct linear and inverse linear magnetostriction effect, that is, $\varsigma(M) \sim M$ and $\Delta M(\varsigma) \sim \varsigma$. The free-energy minimum of the ferromagnet possessing no magnetostriction properties corresponds to the zero partial derivative

$$\frac{\partial F(T,M)}{\partial M} = 2a(T-T_K)M+4bM^3=0,$$ which implies the usual spontaneous magnetization

expression $M_0 = a\sqrt{\dfrac{T_K - T}{2b}}$. An analogous free-energy maximum condition with allowance

for the "mixed" summand $\dfrac{\partial F(T,M)}{\partial M} = 2a(T-T_K)M+4bM^3 +c_1\varsigma=0$ leads to an

expression cubic in M, whose approximate solution includes a correction corresponding to

deformation, $M \approx M_0 + \Delta M(\varsigma) = M_0 +\dfrac{c_1\varsigma}{8a(T_K - T)}$. One can see that the Villari effect

used in the described scheme "increases in inverse proportion" as T approaches the Curie temperature.

A "mixed" summand of the form $c_1 M\varsigma$ corresponds to the linear magnetostriction effects (this is observed in piezomagnets, i.e., antiferromagnets with strain-broken equilibrium of oppositely polarized sublattices) or to "ordinary" quadratic magnetostriction effects where an external magnetic field acts on the sample (the proportionality coefficient then becomes dependent on the external field $c_1=c_1(B_{ext})\sim B_{ext}$) [75]. In the absence of external bias "quadratic" effects should be described by a mixed summand of the form $c_2 M^2\varsigma$ (the tensor character of the coefficients c_1 and c_2 reflecting crystal anisotropy is disregarded here for simplicity [76-78]). Such mixed summand corresponds to the cases of direct quadratic and inverse linear magnetostriction effect, i.e., $\varsigma(M) \sim M^2$ and $\Delta M(\varsigma) \sim M\varsigma$. As before, the magnetic moment corresponding to the minimum ferromagnet free energy with allowance for the summand $c_2 M^2\varsigma$ includes the deformation-dependent correction

$$M \approx M_0 + \delta M(\varsigma) = M_0 -\frac{c_2\varsigma}{4} \sqrt{\frac{2b}{a(T_K - T)}},$$ which is the essence of the inverse

magnetostriction effect. It can however be seen that the temperature dependence of the response to deformation appears to be weaker now: $\Delta M(\varsigma)\sim(T_K-T)^{-1/2}$ instead of $\Delta M(\varsigma)\sim(T_K-T)^{-1}$. At the same time one can see that in both "linear" and "quadratic" cases the Villari effect used in the described SQUID/magnetostrictor gets unlimitedly amplified as $T\to T_K$.

It turns out that to heighten the sensitivity the working temperature of this measuring system should be made as close to T_K as possible. It is a known fact that the Curie temperature of conventional iron- or nickel-based magnetostrictive materials makes up to hundreds of Kelvins, and therefore a sensor using such substances under $T{\to}T_K$ cannot be thought of a as low-noise (and all the more super-low-noise) one. This fact makes topical the search for cryogenic magnetostrictive materials, i.e., magnetostrictors with low (units or fractions of Kelvin) Curie temperatures.

The simplest attempt to consider magnetostriction from the microscopic standpoint immediately confronts the necessity of direct allowance for collective spin-spin interaction effects. If one proceeds from the microscopic mechanism presuming no allowance for collective effects of interaction of each electron with each electron (the spin gas model), then the calculated sensitivity of magnetostriction sensor will appear to be at zero level. The condition of self-consistency of the external magnetizing field B_{ext} and the field B_{int} corresponding to internal spin-system polarization ignoring the spin/spin interaction $B_{ext} + \chi_P B_{int} = B_{int}$ leads   to   the   susceptibility   formula

$$\chi = \frac{\mu_0 m_{int}}{B_{ext}} = \frac{\chi_P}{1-\chi_P} = \frac{\mu_0 \mu_B^2 N_F}{1-\mu_0 \mu_B^2 N_F} \text{ (where } \mu_B \text{ is the Bohr magneton, } N_F = \frac{\partial n}{\partial E}\bigg|_{E=E_F}$$

is the density of states in the region, $n=N/V$, $m_{int}=M_{int}/V=B_{int}/\mu_0$, V is the crystal volume, $\chi_P=\mu_0\mu_B^2 N_F$ – Pauli susceptibility of a non-interacting spin gas). Under deformation, the relative change of the spin separation will be approximately equal to the relative extension of the entire magnetostriction sample, which in turn presumes a possibility of strong dependence of the exchange spin-spin interaction integral J_{SS} on the crystal extension $\frac{\Delta J_{SS}}{J_{SS}} \cong \frac{\Delta \ell_{SS}}{\ell_{SS}} = \frac{\Delta L}{L} = \varsigma$. However, as can be seen from the formula derived with disregard of spin interaction, the exchange integral variation ΔJ_{SS} will in no way affect the system magnetization $\frac{\partial m_{int}}{\partial J_{SS}} = \frac{\partial \chi}{\partial J_{SS}} B_{ext} = 0$, and therefore $\Delta m_{int}(\varsigma)=0$.

The mechanism reinforcing the substance properties that correspond to the Villari effect and, at the same time, the registered response to pressure or extension in the SQUID/magnetostrictor system is closely related to "ferromagnetic collectivism" which is simpler to consider within the phenomenological theory (as has been done above). From the "microscopic" standpoint, collective effects manifest themselves in the self-consistent interaction of each spin with total magnetization of all the other electrons or holes belonging in the corresponding energy range but with an explicit account of the exchange spin/spin interaction. Spin polarization due to exchange interaction and/or the action of external magnetization leads to a split of the original energy band into two subbands. The difference $\Delta n_{\uparrow\downarrow}$ in the number of carriers in the subbands is expressed in terms of the product of the density of states N_F by the difference of the "subband" energies: $N_F \Delta E_{\uparrow\downarrow}= (n_{\uparrow}-n_{\downarrow})/2=\Delta n_{\uparrow\downarrow}$. The subband formation induces an increase in the kinetic energy of the system:

$$\Delta E_{Kin} = \frac{n_\uparrow - n_\downarrow}{2} \Delta E_{\uparrow\downarrow} = \frac{1}{N_F}\left(\frac{n_\uparrow - n_\downarrow}{2}\right)^2 = \frac{(\Delta n_{\uparrow\downarrow})^2}{N_F}.$$

At the same time, the spin-spin interaction energy expressed in terms of the exchange integral J_{SS} will decrease: $\Delta E_{Ex} = J_{Ex} n_\uparrow n_\downarrow - J_{Ex}\left(\frac{n}{2}\right)\left(\frac{n}{2}\right) = J_{Ex} n_\uparrow n_\downarrow - J_{Ex}\left(\frac{n_\uparrow + n_\downarrow}{2}\right)^2 = -J_{Ex}(\Delta n_{\uparrow\downarrow})^2,$

where $J_{Ex}=J_{SS}V.$. Thus, the resultant energy variation with allowance for polarization excited by the external field B_{ext} is written as

$$\Delta E = \Delta E_{Kin} + \Delta E_{Ex} - 2(\mu_B \Delta n_{\uparrow\downarrow})B_{ext} = (1/N_F - J_{Ex})(\Delta n_{\uparrow\downarrow})^2 - 2\mu_B B_{ext}\Delta n_{\uparrow\downarrow},$$ and

the condition of system's energy minimum at $T\approx0$ as $\dfrac{\partial(E_0 + \Delta E)}{\partial(\Delta n_{\uparrow\downarrow})} = 0$ (closeness to $T=0$

allows disregard of the role of entropy here). The minimum condition implies the equation $(1/N_F - J_{Ex})\Delta n_{\uparrow\downarrow} = \mu_B B_{ext}$ whose solution makes it possible to determine the

ferromagnet susceptibility $\chi = \dfrac{\mu_0 m_{int}}{B_{ext}} = \dfrac{\mu_0 \mu_B \Delta n_{\uparrow\downarrow}}{B_{ext}} = \dfrac{\mu_0 \mu_B^2 N_F}{1 - J_{Ex} N_F}$. One can see that at

$T\approx0$ the dependence of magnetization m_{int} on the spin-spin interaction exchange integral increases to the extent of closeness of the Stoner factor $J_{Ex}N_F$ to unity: $\dfrac{\partial m_{int}}{\partial J_{Ex}} = \dfrac{N_F m_{int}}{1 - J_{Ex} N_F}$.

Expressing the ferromagnet susceptibility χ through the Pauli susceptibility χ_P of a non-interacting spin gas, we derive the formula $\chi = \dfrac{\chi_P}{1 - (J_{Ex}/\mu_0\mu_B^2)\chi_P}$ similar in structure to

the well-known expression of the system's amplification factor with allowance for the feedback $K = \dfrac{K_0}{1 - \beta K_0}$. Thus, the exchange interaction $J_{Ex}/(\mu_0\mu_B^2)$ appears to play the role

of the positive feedback coefficient β promoting internal polarization amplification as a response to the effect of the external magnetic field B_{ext}. The amplification factor can be represented as a power series converging for $\beta K_0|<1$:

$$K = \frac{K_0}{1 - \beta K_0} = K_0(1 + \beta K_0 + \beta^2 K_0^2 + \beta^3 K_0^3 + ...).$$

Successive summation of this series corresponds to allowance of the feedback in all higher orders of magnitude: the first summand demonstrates amplification without account of the feedback; the second summand takes into account amplification of part of the already amplified effect withdrawn from the output to the input through the feedback channel with the transfer coefficient β; the third corresponds to amplification of the doubly amplified effect, etc. Such power series can be assigned the diagram series presented in Figure 3.

When the Stoner factor approaches unity, $J_{Ex}N_F\approx1$, the strong dependence of the magnetization m_{int} on the exchange integral

$$\frac{\Delta m_{\text{int}}}{m_{\text{int}}} = \frac{J_{Ex}N_F}{1-J_{Ex}N_F}\frac{\Delta J_{Ex}}{J_{Ex}} \approx \frac{J_{Ex}N_F}{1-J_{Ex}N_F}\frac{\Delta L}{L} \approx \frac{1}{1-J_{Ex}N_F}\frac{\Delta L}{L}$$ makes it possible to attain

high sensitivity during recording the relative extension by the SQUID-magnetostrictor,

$$\Delta\Phi \approx \frac{\chi S B_{ext}}{1-J_{Ex}N_F}\frac{\Delta L}{L},$$ where $\Delta\Phi$ is the SQUID-fixed flux variation and S is the

magnetostrictor cross section. Thus, while within the framework of the phenomenological scheme of the occurrence of ferromagnetism the high magnetostriction sensor sensitivity should be sought near the Curie temperature (see above), from the standpoint of the microscopic theory the highest sensitivity can be reached when the Stoner factor approaches unity (the Stoner factor close to unity is observed, for example, in Pd). The joint action of both mechanisms of Villari effect amplification is obviously possible, and in this case for the magnetostriction sensor a material with Stoner factor close to unity and a low Curie temperature should be sought. Such a sensor should be exploited within the range $T \approx T_K$. The analysis above points out that the best characteristics i.e. the highest sensibility would have a detector operating at temperature in the vicinity of Curie point of its working body possessing a Stoner factor of the order of unit $J_{Ex}N_F \approx 1$. Generally speaking the working body may be both in para- and ferromagnetic states, the first perhaps preferable.

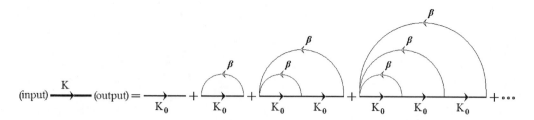

Figure 3. The diagram series demonstrating the formation of the amplification factor in a system with feedback. The straight segments on separate diagrams correspond to an "unperturbed" amplification K_0, the arcs correspond to the feedback channel with transfer coefficient β, the segments are directed from input to output and the arcs – from output to input. Correspondingly, in the case of a magnetic system the straight segments are assigned the Pauli susceptibility χ_P of noninteracting spin gas and the arcs are assigned the spin/spin exchange interaction $J_{Ex}/(\mu_0\mu_B^2)$.

References

[1]   Ryabov V.A., Tsarev V.A., Tskhovrebov A.M. "The search for dark-matter particles", *Usp. Fiz. Nauk*, (2008) 51 1091.

[2]   Golovashkin A.I.,Tskhovrebov A.M., Zherikhina L.N. et al., "Fluctuation limit in measurements of the relative elongation of a magnetostrictive cylinder", *JETP Lett.*, (1994) 60, No. 8, 612.

[3]   Zwicky F., *Helv. Phys. Acta.*, (1933) 6, 110.

[4]   Bertone G., Dan Hooper, Silk J., arXiv:hep-ph/0404175 Aug (2004) 2, 13.

[5]   Chardin G., arXiv:astro-ph/0411503 Feb. (2005)3, 28.

[6]   Riess A G et al., *Astron. J.* (1998) 116 1009.

[7]   Perlmutter S et al. *Astrophys. J.*, (1999), 517 565.

[8]   Lukash V.N., Rubakov V.A. *"Dark energy: myths and reality"* (2008) 51 283 - 290.

[9]   Linde A. *Particle Physics and Inflationary Cosmology*, Chur: Harwood Acad. Publ., 1990.

[10]   Melchiorri A et al. (Boomerang Collab.) *Astrophys. J.* , (2000) 536 L63.

[11]   de Bernardis P. et al. (Boomerang Collab.) *Nature*, (2000) 404 955.

[12]   Lange A E et al. (Boomerang Collab.) *Phys. Rev.* D (2001) 63 042001.

[13]   Balbi A et al. *Astrophys. J.*, (2000) 545 L1-L4.

[14]   Hanany S et al. *Astrophys. J.* , (2000) 545 L5 –L8.

[15]   Jaffe A H et al. (Boomerang Collab.) *Phys. Rev. Lett.*, (2001) 86 3475.

[16]   Spergel D N et al. *Astrophys. J. Suppl.* , (2003) 148 175.

[17]   Spergel D N et al., *Astrophys. J. Suppl.*, (2007) 170 377.

[18]   Hinshaw G et al., *Astrophys. J. Suppl.* , (2007) 170 288.

[19]   Nacelckiy P.D., Novikov D.I., Novikov I.D., *The relict Radiation of the Universe*, Voscow, *Publ. Science,* (2003).

[20]   Sazhin M V. *Phys. Usp.*, (2004) 47 187.

[21]   Zeldovich Ya.B., Myshkis A.D., *The elements of mathematical Physics,* Moscow, Pupl. Science, (1973).

[22]   Bender P et al. LISA: *Pre-Phase* a Report 2nd ed. (Garching, 1998).

[23]   Jaffe R. L., The Casimir Effect and the Quantum Vacuum *Phys.Rev. D* (2005) 72 021301.

[24]   Casimir H.B.G., *Kon. Ned. Akad.Wetensch. Proc.,* (1948) 51 793.

[25]   Levin F.S. and Micha D.A., P*lenum, Long-Range Casimir Forces,* eds. , New York, (1993).

[26]   .Lamoreaux S.K, *Phys. Rev. Lett.,* (1997) 78, 5.

[27]   Brax P., van de Bruck C., Davis A.-C., Mota D.F., Shaw D., *Phys. Rev. D* (2007) 76, 124034.

[28]   Khoury J. and Weltman A., *Phys. Rev. Lett.,* (2004) 93, 171104.

[29]   Khoury J. and Weltman A., *Phys. Rev. D,* (2004) 69, 044026.

[30]   Gubser S.S. and Khoury J., *Phys. Rev. D,* (2004) 70, 104001.

[31]   Brax P., van de Bruck C., Davis A.-C., Khoury J., and Weltman A., *Phys. Rev. D,* (2004) 70, 123518.

[32]   Upadhye A., Gubser S.S. and Khoury J., *Phys. Rev. D,* (2006) 74, 104024.

[33]   Adelberger E.G. et al., *Phys. Rev. Lett.,* (2007) 98 131104.

[34]   Burrage C., Davis A.-C. and Shaw D. J. *Phys. Rev. D,* (2009) 79 044028.

[35]   Mota D. F. and Shaw D.J., *Phys. Rev. Lett.,* (2006) 97, 151102.

[36]   Mota D.F. and Shaw D.J., *Phys. Rev. D,* (2007) 75, 063501.

[37]   Chou A.S., Wester W., Baumbaugh A., Gustafson H.R., Irizarry-Valle Y., Mazur P.O., Steffen J.H., Tomlin R., Upadhye A., Weltman A., Yang X., Yoo J., *Phys. Rev. Lett.,* (2009) 102, 030402.

[38]   Mendeleev D.I. An Attempt of Chemical Interpretation of the Universal Ether, SPb, Publ. M.P.Frolova, (1905).

[39]   Chernin A D "Dark energy and universal antigravitation" *Phys. Usp.,* (2008) 51 253.

[40]   Peebles P J E, *Ratra B Astrophys. J.,* (1988) 325 L17.

[41]   Zlatev I, Wang L, Steihardt P, *J Phys. Rev. Lett.,* (1999) 82 896.

[42]   Steinhardt P J, Wang L, Zlatev I, *Phys. Rev. D,* (1999) 59 123504.

[43]   Gorini V et al., arXiv:0711.4242.

[44]   Alam U, Sahni V, Starobinsky A A, JCAP, (2007) (02) 011.

[45]   Gannouji R et al., astro-ph/0701650.

[46]   Saini T D et al. *Phys. Rev. Lett.,* (2002) 85 1162.

[47]   Sereno M, Piedipalumbo E, Sazhin M *V Mon. Not. R. Astron. Soc.,* (2002) 335 1061.

[48]   Rubakov V A *Phys. Rev.* D, (2000) 61 061501.

[49]   Chernin A D, Santiago D I, Silbergleit A S, *Phys. Lett. A,* (2002) 294 79.

[50]   Dabrowski M P, Stachowiak T *Ann. Phys.* (New York), (2006) 321 771.

[51]   Sami M, Toporensky A *Mod. Phys. Lett. A,* (2004) 19 1509.

[52]   Libanov Met al. *JCAP,* (2007) (08) 010.

[53]   Superconductor Applications: *SQUIDs and Machines.* Eds. Schwartz Br.B. and Foner S., Plenum Press, New York (1977).

[54]   A.Barone, G.Paterno. *Physics and applications of Josephson Effect.* A Wiley-Interscience Publication John Wiley & Sons. New York – Chichester – Brisban – Toronto – Singapore (1982).

[55]   Falferi, Class. *Quantum Grav.,* (2004) 21 S973-S976.

[56]   Gottardi L., Podt M., Bassan M. *et.al., Class. Quantum Grav.,* (2004) 21 S1191-S1196.

[57]   Voss R.F., Laibowitz R.B., Raider S.I., and Clarke J., *J. Appl. Phys.,* (1980) 51, 2306.

[58]   Ketchen M.B. and Voss F., *Appl. Phys. Lett.,* (1979) 35 812.

[59]   Clarke J., IEEE Trans. *Electron Dev.* ED-27, (1980) 1896.

[60]   Ketchen M.B. and Jaycox J.M., *Appl.Phys.Lett.,* (1982) 40 736.

[61]   Clarke J. *Physics Today*, March, (1986) 39 №3.36.

[62]   Voss R.F.et al., *Second Int. Conf. on Supercond. Quantum Devices,* Berlin, May (1980) 94.

[63]   Braginsky V.B., Vorontsov Y.I., Thorne K.S. *Quantum nondemolition measurements. Science,* (1980) 209 547-557.

[64]   Likharev K. K., Ulrikh B. T., *Systems with Josephson contacts. Moscow Univ.* Press (1978).

[65]   Golovashkin A. I., Kuleshova G. V., Tskhovrebov A. M., Zherikhina L.N., FPS'08, Moscow/Zvenigorod (2008) 292-293.

[66]   Pol P.B., The Doctrine of Electricity, Moscow, *Publ. Phys.-Math. Lit.,*(1962).

[67]   Belov K.P., *The Magnetostriction phenomena and their technical applications,* Moscow, Publ. Science, (1987).

[68]   Putilov K.P., *The Course of Physics,* v.2, Moscow, Publ. Phys.-Math. Lit.,(1963).

[69]   Sverdlin G.M., *Applied Hydroacoustics,* Leningrad, Shipbuilding,(1990).

[70]   Kundys B., Bukhantsev Y., Szymczak H., Gibbs M.R.J., Zuberek R., *J.Phys.D: Appl.Phys.,* (2002) 35, 1095-1098.

[71]   Cohen D., *Science,* (1968) 161 784.

[72]   O.V.Lounasmaa. *Experimental principles and methods below 1K.* London and New York, Academic Press (1974).

[73]   Karuzskiy A., Kuleshova G., Tshovrebov A., Zherikhina L., Quantum interferometers on multichain of josephson junctions , *International Conference "Micro- and nanoelectronics – 2009"* Moscow-Zvenigorod, October 5-9, (2009) P1-41.

[74]   Landau L.D., Lifshits E.M., *Statistical Physics.Part 1. Theory Physics,* v.5, Moscow, Publ. Phys.-Math. Lit.,(1976).

[75]   Landau L.D., Lifshits E.M., *Electrodynamics of continuous medium. Theory Physics,* v.7, Moscow, Publ. Phys.-Math. Lit.,(1982).

[76] Akulov N.S., *Ferromagnetism,* ONTI, (1939).

[77] Vonsovskiy S.V., Modern doctrine of magnetism, *Physics Uspekhi,* (1949) №1 1-64.

[78] Krinchik G.S., *Physics of magnetic phenomena,* Moscow, Publ. MSU (1985).

In: Dark Energy: Theories, Developments and Implications   ISBN 978-1-61668-271-2
Editors: K. Lefebvre and R. Garcia, pp. 21-32   © 2010 Nova Science Publishers, Inc.

Chapter 2

A REVIEW ON DARK ENERGY OBJECTS

P. Rocha[1,2,][*]***, M.F.A. da Silva***[2,][†]***and R. Chan***[3,‡]
[1] Gerência de Tecnologia da Informação, ACERP, TV Brasil,
Rádios Nacional e MEC,
Rua da Relação 18, Lapa, CEP 20231-110,
Rio de Janeiro, RJ, Brazil,
[2] Departamento de Fisica Teórica,
Instituto de Fisica, Universidade do Estado do Rio de Janeiro,
Rua São Francisco Xavier 524, Maracanã,
CEP 20550-900, Rio de Janeiro, RJ, Brazil,
[3] Coordenação de Astronomia e Astrofisica, Observatório Nacional,
Rua General José Cristino, 77, São Cristóvão, CEP 20921-400,
Rio de Janeiro, RJ, Brazil,

Abstract

Since it was discovered that the galaxies are moving away from each other accelerately, researches have considered the hypothesis that the universe is predominately constituted by an exotic fluid, called dark energy, in order to preserve the general relativity theory. In this context, it is natural to ask about the connection between dark energy and the structures formation, as stars, black holes, naked singularities and including the possibility of the formation of new structures as the gravastars of dark energy. Here we present a review of the main results on this last troublemaker subject.

1. Introduction

Nowadays, several kinds of observational data indicate that our universe is in accelerated expansion. In Einstein's general relativity, in order to have such an acceleration, one needs to introduce a component to the matter distribution of the universe with a large negative pressure. This component is usually referred as dark energy. Astronomical observations

---

[*]E-mail address: pedrosennarocha@gmail.com.

[†]E-mail address: mfasnic@gmail.com.

[‡]E-mail address: chan@on.br.

indicate that our universe is flat and currently consists of approximately 2/3 dark energy and 1/3 dark matter. The nature of dark energy as well as dark matter is unknown, and many radically different models have been proposed, such as, a tiny positive cosmological constant. Based on this fact, we would like to ask how the picture of the structures formation is affected by the presence of dark energy in the universe.

In another direction, although we have strong theoretical and experimental evidences in favor of the existence of black holes, lots of paradoxical problems about them do exist[1]. Besides, it was shown recently that observational data can give strong arguments in favor of the existence of event horizons, but they can not prove it in a direct fundamental form[2].

As pointed out by Chapline [3] and other researchers, the picture of gravitational collapse provided by Einstein's general relativity can not be completely correct since in the final stages of collapse quantum effects must be taken into account at high curvature values, or short distances, compared with the Planck length scale.

These facts frequently motivated authors to try to find new alternatives for the final state of a collapsing star without horizons. Among these models we can mention gravastars [4, 5, 6, 7], Bose superfluid[8], dark stars [9] and holostars [10, 11, 12, 13, 14, 15], some of these connecting the supposed new contents of the universe with the structures' formation. As an example, the concept of a dark energy star, as proposed by Chapline [16], was motivated by the fact that the appearance of trapped surfaces makes it impossible to everywhere synchronize atomic clocks. Thus, Chapline has built a model of dark energy stars based on the analogy between a superfluid condensate near its critical point with the neighborhood of an event horizon. His dark energy stars have a de Sitter spacetime matched with a Schwarzschild exterior spacetime.

There are many others additional models proposed as black holes mimickers, but among them, alternative models to these compact objects, the gravitational vacuum stars have received special attention [17, 18, 19, 20, 21, 22, 23, 24, 25]. It was firstly proposed by Mazur and Mottola (MM) [4, 5, 6, 7]. In that model, and in the work related developed by Laughlin *et al.* [8], the quantum vacuum itself suffers a phase transition near or at the place where the event horizon was expected to form.

In a more detailed description, a gravastar is a compact object with a de Sitter condensate in the interior described by an equation of state $p = -\rho$, where p is the isotropic pressure and ρ is the energy density, matched to a Schwarzschild exterior spacetime by a dynamical infinitely thin shell of stiff fluid with equation of state $p = \sigma$, where σ is the superficial energy density. This thin shell replaces both the event horizon and the cosmological event horizon. This model has not either singularity or event horizon since its rigid surface is localized in a radius slightly greater than the Schwarzschild's radius and smaller than the de Sitter horizon. MM have shown that the gravastar model is termodinamically stable. Related models analyzed in a different context were also considered by Dymnikova [26, 27, 28, 29, 30].

In fact, in the original model [4], gravastars consisted of five layers: an internal core $0 < r < r_1$, described by the de Sitter universe, an intermediate thin layer of stiff fluid $r_1 < r < r_2$, an external region $r > r_2$, described by the Schwarzschild solution, and two infinitely thin shells, appearing, respectively, on the hypersurfaces $r = r_1$ and $r = r_2$. By properly choosing the free parameters involved, one can show that the two shells can only have tensions but with opposite signs[4]. Visser and Wiltshire (VW) [31] have argued that such five-layer models from MM could be simplified to three layer ones. The main motivation of VW was simplification of the model for the possibility of a full dynamical analysis. They have achieved that replacing the two infinitely thin shells and the intermediate region by one infinitely thin shell, so that the function $f(r)$ in the Schwarzschild's metric and in the de Sitter metric are, respectively given by

$$f(r) = 1 - \frac{2m}{r}, \quad r > a(\tau), \tag{1}$$

or

$$f(r) = 1 - \left(\frac{r}{l}\right)^2, \quad r < a(\tau), \tag{2}$$

where m is the total mass of the system, $r = a(\tau)$ is a time-like hypersurface, at which the infinitely thin shell is located, the parameter $l = \sqrt{3/\Lambda}$ (where Λ is the cosmological constant) and τ denotes the proper time. On the hypersurface $r = a(\tau)$ Israel junction conditions yield:

$$\frac{1}{2}\dot{a}^2 + V(a) = 0, \tag{3}$$

where an overdot denotes the derivative with respect to the proper time τ of the thin shell.

Therefore, in the region $r > a(\tau)$ the space-time is locally Schwarzschild, while in the region $r < a(\tau)$ is locally de Sitter. These two different regions are connected through a dynamical infinitely thin shell located at $r = a(\tau)$ to form a new spacetime of the gravastar.

To study the dynamics of equation (3), one can follow two different approaches: one is to assume a potential $V(a)$ and leave the equation of state of the shell as derived, and the other is to assume an equation of state and leave the potential $V(a)$ as derived. VW have followed the first approach, and studied in detail the case where

$$V(a_0) = 0, \quad V'(a_0) = 0, \quad V''(a_0) > 0, \tag{4}$$

where the prime denotes the ordinary differentiation with respect to the indicated argument. If and only if there exists such an a_0 for which the above conditions are satisfied, the model is said to be stable. Among other things, VW have found that there are many equations of state for which the gravastar configurations are stable, while others are not [31]. Carter [32] has studied the same problem and he has found new equations of state for which the gravastar is stable, while De Benedictis et al. [33] and Chirenti e Rezzolla [34] have investigated the stability of the original model of MM against axial perturbations, and they have found that gravastars are stable to these perturbations. Chirenti and Rezzolla [34] have also shown that their quasi-normal modes differ from those of a black hole of the same mass, and thus can be used to discern a gravastar from a black hole.

As VW noticed, there is a less stringent notion of stability, the so called "bounded excursion" models, in which there exist two radii a_1 and a_2 such that

$$V(a_1) = 0, \quad V'(a_1) \leq 0, \quad V(a_2) = 0, \quad V'(a_2) \geq 0, \tag{5}$$

with $V(a) < 0$ for $a \in (a_1, a_2)$, where $a_2 > a_1$.

Physical configurations with a de Sitter interior have been reviewed and it can be found in the work of Dymnikova and Galaktionov [35].

The tight connection between the cosmological constant and our accelerated expanding universe [36, 37], gave rise to consider the possibility of construction of other structures constituted by other kind of dark energy. Thus, some alternative models to the MM gravastar were also proposed. Among them, we can find a Chaplygin dark star [38], a gravastar supported by non-linear electrodynamics[21] and a gravastar with continuous anisotropic pressure[25]. Additionally, Lobo [20] has studied two models for a dark energy fluid. One of them describes a homogeneous energy density and the other describes an ad-hoc monotonically decreasing energy density, although both of them are with anisotropic pressure. In order to match an exterior Schwarzschild spacetime he has introduced a thin shell between the interior and the exterior spacetimes. Some of us have constructed a model consisting of four parts [39]: (i) a homogeneous inner core with anisotropic pressure (ii) an infinitesimal thin shell separating the core and the envelope; (iii) an envelope of inhomogeneous density and isotropic pressure; (iv) an infinitesimal thin shell matching the envelope boundary and the exterior Schwarzschild spacetime. There, it was considered the possibility of gravitational trapping of dark energy by standard energy, i.e., not dark energy. It was found that, in order to have static solutions, at least one of the regions must be constituted by dark energy. Then, in this context we could have the dark energy sustaining the collapse of the standard energy, while the standard matter would trap the dark energy.

The consideration of this new component of energy, called dark energy, in the context of the gravitational collapse and star formation, where anisotropies are particularly important, imposes a careful analysis of the energy conditions and of the roles of the components of the pressure. The standard cosmological model considers isotropy of the pressure and assumes an equation of state $p = \omega\rho$, relating the pressure p and the energy density ρ, where ω is a constant. The interval of the parameter ω defines the kind of matter of the universe, related to the fulfillment, or not, of the energy conditions of the fluid. In another recent work some of us have pointed out this aspect [40].

2. Gravastars Can Exist but They Do Not Exclude the Existence of Black Holes

Recently we have studied the problem of the stability of gravastars[41] . Firstly, we have taken a model which consisted of an internal de Sitter spacetime, a dynamical infinitely thin shell of stiff fluid, and an external Schwarzschild spacetime, in analogy to the VW's model. We have shown explicitly that the final output can be a black hole, a "bounded excursion" stable gravastar, a Minkowski, or a de Sitter spacetime, depending on the total mass m of the system, the cosmological constant Λ, and the initial position of the dynamical shell. Therefore, we have established, for the first time in the literature, that although it does exist a region of the space of the initial parameters where it is always formed stable gravastars, it also exist a large region of this space where we can still find black hole formation. Then, gravastar is not an alternative model to black hole as it was originally proposed by MM and VW models. Nevertheless, it is a very interesting theoretical dark energy object, which could coexist with the black holes.

After that [42], we have extended our previous results considering in that time an equation of state $p = (1 - \gamma)\sigma$ for the shell, where σ is the superficial energy density and γ is a constant, instead of only using a stiff fluid ($\gamma = 0$). We have found that stable gravastars can be formed even for $\gamma \neq 0$, since $\gamma < 1$, generalizing the gravastar models proposed until then. In addition, black holes can be also formed, depending on the initial parameters. We also consider the cases of null Schwarzschild mass, where the shell has and has not charge [43][44].

There are some works which have included the rotation in the analysis of the stability of the gravastars. In one of them, Cardoso *et al.* [45] have studied the problem of stability of spinning gravastars and black hole mimickers such as, for example, boson stars and they have found that these ultra-compact objects develop a strong ergoregion of instability when rapidly spinning. Instability timescales can be of the order of 0.1 seconds to one week for objects with mass $m = 1 - 10^6$ solar masses and angular momentum $J > 0.4m^2$. Their conclusion was that their results provide a strong indication that ultra-compact objects with large rotation are black holes. Chirenti and Rezzolla[46] have also analyzed in details the problem of the ergoregion instability in rotating gravastars and they have arrived to conclusions less restrictive than Cardoso *et al.* [45] noticing that not all rotating gravastars are unstable to the ergoregion instability. They have also found that stable models of rotating gravastars without an ergoregion can be constructed also for extreme rotation rates, namely, for models with $J/m^2 \geq 1$, where J and m are the gravastar's angular momentum and mass, respectively. As an important conclusion, they have reported in their work that not all ultra-compact astrophysical objects rotating with $J/m^2 \leq 1$ are to be considered necessarily black holes, but in spite of these conclusion Chirenti and Rezzolla [46] have confirmed the results of the work of Cardoso *et al.* [45] for the models considered in their work and it has opened a door to the important effort to the possibility of observationally distinguish gravastars from black holes.

The possibility of distinguishing observationally gravastars from black holes has constituted another important branch in this subject. This is analyzed in the work Harko *et al.*[47] considering accretion disks around slowly rotating gravastars, with all the metric tensor components estimated up to the second order in the angular velocity. Due to the differences in the exterior geometry, the thermodynamic and electromagnetic properties of the disks (energy flux, temperature distribution and equilibrium radiation spectrum) are different for these two classes of compact objects, consequently giving clear observational signatures that could distinguish gravastars from black holes. In addition to this, it was also shown that the conversion efficiency of the accreting mass into radiation is always smaller than the conversion efficiency for black holes, i.e., gravastars provide a less efficient mechanism for converting mass to radiation than black holes. Thus, these observational signatures provide the possibility of clearly distinguishing rotating gravastars from Kerr-type black holes.

3. The Connection between Acceleration of the Universe and Star-Black Hole Formation

Narrowing the distance between the gravastar models with the dark energy necessary to accelerate the universe, it would be natural to ask how the picture of its formation and evolution could be influenced by an exterior spacetime with a positive cosmological constant. Following these ideas, Carter [32] has studied spherically symmetric gravastar solutions which possess a (anti) de Sitter and a (anti) de Sitter-Schwarzschild or Reissner-Nordstrom exterior. He has followed the same approach of VW assuming a potential $V(a)$ and, then, finding the equation of state of the shell. He has found a wide range of parameters which allows stable gravastar solutions, and he has presented the different qualitative behaviors of the equation of state for these parameters. In a recent work [48], we have generalized our second work on gravastars by considering now an external de Sitter-Schwarzschild space-time. There, a different approach, in comparison with the Carter's [32] work, to study how the cosmological constant affects the gravastar formation was used and it is the same approach we have utilized in our previous works [41, 42]. We have first assumed an equation of state, $p = (1 - \gamma)\sigma$, and, using Israel conditions, we have derived a potential depending on the parameters of the interior, the shell and the exterior of the gravastar's prototype. We have, then, studied all types of compact objects that can be generated according to this potential, to the parameters related to the cosmological constants and to the masses of our model and we found that gravastars and black holes can be formed for standard energy shells, satisfying or not the dominant energy condition. On the other hand, for dark energy shells and for phantom energy shells no gravastars can be formed. In fact, stable gravastars were found only for an interval of the external cosmological constant (Λ_e), which depends on the values of interior cosmological constant (Λ_i). We have also concluded that the smaller is Λ_i (for $\Lambda_e = 0$) the bigger is the tendency to the collapse of the shell, forming a "bounded excursion" gravastar or a black hole. Moreover, for a given Λ_i, the formation of gravastars depends on the value of Λ_e in a such way that, instead of what occurs for Λ_i, the bigger is Λ_e the bigger is the tendency to the collapse. These conclusions are in agreement to the gravastar requirement proposed by Horvat and Ilijic [17]. The reason is that the dark energy density inside the gravastar have to be greater than the surrounding spacetime, i.e., $\Lambda_i > \Lambda_e$, in order to sustain the collapse of the shell.

Another natural question is how dark energy, abundantly presents in the universe, would participate of the process of the gravitational collapse or the star formation. Since the dark energy exerts a repulsive force on its surroundings, this force may prevent the star from collapse. Thus, we could speculate that a massive star does not simply collapse to form a black hole, instead, to form a star that contains dark energy, as it was supposed in the origins of the gravastars models. Besides, we can ask how dark energy influences already formed black hole, since it was shown that the mass of a black hole decreases due to phantom energy accretion and tends to zero when the Big Rip approaches [49, 50]. Gravitational collapse and formation of black holes in the presence of dark energy were first considered by several works [51, 52]. Chakraborty and Bandyopadhyay [53, 54] studied the gravitational collapse of an inhomogeneous spherically and quasi-spherically symmetric star, whose matter inside contains a combination of dark matter (dust) and dark energy (anisotropic), in order to investigate what would be the effects of the dark energy on the collapse process, including its effects on the Cosmic Censorship Conjecture. They have found that, depending on the tangential pressure (of the dark energy component, with zero radial pressure) and on the total matter density, the collapse could result in a black hole or in a naked singularity. In a more recent work, Ghezzi [55] has considered a model where the star is isotropic in its interior but it has an anisotropic thin shell. This model share many similar features with the gravastars. The hydrodynamic stability reveals that there is a maximum mass, implying that the existence of black holes is unavoidable. Moreover, he has shown that exists a critical coupling parameter (the rate between the dark energy density and the total energy density), which separates dark energy stars from regular black hole solutions.

A generalization of a gravastar model, adopting a physical dark energy fluid in its interior, was considered by us [56]. That model consists of an internal phantom fluid, a dynamical infinitely thin shell of perfect fluid with the equation of state $p = (1 - \gamma)\sigma$, and an external Schwarzschild spacetime. We have studied the problem of the stability of gravastars and the final output can be a black hole, a "bounded excursion" stable gravastar, a Minkowski, or a phantom spacetime, depending on the total mass m of the system, the parameter ω, the constant b (that is proportional to the central energy density), the parameter γ and the initial position of the dynamical shell. An interesting result is that we can have black hole formation even with an interior phantom energy for any given γ.

We have also proposed another alternative model of a dark energy star, like a gravastar model [39]. In that model the mass function is a natural consequence of the Einstein's field equations and the energy density as well as the pressure decreases with the radial coordinate (envelope), as expected for known stellar models. In order to eliminate the central singularity present in this model, we have considered a core with a homogeneous energy density, described by the Lobo's [20] first solution. We could conclude that we always have the presence of the dark energy, at least, in one of the thin shell or in the core. Besides, there is no physical reason to have a superior limit for the mass of these objects, but for the ratio of mass and radius, in order to find out which one is made of dark energy.

4. Conclusion

Our purpose here is to present a brief review of the main results, found until now, on the structure formation processes, such as stars, gravastars, black holes and even naked singularities, considering the dark energy effects. Among them, gravastars have received a special attention since they were proposed as an alternative to another open problem, which is the proper existence of black holes.

Gravastars were proposed as an alternative model to black holes by MM. They argued in favor of the thermodynamic stability of these compact objects. In order to analyze the dynamical stability, VW simplified the five layer gravastar model of MM to three layer one. Later, Lobo [20] generalized the idea of gravastar replacing the de Sitter's interior by an anisotropic dark energy fluid. Prescribing first an equation of state, using Israel conditions and finding a potential to be analyzed, we have shown explicitly that the final output of the collapse of a three layer gravastar's prototype can be a black hole, a "bounded excursion" stable gravastar, a stable gravastar, or a de Sitter spacetime, depending on the parameters of the model considered and the initial position of the dynamical shell in all our works. All these possibilities have non-zero measurements in the space of the parameters considered in each work, although the region of gravastars is very small in comparison with that of black holes.

All the results shown in this review forecloses gravastars from being a serious alternative to the black hole model although they cannot exclude gravastar for being a real astrophysical object. It is worth saying, in order to to conclude this work, that there are not yet any known experimental result that can exclude gravastars or black holes from being the final state of the stellar collapse and only much more work on the subject (theoretical and experimental) will make this clearer to us. Maybe it will only be achieved with a full comprehension of a quantum theory of gravity.

Although the gravastars have been created based on a phase transitions of quantum vacuum fluctuations, without any connection with the cosmological dark energy, their generalization to other kind of dark energy was unavoidable, admitting the possibility of the dark energy object formation.

We cannot exclude the hypothesis that our universe is not actually accelerated, but this might be an effect of wrong suppositions. Thus, as an alternative to admitting the existence of dark energy, a review of the postulates which imply in its introduction[57] has been done. In particular, some researchers are proposing a violation of one of the two fundamental principles in cosmology, i.e., that the Universe is neither homogeneous nor isotropic. According to them, explanations to type Ia supernovae and the WMAP (Wilkinson Microwave Anisotropy Probe) data can be made if our local environment were emptier than the surrounding Universe. It means that we could be living in a void, rejecting the Copernican principle and it could explain data observed without the necessity of the existence of exotic substances, such as dark energy, extra dimensions and modifications to gravity. But, as pointed out by a recent work, in order to the void model to be taken seriously, several key issues have to be addressed[58]. Only in coming years more observational studies will make possible to experimentally distinguish among these scenarios, excluding or not the dark energy as a necessary ingredient in the Universe.

Furthermore, there are some researchers who work with other alternative theories, modifying the Einstein's theory of General Relativity. There are numerous ways to deviate from General Relativity such as scalar-tensor theory, DGP gravity, brane-world gravity, $f(R)$ theories of gravity and many others[59]. In all these theories it is very important the issue of gravitational collapse and the types of astrophysical objects that can be found when considering each one of them.

References

[1] R. M. Wald, " The Thermodynamics of Black Holes", *Living Rev. Rel.* **4**, 6 (2001) [arXiv:gr-qc/9912119].

[2] M.A. Abramowicz, W. Kluźniak and J.-P. Lasota, " No observational proof of the black-hole event-horizon", *Astron. & Astrophys,* **396**, L31 (2002).

[3] G. Chapline, *"Dark Energy Stars and AdS/CFT"*, Talk at the 12th Marcel Grossman Meeting, (2009) [arXiv:gr-qc/0907.4397].

[4] P.O. Mazur and E. Mottola, *"Gravitational Condensate Stars: An Alternative to Black Holes"*, (2001) [arXiv:gr-qc/0109035].

[5] P.O. Mazur and E. Mottola, *Proc. Nat. Acad. Sci.* **101**, 9545 (2004) [arXiv:gr-qc/0407075].

[6] P. O. Mazur and E. Mottola, *"Dark energy and condensate stars: Casimir energy in the large"*, (2004) [arXiv:gr-qc/0405111].

[7] P. O. Mazur and E. Mottola, " Gravitational Vacuum Condensate Stars", *Proc. Nat. Acad. Sci.* **111**, 9545 (2004) [arXiv:gr-qc/0407075].

[8] G. Chapline, E. Hohlfeld, R.B. Laughlin and D.I. Santiago, " Quantum Phase Transitions and the Breakdown of Classical General Relativity", *Int. J. Mod. Phys.* A**18**, 3587 (2003) [arXiv:gr-qc/0012094].

[9] T. Vachaspati, *"Black Stars and Gamma Ray Bursts"*, (2007) [arXiv:gr-qc/07061203].

[10] M. Petri, *"Charged holostars"*, (2003) [arXiv:gr-qc/0306068].

[11] M. Petri, *"Compact anisotropic stars with membrane - a new class of exact solutions to the Einstein field equations"*, (2003) [arXiv:gr-qc/0306063].

[12] M. Petri, *"The holographic solution - Why general relativity must be understood in terms of strings"*, (2004) [arXiv:gr-qc/0405007].

[13] M. Petri, *"Holostar thermodynamics"*, (2003) [arXiv:gr-qc/0306067].

[14] M. Petri, *"Are we living in a string dominated universe?"*, (2004) [arXiv:gr-qc/0405011].

[15] M. Petri, *"On the origin of the matter-antimatter asymmetry in self-gravitating systems at ultra-high temperatures"*, (2004) [arXiv:gr-qc/0405010].

[16] G. Chapline, *"Dark Energy Stars"*, Proceedings of the Texas Conference on Relativistic Astrophysics, Stanford, CA (2004) [arXiv:astro-ph/0503200].

[17] D. Horvat and S. Ilijic, *"Gravastar energy conditions revisited"*, (2007) [arXiv:igr-qc/07071636].

[18] P. Marecki, *"On quantum effects in the vicinity of would-be horizons"*, (2006) [arXiv:gr-qc/0612178].

[19] F.S.N. Lobo, " Stable dark energy stars: An alternative to black holes?", *Phys. Rev.* D**75**, 024023 (2007) [arXiv:gr-qc/0612030].

[20] F.S.N. Lobo, " Stable dark energy stars", *Class. Quant. Grav.* **23**, 1525 (2006).

[21] F.S.N. Lobo and A.V.B. Arellano, " Gravastars supported by nonlinear electrodynamics", *Class. Quant. Grav.* **24**, 1069 (2007) [arXiv:gr-qc/0611083].

[22] T. Faber, *"Galactic halos and gravastars: static spherically symmetric spacetimes in modern general relativity and astrophysics"*, (2006) [arXiv:gr-qc/0607029].

[23] C. Cattoen, *"Cosmological milestones and gravastars - topics in general relativity"*, (2006) [arXiv:gr-qc/0606011].

[24] O.B. Zaslavskii, " N-spheres in general relativity: Regular black holes without apparent horizons, static wormholes with event horizons and gravastars with a tube-like core", *Phys. Lett.* B**634**, 111 (2006).

[25] C. Cattoen, T. Faber, and M. Visser, " Gravastars must have anisotropic pressures", *Class. Quant. Grav.* **22**, 4189 (2005).

[26] I. Dymnikova, " Vacuum nonsingular black hole", *Gen. Rel. Grav.* **24**, 235 (1992).

[27] I. Dymnikova, " The algebraic structure of a cosmological term in spherically symmetric solutions", *Phys. Lett.* **B472**, 33 (2000) [arXiv:gr-qc/9912116].

[28] I. Dymnikova, " Cosmological term as a source of mass", *Class. Quant. Grav.* **19** 725 (2002) [arXiv:gr-qc/0112052].

[29] I. Dymnikova, " Spherically symmetric space-time with the regular de Sitter center", *Int. J. Mod. Phys.* **D 12**, 1015 (2003) [arXiv:gr-qc/0304110].

[30] I. Dymnikova and E. Galaktionov, " Stability of a vacuum nonsingular black hole", *Class. Quant. Grav.* **22**, 2331 (2005) [arXiv:gr-qc/0409049].

[31] M. Visser and D.L. Wiltshire, " Stable gravastars - an alternative to black holes?", *Class. Quant. Grav.* **21**, 1135 (2004) [arXiv:gr-qc/0310107].

[32] B.M.N. Carter, *"Stable gravastars with generalised exteriors"*, *Class. Quant. Grav.* **22**, 4551 (2005) [arXiv:gr-qc/0509087].

[33] A. DeBenedictis, D. Horvat, S. Ilijic, S. Kloster and K. S. Viswanathan, " Gravastar Solutions with Continuous Pressures and Equation of State", *Class. Quant. Grav.* **23**, 2303 (2006) [arXiv:gr-qc/0511097].

[34] C.B.M.H. Chirenti and L. Rezzolla, *"How to tell a gravastar from a black hole"*, (2007) [arXiv:igr-qc/07061513].

[35] I. Dymnikova and E. Galaktionov, " Vacuum Dark Fluid", *Physics Letters B* **645**, 358 (2007).

[36] E.J. Copeland, M. Sami and S. Tsujikawa, " Dynamics of Dark Energy", *Int. J. Mod. Phys.* **D15**, 1753 (2006);

[37] T. Padmanabhan, *"Dark Energy and Gravity"*, (2007) [arXiv:gr-qc/07052533].

[38] O. Bertolami and J. Páramos, " The Chaplygin dark star", *Phys. Rev. D* **72**, 123512 (2005) [arXiv:astro-ph/0509547].

[39] R. Chan, M.F.A. da Silva and J.F. Villas da Rocha, " Star Models with Dark Energy, *Gen. Rel. Grav.* **41**, 1835 (2009) [arXiv:gr-qc/08033064].

[40] R. Chan, M.F.A. da Silva and J.F. Villas da Rocha, " On Anisotropic Dark Energy", *Mod. Phys. Lett. A* **24**, 1137 (2009) [arXiv:gr-qc/08032508].

[41] P. Rocha, A.Y. Miguelote, R. Chan, M.F.A. da Silva, N.O. Santos and Anzhong Wang, " Bounded excursion stable gravastars and black holes", *J. Cosmol. Astropart. Phys.* **6**, 25 (2008) [arXiv:gr-qc/08034200].

[42] P. Rocha, R. Chan, M.F.A. da Silva and Anzhong Wang, " Stable and "bounded excursion" gravastars, and black holes in Einstein's theory of gravity", *J. Cosmol. Astropart. Phys.* **11**, 10 (2008) [arXiv:gr-qc/08094879].

[43] R. Chan, M.F.A. da Silva and Jaime F. Villas da Rocha, *"Massive Non-gravitational Object from Gravastars"*, (2010) [arXiv:gr-qc/10042906]

[44] R. Chan and M.F.A. da Silva, *"How the Charge Can Affect the Formation of Gravastars"*, (2010) [arXiv:gr-qc/10053703]

[45] V. Cardoso, P. Pani, M. Cadoni and M. Cavaglia, " Ergoregion instability of ultracompact astrophysical objects", *Phys. Rev. D* **77**, 124044 (2008).

[46] C.B.M.H. Chirenti and L. Rezzolla, " On the ergoregion instability in rotating gravastars", *Phys.Rev.D* **78**, 084011 (2008) [arXiv:gr-qc/80804080].

[47] T. Harko, Z. Kovács and F.S.N. Lobo, " Can accretion disk properties distinguish gravastars from black holes?", *Class. Quant. Grav.* **26**, 215006 (2009) [arXiv:gr-qc/09051355].

[48] R. Chan, M.F.A. da Silva and P. Rocha, " How the Cosmological Constant Affects the Gravastar Formation", *J. Cosmol. Astropart. Phys.* **12**, 17 (2009) [arXiv:gr-qc/09102054].

[49] E. Babichev, V. Dokuchaev, and Y. Eroshenko, " Black Hole Mass Decreasing due to Phantom Energy Accretion", *Phys. Rev. Lett.* **93**, 021102 (2004).

[50] S. Nojiri and S.D. Odintsov, " The final state and thermodynamics of dark energy universe", *Phys. Rev.* **D70**, 103522 (2004) [arXiv:hep-th/0408170].

[51] Z.-H. Li and Anzhong Wang, " Existence of black holes in Friedmann-Robertson-Walker universe dominated by dark energy", *Mod. Phys. Lett.* **A22**, 1663 (2007) [arXiv:astro-ph/0607554].

[52] U. Debnath and S. Chakraborty, *"Role of Modified Chaplygin Gas as a Dark Energy Model in Collapsing Spherically Symmetric Cloud*, (2006) [arXiv:gr-qc/0601049].

[53] S. Chakraborty and T. Bandyopadhyay, *"Collapse Dynamics of a Star of Dark Matter and Dark Energy"*, (2006) [arXiv:gr-qc/0609038].

[54] T. Bandyopadhyay and S. Chakraborty, " Quasi-Spherical Star of Dark Matter and Dark Energy and a Study of its Collapse Dynamics", *Mod. Phys. Lett. A,* **22**, 2839 (2007).

[55] C.R. Ghezzi, *"Anisotropic dark energy stars"*, (2009) [arXiv:gr-qc/09080779].

[56] R. Chan, M.F.A. da Silva, P. Rocha, and Anzhong Wang, " Stable Gravastars of Anisotropic Dark Energy", *J. Cosmol. Astropart. Phys.* **3**, 10 (2009) [arXiv:gr-qc/08124924].

[57] T. Clifton, P. G. Ferreira, and K. Land, " Living in a void: Testing the Copernican Principle with Distant Supernovae", *Phys. Rev. Lett.* **101**, 131302 (2008).

[58] S. Alexander, T. Biswas, A. Notari, and D. Vaid, " Local void vs dark energy: confrontation with WMAP and type Ia supernovae", *J. Cosmol. Astropart. Phys.* **9**, 25 (2009) [arXiv:gr-qc/07120370].

[59] T. Sotiriou, and V. Faraoni, "$f(R)$ *theories of gravity*", (2008) [arXiv:gr-qc/08051726v3]

In: Dark Energy: Theories, Developments and Implications   ISBN 978-1-61668-271-2
Editors: K. Lefebvre and R. Garcia, pp. 33-48   © 2010 Nova Science Publishers, Inc.

Chapter 3

The Dark Energy Scale in Superconductors: Innovative Theoretical and Experimental Concepts

Christian Beck[1],*and Clovis Jacinto de Matos*[2],†
[1] School of Mathematical Sciences, Queen
Mary, University of London, London E1 4NS, UK,
[2]ESA-HQ, European Space Agency, 8-10 rue Mario Nikis,
75015 Paris, France

Abstract

We revisit the cosmological constant problem using the viewpoint that the observed value of dark energy density in the universe actually represents a rather natural value arising as the geometric mean of two vacuum energy densities, one being extremely large and the other one being extremely small. The corresponding mean energy scale is the Planck Einstein scale $l_{PE} = \sqrt{l_{Pl}l_E} = (\hbar G/c^3 \Lambda)^{1/4} \sim 0.037mm$, a natural scale both for dark energy and the physics of superconductors. We deal with the statistics of quantum fluctuations underlying dark energy in superconductors and consider a scale transformation from the Planck scale to the Planck-Einstein scale which leaves the quantum physics invariant. Our approach unifies various experimentally confirmed or conjectured effects in superconductors into a common framework: Cut-off of vacuum fluctuation spectra, formation of Tao balls, anomalous gravitomagnetic fields, non-classical inertia, and time uncertainties in radioactive superconductors. We propose several new experiments which may further elucidate the role of the Planck-Einstein scale in superconductors.

1. Introduction

Observational cosmology of the last decade has revealed a composition of the universe that poses one of the greatest challenges theoretical physics has ever faced: We are living in

---

*E-mail address: c.beck@qmul.ac.uk
†E-mail address: Clovis.de.Matos@esa.int

a universe that exhibits accelerating expansion in (approximately) four space-time dimensions, with a de Sitter spacetime metric with cosmological constant $\Lambda = 1.29 \times 10^{-52} [m^{-2}]$ [1, 2, 3, 4, 5]. A small cosmological constant is equivalent to a small vacuum energy density (dark energy density with equation of state $w = -1$) given by

$$\rho_{vac\Lambda} = \frac{c^4 \Lambda}{8\pi G} = 6.21 \times 10^{-10} [J/m^3]. \tag{1}$$

In quantum field theory, the cosmological constant counts the degrees of freedom of the vacuum. Heuristically, we may sum the zero-point energies of harmonic oscillators and write:

$$E_{vac} = \sum_i \left(\frac{1}{2} \hbar \omega_i \right) \tag{2}$$

The sum is manifestly divergent. The natural maximum cutoff to impose on a quantum theory of gravity is the Planck frequency, leading to the Planck energy density of the vacuum: $\rho_P = E_P / l_P^3 = \frac{c^7}{G^2 \hbar} = 4.6 \times 10^{113} [J/m^3]$. This prescription yields an ultraviolet enumeration of the zero-point energy that is 122 orders of magnitude above the measured value, eq.(1). This is the well-known cosmological constant problem. There is also a natural minimum cutoff that we can impose on the sum eq.(2), which is the minimum energy, $E_E = \sqrt{c^2 \hbar^2 \Lambda}$ associated with the cosmological length scale (or Einstein length scale), $l_E := \Lambda^{-1/2}$. This leads to the Einstein energy density for the vacuum, $\rho_E = c\hbar\Lambda^2 = 5.26 \times 10^{-130} [J/m^3]$, which is 121 orders of magnitude below the measured value of the dark energy density, eq.(1). Apparently, the correct value of vacuum energy density in the universe as seen from WMAP [1] observations is approximately the geometric mean of both values (see also [6, 7, 8] for related work).

We may therefore ask: What is the domain of physics and phenomenology that provides a natural scale for dark energy? In fact, the observed dark energy scale is very similar to typical energy scales that occur in solid state physics in the physics of superconductors. The principal idea discussed in recent papers [9, 10, 11] is that this is not a random coincidence, but that there is a deep physical reason for this coming from a similar structure of the theory of superconductors and that of dark energy. A possible model in this direction is the Ginzburg-Landau theory of dark energy as developed in [10]. In this model vacuum fluctuations exhibit a phase transition from a gravitationally active to a gravitationally inactive state at a critical frequency, similar to the superconductive phase transition at a critical temperature. On the experimental side, there is an interesting possibility arising out of this approach, namely that superconductors could be used as suitable detectors for quantum fluctuations underlying dark energy [10, 11]. In the following we will further work out this concept. In particular, we will deal with uncertainty relations for vacuum energy densities and space-time volumes, and deal with the corresponding fluctuation statistics in superconductors. We will suggest several new laboratory experiments which could be performed to test the theoretical concepts.

We will re-investigate the cosmological constant problem by considering an uncertainty relation between vacuum energy density and the four-dimensional volume of the universe. We start from the Einstein-Hilbert action and show that in this approach one actually obtains an 'inverse' cosmological constant problem: The cosmological constant comes out *too small* by 120 orders of magnitude! This naturally suggests to regard the observed dark energy density of the universe as the geometric mean of two values of vacuum energy, one being 120 orders of magnitude too large and the other one being 120 orders of magnitude too small. The corresponding mean energy scale is the Planck-Einstein scale, corresponding to lengths of about 0.037mm, a natural scale both for dark energy and superconductive materials.

Quite interesting phenomena may arise out of the fact that the relevant length scale for quantum fluctuations is the Planck-Einstein scale in superconductive materials. Basically, our model for quantum fluctuations in the superconductor is like a model of quantum gravity where the Planck mass $m_P = (\hbar c/G)^{1/2}$ is replaced by a much smaller value, the Planck-Einstein mass $m_{PE} = (\hbar^3 \Lambda/cG)^{1/4}$. Formally replacing the Planck mass by much smaller values has also been discussed in the context of extra dimensions [12]. Our approach here does not require extra dimensions but just a superconducting environment. We will analyse the observed formation of large sperical clusters of superconductive particles, so-called Tao balls [13, 14, 15, 16] in this context, a phenomenon that is so far unexplained in the usual theory of superconductors, but which can be understood using our current approach. We will also suggest to measure the formal fluctuations of space-time in the superconductor by looking at the lifetime of radioactive superconductors, as well as by comparing two clocks, one located inside the superconductive cavity, and the other being located outside.

This paper is organized as follows. In section 2 we sketch the inverse cosmological problem and describe how to obtain a natural scale of dark energy, the Planck-Einstein scale. Properties of this scale are summarized in section 3. In section 4 we deal with scale transformations from the Planck scale (relevant in a non-superconducting environment) to the Planck-Einstein scale (relevant in a superconducting environment). We discuss several interesting phenomena that may produce measurable effects in this context: Cutoffs of quantum noise spectra, formation of Tao balls, fundamental time uncertainties in radioactive superconductors and non-classical inertia. A more detailed model for the formation of Tao balls is discussed in section 5. Finally, in section 6 we list some suggestions for future experiments that could further clarify the role of dark energy and gravity in superconductors.

2.   Inverse Cosmological Constant Problem and the Uncertainty Principle

The cosmological constant problem is the problem that the typical value of vacuum energy density predicted by quantum field theory is *too large* by a factor of 10^{120} as compared to astronomical observations. Here we show that using a different argument, one can actually get a value that is *too small* by a factor of order 10^{-120}. Hence the observed value of the cosmological constant in nature seems not that unnatural at all, being the geometric mean of both values.

We start from the Einstein Hilbert action

$$S = \int [k(R - 2\Lambda) + L_M]\sqrt{-g}d^4x, \tag{3}$$

where R is the Ricci scalar, g is the determinant of the space-time Lorentz metric, L_M is the matter Lagrangian, and the constant k is

$$k = \frac{c^4}{16\pi G},\tag{4}$$

where G is the gravitational constant. Einstein's field equations are invariant under complex transformations of the space-time coordinates $x^\mu \to iy^\mu$ except for the cosmological constant term [17], which is the only relevant term for our approach in the following. The part S_Λ of the action corresponding to the cosmological constant Λ can be written as a product between the vacuum energy density $\rho_{vac\Lambda}[J/m^3]$ and the four-dimensional volume $V[m^4]$

$$S_\Lambda = -\rho_{vac\Lambda}V,\tag{5}$$

where the four-dimensional volume is expressed in its covariant form as

$$V = \int d^4x\sqrt{-g},\tag{6}$$

and the vacuum energy density is given by

$$\rho_{vac\Lambda} = \frac{c^4\Lambda}{8\pi G}.\tag{7}$$

We may now regard $\rho_{vac\Lambda}$ and V occuring in eq. (5) as canonically conjugated quantities, as previously suggested in [18, 19]. In a quantum theory of gravity, we expect that the fluctuations in one observable are related to fluctuations in its conjugate, according to Heisenberg's uncertainty relation. Thus

$$\Delta\rho_{vac\Lambda}\Delta V \sim \hbar c.\tag{8}$$

This resembles a kind of uncertainty relation in 4-dimensional spacetime. Substituting eq.(7) into eq.(8), we can write eq.(8) as

$$\Delta\Lambda\Delta V \sim 8\pi l_P^2\tag{9}$$

where $l_P = \sqrt{G\hbar/c^3} = 1.61 \times 10^{-35}[m]$ is the Planck length.

Let us now assume that the universe has a finite lifetime τ. τ is bounded from below by the current age of the universe. From the measured Hubble constant, $H_0 = 2.3 \times 10^{-18}[s^{-1}]$, which is of the order of the inverse age of the universe, the four-volume of the universe can be estimated:

$$\Delta V \sim V \sim \frac{4}{3}\pi(c\tau)^4 \sim \frac{4}{3}\pi\left(\frac{c}{H_0}\right)^4 \sim 1.2 \times 10^{105}[m^4]\tag{10}$$

Substituting eq.(10) into eq.(9) a typical order of magnitude of the cosmological constant can be computed, regarding the present value as a quantum fluctuation:

$$\Delta\Lambda \sim \Lambda = 5.44 \times 10^{-174}[1/m^2]\tag{11}$$

This value is in total disagreement with the experimental results, being 121 orders of magnitude *below* the value measured by WMAP:

$$\Lambda = 1.29 \times 10^{-52}[1/m^2] \tag{12}$$

Apparently we get by this formal approach a different cosmological constant problem, which we may call the *inverse* (or infrared) cosmological problem. The typical order of magnitude of the cosmological constant, as derived from this formal approach, turns out to be 120 orders of magnitude *too small* as compared to the astronomical observations. If the universe lives much longer, the estimated value of the typical order of magnitude of Λ would even further decrease.

Our conclusion is that the observed value of the cosmological constant is not so unnatural after all. It's just given by the geometric mean of both approaches, the one starting from zeropoint energies in quantum field theories and the one starting from an uncertainty relation for S_Λ. Full symmetry between both approaches is obtained if $\tau \sim H_0^{-1}$.

Let us now try to reconcile both approaches. Following Sorkin's work [20, 21] in causal set theory, fluctuations in Λ are inversely related to fluctuations in V. The fluctuations of relevance to us are in the number n_{cells} of Planck sized cells that fill up the four-dimensional spacetime of the universe:

$$n_{cells} \sim \frac{V}{l_P^4} \Rightarrow \Delta n_{cells} \sim \sqrt{n_{cells}} \Rightarrow \Delta V \sim \sqrt{V} l_P^2, \tag{13}$$

Substituting eq.(13) into eq.(9), we obtain:

$$\Delta \Lambda \sqrt{V} \sim 8\pi \tag{14}$$

Substituting the value of the four-volume of the universe of eq.(10) into eq.(14) we find a value of the cosmological constant in agreement with the experimentally measured value eq.(12):

$$\Delta \Lambda \sim \Lambda \sim 10^{-52}[1/m^2] \tag{15}$$

In summary we see that the quantization of the universe's spacetime volume with Planck sized four-dimensional cells can solve the cosmological constant problem if we interpret the value of cosmological constant as being due to statistical fluctuations of the total number of cells making up this volume, according to the uncertainty principle eq.(9).

3.   The Planck-Einstein Scale

The Planck-Einstein scale corresponds to the geometric mean value between the Planck scale, l_P, which determines the highest possible energy density in the universe, and the cosmological length scale, or Einstein scale, $l_E = \Lambda^{-1/2}$, which determines the lowest possible energy in the universe [22, 23]. The Planck-Einstein energy density is the geometric mean $\rho_{PE} = \sqrt{\rho_P \rho_E}$ between the two energy densities, and the Planck-Einstein length $l_{PE} = \sqrt{l_P l_E}$ is the geometric mean of the two length scales in the universe [6, 7, 8]. In the following table we list side by side the relevant quantities and their numerical values:

	Einstein scale	Planck-Einstein Scale	Planck scale
	Λ, $\hbar$, c, k	Λ, $\hbar$, c, k G	c, $\hbar$, k, G
Temperature [K]	$T_E = \frac{1}{k}\sqrt{c^2\hbar^2\Lambda}$	$T_{PE} = \sqrt{T_E T_P}$	$T_P = \frac{1}{k}\sqrt{\frac{\hbar c^5}{G}}$
	2.95×10^{-55}	60.71	1.42×10^{32}
Time [s]	$t_E = \sqrt{\frac{1}{c^2\Lambda}}$	$t_{PE} = \sqrt{t_E t_P}$	$t_P = \sqrt{\frac{\hbar G}{c^5}}$
	2.58×10^{43}	1.26×10^{-13}	5.38×10^{-44}
Length [m]	$l_E = \sqrt{\frac{1}{\Lambda}}$	$l_{PE} = \sqrt{l_E l_P}$	$l_P = \sqrt{\frac{\hbar G}{c^3}}$
	8.8×10^{25}	3.77×10^{-5}	1.61×10^{-35}
Mass [Kg]	$m_E = \sqrt{\frac{\hbar^2\Lambda}{c^2}}$	$m_{PE} = \sqrt{M_E M_P}$	$m_P = \sqrt{\frac{\hbar c}{G}}$
	5.53×10^{-95}	9.32×10^{-39}	2.17×10^{-8}
Energy [J]	$E_E = \sqrt{c^2\hbar^2\Lambda}$	$E_{PE} = \sqrt{E_E E_P}$	$E_P = \sqrt{\frac{\hbar c^5}{G}}$
	4.07×10^{-78}	8.38×10^{-22}	1.96×10^9
Energy density [J/m^3]	$\rho_E = \sqrt{c^2\hbar^2\Lambda^4}$	$\rho_{PE} = \sqrt{\rho_E \rho_P}$	$\rho_P = \sqrt{\frac{c^{14}}{G^4\hbar^2}}$
	5.26×10^{-130}	3.73×10^{-9}	4.6×10^{113}

Explicitly one has the following formulas at the Planck-Einstein scale:

$$E_{PE} = kT_{PE} = \left(\frac{c^7\hbar^3\Lambda}{G}\right)^{1/4} = 5.25[meV] \tag{16}$$

$$m_{PE} = \frac{E_{PE}}{c^2} = \left(\frac{\hbar^3\Lambda}{cG}\right)^{1/4} = 9.32 \times 10^{-39}[Kg] \tag{17}$$

$$l_{PE} = \frac{\hbar}{M_{PE}c} = \left(\frac{\hbar G}{c^3\Lambda}\right)^{1/4} = 0.037[mm] \tag{18}$$

$$t_{PE} = \frac{l_{PE}}{c} = \left(\frac{\hbar G}{c^7\Lambda}\right)^{1/4} = 1.26 \times 10^{-13}[s] \tag{19}$$

$$\rho_{PE} = \frac{E_{PE}}{l_{PE}^3} = \frac{c^4\Lambda}{G} = 104[eV/mm^3] \tag{20}$$

One readily notices that the numerical values of Planck-Einstein quantities correspond to typical time, length or energy scales in superconductor physics, as well as to typical energy scales for dark energy. In previous papers it has been pointed out [9, 10, 11] that there could be a deeper reason for this coincidence: It is possible to construct theories of dark energy that bear striking similarities with the physics of superconductors. In these theories the Planck-Einstein scale replaces the Planck scale as a suitable cutoff for vacuum fluctuations.

4. Scale Transformation in Superconductors

Our main hypothesis in this paper, to be worked out in the following, is that when proceeding from a normal to a superconducting environment it makes sense to consider a scale transformation from the Planck length to the Planck-Einstein length, which keeps many features of the quantum physics invariant. We will give many examples below. The scale transformation may induce new interesting observable phenomena in superconductors. Quantum gravity phenomena that normally happen at the Planck scale l_P only could possibly induce related phenomena at the Planck-Einstein scale in a superconducting environment. Due to our scale transformation, the gravitational constant $G = \hbar c / m_P^2$ formally becomes much stronger in a superconductor if m_P is replaced by the much smaller value m_{PE}. Similarly, frame dragging effects and gravitomagnetic fields could become much stronger as well, in line with recent experimental observations [24, 25, 26]. The vacuum energy density of vacuum fluctuations would become much smaller as well (of the order of dark energy density). In the following subsections we will investigate the consequences of our scale transformation hypothesis and show that the hypothesis is consistent with some recent experimental observations for superconducting materials. Moreover, we will predict some new phenomena as a consequence of the scale transformation that could be experimentally tested.

4.1. Cutoff for Vacuum Fluctuations in Superconductors

As mentioned before, in quantum field theories the natural cutoff frequency ω_c for vacuum fluctuations is given by $\hbar \omega_c \sim m_P c^2$. This leads to the cosmological constant problem, since the corresponding vacuum energy obtained by integrating over all frequencies up to ω_c is much too large. In Josephson junctions (two superconductors separated by a thin insulator) vacuum fluctuations of the electromagnetic field can lead to measurable noise spectra [27, 28]. This measurability is due to the Josephson effect and the fluctuation dissipation theorem (details in [9, 29]). However, the maximum Josephson frequency that can be reached with a given superconductor (and thus the cutoff frequency of measurable vacuum fluctuations) is determined by the gap energy of the superconductor. This gap energy of a superconductor is proportional to kT_c, where T_c is the critical temperature of the superconductor under consideration (in the BCS theory, the proportionality factor is given by 3.5). Thus measurable noise spectra induced by vacuum fluctuations can only be measured in superconducting Josephson junctions up to a critical value of the order $\hbar \omega_c \sim kT_c$.

Re-interpreted in terms of our scale transformation, this means that for superconductors the Planck scale m_P as a cutoff for vacuum fluctuations is formally replaced by something of the order of the Planck-Einstein scale, since the critical temperaure $T_c = 1...140K$ of normal and high-T_c superconductors is of the same order of magnitude as the Planck-Einstein temperature $T_{PE} = 60.7K$. Hence our scale transformation hypothesis makes sense for vacuum fluctuations and vacuum energy as observed in a superconducting environment. The relevant scale factor is of the order $l_{PE}/l_P \sim 10^{30}$.

4.2. Formation of Tao Balls

When a strong electric field is applied to a mixture of superconducting and non-superconducting particles, a remarkable effect is observed [13, 14, 15, 16]: Millions of superconducting micropartices of μm size spontaneously aggregate into spherical balls of mm size. The normal particles in the mixture do not show this behavior, only the superconducting ones. The effect has not been explained within the conventional theory of superconductors so far. In fact, within the conventional theory of superconductors one expects that normal particles respond to electrostatic fields in just the same way as superconducting ones do. Hence the Tao effect represents an unsolved puzzle: Superconducting and non-superconducting matter behave in a fundamentally different way. Assuming that the superconducting and non-superconducting particles differ in no other way, one may even regard this effect as pointing towards a violation of the equivalence principle, or more generally the principle of general covariance.

In a sense the formation of Tao balls reminds us of a kind of 'planet formation' on a scale that is much smaller than the solar system, which is possible for superconducting matter only. When working out this analogy, again relevant scale factors of the order $10^{30} \sim l_{PE}/l_P \sim l_E/l_{PE}$ arise: Tao balls have a size of order $10^{-3}m$, whereas typical planets such as the earth have a size of order $l_E \sim 10^7 m$. Hence the volume of a Tao ball is smaller than the volume of a typical planet by a factor 10^{-30}, and so is the mass of the Tao ball. The possible role of gravitational forces in the formation process of Tao balls is further discussed in section 5.

4.3. Fundamental Space-Time Uncertainty in a Radioactive Superconductor

In the following we predict a new effect for radioactive superconductors, which arises out of the scale transformation and which could possibly be confirmed in future experiments. We start from an effective Planck length comparable to the Planck-Einstein length in superconductors. We are lead to envisage that the spacetime volume of a superconductor is made of Planck-Einstein sized cells, l_{PE}^4, which will statistically fluctuate according to eq.(13):

$$\Delta V \sim \sqrt{V} l_{PE}^2 \qquad (21)$$

What should we now take for the space-time volume of a superconductor? The problem is well-defined if we consider a superconducting material with a finite life time, a radioactive superconductor. Let v denote the volume of the superconducting material and τ the mean life time of the radioactive material. We then choose the 4-volume as

$$V = vc\tau. \qquad (22)$$

This is similar to the approach in section 2, where we considered a universe with a finite life time τ of order H_0^{-1}. Since the 3-volume v is fixed, for a superconductor in the laboratory there can only be a time uncertainty Δt given by

$$\Delta V = vc\Delta t. \qquad (23)$$

Putting eq. (22) and (23) into (21) we obtain an equation for the order of magnitude of a fundamental time uncertainty in radioactive superconductors:

$$\Delta t \sim \sqrt{\frac{\tau}{cv}} l_{PE}^2 \tag{24}$$

To estimate some numbers, let us consider the metastable state of a Nb^{90m} superconductor with a mean life time of $\tau = 34.6[s]$. The volume of the superconductor in Tate's experiment [30] (just as an example) is $v = 1.28 \times 10^{-13}[m^3]$. From this we get

$$\Delta t \sim 1.3 \times 10^{-6}[s]. \tag{25}$$

Fundamental time uncertainties of the above kind should create a broadening of the decay energy line width $\Gamma = \hbar/\tau$:

$$\frac{\Delta t}{\tau} \sim \frac{\Delta \Gamma}{\Gamma}. \tag{26}$$

For the above example we get

$$\frac{\Delta \Gamma}{\Gamma} \sim 10^{-8} \tag{27}$$

which is challenging to measure. The smallness of the above number might explain why this effect has not been revealed by the experiments of Mazaki [31] on the search for a superconducting effect on the decay of Technetium$-99m$. However the possibility of time fluctuations in radioactive superconductors offers a new perspective to interpret the positive results from Olin [32] on the influence of superconductivity on the lifetime of Niobium$-90m$.

4.4. Uncertainty Principle and Non-classical Inertia in Superconductors

Tajmar et al. [24, 25, 26] have measured anomalous acceleration signals around isolated accelerated superconductors, as well as anomalous gyroscope signals around constantly rotating superconductors. These signals can be interpreted in terms of an anomalous gravitomagnetic field that is about 30 orders of magnitude larger than expected from normal gravity. We note again that $l_{PE}/l_P \sim l_E/l_{PE} \sim 10^{30}$, thus the effect could again stand in relation to a scale transformation in superconductors.

A recent paper [11] connects the anomalous gravitomagnetic fields and non-classical inertial properties of superconductive cavities [33, 34] with the electromagnetic model of dark energy of Beck and Mackey [10]. In this approach the vacuum energy stored in a given superconductor is given by

$$\rho_{vac} = \frac{\pi (\ln 3)^4}{2} \frac{k^4}{(ch)^3} T_c^4 \tag{28}$$

One defines an dimensionless parameter χ by

$$\chi = \frac{B_g}{\omega} = -2\frac{g}{a} \tag{29}$$

Here B_g is the gravitomagnetic field created by a rotating superconductor, ω the angular velocity of the rotating superconductor, g is the acceleration measured inside the superconductive cavity and a the acceleration communicated to the superconductive cavity [35, 36]:

For a cavity made of normal matter $\chi = 2$, which means that the gravitational Larmor theorem, $B_g = 2\omega$ [37], and the principle of general covariance, $g = -a$, are verified. For a superconductive cavity χ turns out to be a function of the ratio between the electromagnetic vacuum energy density contained in the superconductor, ρ_{vac} as given by eq.(28), and the cosmological vacuum energy density, $\rho_{vac\Lambda}$ as given by eq.(7):

$$\chi = \frac{3}{2} \frac{\rho_{vac}}{\rho_{vac\Lambda}} \tag{30}$$

Substituting eq.(28) and eq.(21) into eq.(8) we obtain for the typical size of fluctuations in χ the value

$$\Delta\chi\sqrt{V} \sim \frac{2\pi^2}{3}l_{PE}^2. \tag{31}$$

This can be interpreted in the sense that the inertia inside a superconductive cavity change with respect to their classical laws due to the superconductor's spacetime volume fluctuations. Again the relevant length scale is the Planck-Einstein length, rather than the Planck length.

5.   Gravitational Surface Tension of Tao Balls

As already mentioned in section 4.2, a Tao ball is made up of many superconductive microparticles of size $r \sim 1\mu m$ and its radius is of the order $a \sim 1mm$. Therefore a Tao Ball consists of roughly

$$\left(\frac{a}{r}\right)^3 \sim 10^6 microparticles. \tag{32}$$

Tao et al. [13] proposed to explain the strong cohesion into spherical balls by a new type of surface tension σ. Also Hirsch [16] emphasizes that the formation of Tao balls cannot be understood by conventional superconductor physics.

The Tao ball surface energy ϵ_a is the product of the Tao ball surface $4\pi a^2$ and the surface tension σ:

$$\epsilon_a = 4\pi a^2 \sigma \tag{33}$$

If there are $(a/r)^3$ separated spherical particles then the total surface energy ϵ_{tot} is

$$\epsilon_{tot} = 4\pi r^2 \sigma \left(\frac{a}{r}\right)^3 \sim \epsilon_a \frac{a}{r} \sim 100\epsilon_a. \tag{34}$$

Therefore the surface energy ϵ_a of a Tao ball is just 1 % of the total surface energy ϵ_{tot} of the separated superconductive microparticles it consists of. This makes it plausible why macroscopic objects form. However, the question is what type of force creates the surface tension. It must be a force that is strong for superconducting matter only, and it must allow for the spherical symmetry of the objects formed.

Let us consider a homogeneous spherical body of mass m, uniform density ρ, and radius a, generating a Newtonian gravitational field. We may formally define a gravitational surface energy ϵ_g (also called surface pressure) [38] by

$$\epsilon_g = G\frac{m^2}{a} \tag{35}$$

where G is the gravitational constant. The gravitational surface tension σ_g is then given by

$$\epsilon_g = 4\pi a^2 \sigma_g, \tag{36}$$

or

$$\sigma_g = \frac{1}{3}Gm\rho = \frac{mg_0}{4\pi a}, \tag{37}$$

where $g_0 = Gm/a^2$ is the gravitational acceleration at the surface of the body. Is it consistent to explain the strong cohesion of Tao balls via a gravitational surface tension of the type of eq.(37)?

To answer this question let us assume that the gravitational acceleration g_0 responsible for this hypothetical surface tension σ_g, eq.(37), is generated from the acceleration A communicated to the Tao ball by the electric field E_0 via its electric charge $|q|$, according to the law of non-classical inertia in superconductors, eq.(29):

$$|g_0| = \frac{\chi}{2}|A| \tag{38}$$

Thus our hypothesis is that the strong applied electric field in a Tao cell triggers the creation of gravitational fields that are much stronger than usual, via a suitable scale transformation $G \to G'$. From Newton's law we calculate the Tao ball acceleration as

$$|A| = \frac{|q|E_0}{m_{TB}}, \tag{39}$$

where m_{TB} is the Tao ball mass. The electric charge $|q|$ acquired by a Tao ball while bouncing between the electrodes is given by [13]

$$|q| = 4\pi\epsilon_0 k_L E_0 a^2, \tag{40}$$

where a is the radius of the Tao ball, $k_L = 1.44$ is the dielectric constant of liquid nitrogen, and ϵ_0 is the vacuum permittivity in SI units. Substituting equations (38), (39) and (40) into eq.(37) we obtain the gravitational surface tension of a Tao ball as

$$\sigma_g = \frac{\chi}{2}k_L\epsilon_0 E_0^2 a. \tag{41}$$

Substituting the law defining χ, eq.(30), into eq.(41), we see that this gravitational surface tension is proportional to the fourth power of the critical transition temperature T_c of the superconductive material:

$$\sigma_g = \frac{3}{2}\frac{(\ln 3)^4}{4\pi}\left(\frac{T_c}{T_{PE}}\right)^4 k_L\epsilon_0 E_0^2 a \tag{42}$$

As T_c increases the gravitational surface tension increases. This is in qualitative agreement with the experimental observation: According to [14], the Tao balls formed by low-T_c

superconductors are weaker and easier to break than those formed by high-T_c superconductors.

NdBCO high-T_c superconductors have a critical temperature of $T_c = 94K$. In that case case the numerical prefactor in eq. (42) reduces to 1, and our theory for σ_g then reproduces the same result that was derived in [13] for the surface tension using a different model:

$$\sigma_g^{NdBCO} = k_L \epsilon_0 E_0^2 a \sim 2 \times 10^{-3} [N/m] \tag{43}$$

It is interesting to note that if we substitute the spacetime volume of a Tao ball, $V = \frac{4}{3}\pi a^3 c \Delta t$, into the space-time uncertainty relation in a superconducting environment, eq.(31), we find the following expression for the typical radius of a Tao ball:

$$a \sim \pi \left(\frac{1}{3\chi^2} \frac{l_{PE}^4}{c\Delta t} \right)^{1/3} \tag{44}$$

Assuming that the temporal length $c\Delta t$ in eq.(44) is equal to the size $c\Delta t \sim 1\mu m$ of the microparticles forming the Tao ball, we obtain the correct order of magnitude for the radius of Tao balls, i.e., $a \sim 0.17 [mm]$ for the case of NdBCO (with $\chi^{NdBCO} = 2$). This means that the coarse-grained microstructure of Tao balls consisting of many smaller particles of μm size is correctly described. Tao's experiments could be seen as the spatial counterpart of the experiments with radioactive superconductors discussed above, which deal with the temporal aspects of the space-time volume fluctuations.

6. Further Experimental Suggestions

The scale transformation of section 4 strongly enhances gravitational effects in a superconducting environment. At the same time, it strongly suppresses unwanted vacuum energy. Effects induced by the scale transformation should be measurable. A couple of interesting laboratory tests can then be performed with superconductors. In the following, we list a few proposals in this direction:

1. Measuring the cutoff frequency of quantum noise spectra in superconductors. This experiment is currently performed in London and Cambridge (UK), extending previous work of Koch et al. [39].

2. Measuring gravitomagnetic fields and frame dragging effects in the vicinity of rotating superconductors. These experiments are currently performed in Seibersdorf (Austria) and Canterbury (NZ) [26, 40]. Performing similar experiments with rotating supersolids would also be a valuable concept [41].

3. Measuring time with high precision inside and outside superconductive cavities. Since some inertial-like effects of rotating superconductive rings seem to propagate outside the ring [24, 25, 26], it would also be interesting to probe for temporal statistical fluctuations in the neighborhood of a superconductive material. In order to carry out this investigation we propose to compare the measurement of time intervals

measured by two synchronized identical clocks located inside and outside a superconductive cavity. The clocks would be synchronized before the cavity is made superconductive, and at constant intervals of time, the time indicated by the clocks would be regularly compared, in order to probe for any difference between time statistical fluctuations inside and outside the cavity.

4. Investigating broadening phenomena of the decay energy line width in radioactive superconductors.

5. Measuring Coriolis forces on test masses moving inside rotating superconductive cavities. This would basically correspond to carrying out Foucault-type pendulum experiments inside rotating superconductive cavities.

6. Comparing the measurement of acceleration exerted on masses inside accelerated superconductive cavities with similar accelerations detected inside cavities made of normal materials.

7. Carrying out the famous Einstein *Gedanken* elevator experiment, by comparing the measurement of the acceleration inside a superconductive cavity falling under the sole influence of the earth's gravitational field with that of a cavity made of normal materials, which would also be in a state of free fall.

8. Repeating the small-scale tests of the gravitational inverse square law as performed by Adelberger et al. [42, 43] in a superconducting environment.

9. Investigating the formation of Tao balls [13] in more detail. Is their formation connected with anomalous acceleration and gyroscope signals? Place accelerometers and gyroscopes near to the Tao cell, similar as in Tajmar's experiments [25].

10. Investigating a rotating Tao cell. How do Tao balls form in a rotating environment? Compare with planetary aggregation models where G is replaced by a rescaled G'. Similar questions could be dealt with when a magnetic field is applied to the Tao cell [14, 15].

11. Carrying out Hertz-like experiments with Tao balls and checking for gravitational radiation, in line with a similar proposal of Chiao[44] for electrically charged superfluid Helium droplets.

12. Performing high-precision measurements of force fields in superconductors using SQUIDS and Josephson junction arrays, in line with a suggestion of Fischer et al. [45]

7. Conclusion

In this paper we were turning the cosmological constant problem around, to argue that there is also an inverse cosmological constant problem where formally the cosmological constant comes out 120 orders of magnitude *too small*. For the inverse cosmological constant problem, one starts from the Einstein Hilbert action and considers an uncertainty relation for

4-dimensional spacetime. The true value of the cosmological constant, as observed by WMAP, is given by the geometric mean of both approaches, the quantum field theoretical one predicting a value 120 orders of magnitude too large and and the one starting from the Einstein-Hilbert action, predicting a value 120 orders of magnitude too small. This intermediate value represents the Planck-Einstein scale, a natural scale for dark energy, superconductors, and solid state physics in general.

We have formulated the hypothesis that in a superconducting environment it makes sense to formally consider a scale transformation from the Planck scale by the Planck-Einstein scale. This scale transformation leads to a strong suppression of certain quantum mechanical observables (such as vacuum energy density) and strong enhancement of others (such as gravitomagnetic fields). We have shown that these suppression and enhancement effects are consistent with some recent experimental observations. The formation of large superconducting spherical balls (the Tao effect) can also be understood in this context. A fundamental space-time uncertainty in a superconducting environment is predicted, which can be checked by future experiments. Ultimately non-classical inertia in superconductive cavities can be related to fluctuations in the number of relevant Planck-Einstein sized space-time cells in a superconductor. We believe that it is important to perform further precision experiments with superconductors, to fully explore the Planck-Einstein scale and to further investigate the connection between dark energy, gravity, and the physics of superconductors.

Acknowledgement

C.B.'s research has been supported by a Springboard fellowship of EPSRC.

References

[1] D. N. Spergel et al., *Astrophys. J. Suppl.* **148**, 175 (2003)

[2] P. J. E. Peebles, B. Ratra, *Rev. Mod. Phys.* **75**, 559 (2003)

[3] E.J. Copeland, M. Sami, and S. Tsujikawa, *Int. J. Mod. Phys. D* **15**, 1753 (2006)

[4] T. Padmanabhan, *Phys. Rep.* **380**, 235 (2005)

[5] T. Padmanabhan, "Dark Energy: Mystery of the Millenium", astro-ph/0603114

[6] S. Hsu, A. Zee, *Mod. Phys. Lett. A* **20**, 2699 (2005) [hep-th/0406142]

[7] T. Padmanabhan, *Class. Quant. Grav.* **22**, L107 (2005) [hep-th/0406060]

[8] C. Beck, *Phys. Rev. D* **69**, 123515 (2004) [astro-ph/0310479]

[9] C. Beck, M. C. Mackey, *Physica A* **379**, 101 (2007)

[10] C. Beck, M. C. Mackey, *Int. J. Mod. Phys. D* **17**, 71 (2008) [astro-ph/0703364]

[11] C. J. de Matos, C. Beck, "Possible measurable effects of dark energy in rotating superconductors", arXiv: 0707.1797, to appear in Adv. Astron. (2009)

[12] N. Arkani-Hamed, S. Dimopoulis, G.R. Dvali, *Phys. Lett. B* **436**, 257 (1998)

[13] R. Tao, X. Zhang, X. Tang, P.W. Anderson, *Phys. Rev. Lett.* **83**, 5575 (1999)

[14] R. Tao, X. Xu, Y.C. Lan, Y. Shiroyanagi, *Physica C* **377**, 357 (2002)

[15] R. Tao, X. Xu, E. Amr, *Physica C* **398**, 78 (2003)

[16] J.E. Hirsch, *Phys. Rev. Lett.* **94**, 187001 (2005)

[17] G. 't Hooft, S. Nobbenhuis, Class. Quant. Grav. **23**, 3819 (2006) [gr-qc/0602076]

[18] V. Jejjala, M. Kavic, and D. Minic, "Fine structure of dark energy and new physics", arXiv:0705.4581

[19] V. Jejjala, D. Minic, "Why is there something so close to nothing: towards a fundamental theory of the cosmological constant", hep-th/0605105

[20] R.D. Sorkin, "Causal Sets: Discrete Gravity", proceedings of the 2002 Valdivia Summer School, eds. A. Gomberoff and D. Marolf, p. 305-328 (2005) [gr-qc/0309009]

[21] A. Maqbool, S. Dodelson, P.B. Green, R. Sorkin, *Phys. Rev. D* **69**, 103523 (2004) [astro-ph/0209274]

[22] C. Sivaram, "What is special about the Planck mass", arXiv:0707.0058

[23] C. Sivaram, *Mod. Phys. Lett.* **14**, 2363 (1999)

[24] M. Tajmar, F. Plesescu, K. Marhold, C. J. de Matos, "Experimental Detection of the Gravitomagnetic London Moment", gr-qc/0603033

[25] M. Tajmar, F. Plesescu, B. Seifert, K. Marhold, "Measurement of gravitomagnetic and acceleration fields around a rotating superconductor", *AIP Conf. Proc.* **880**, 1071 (2007) [gr-qc/0610015]

[26] M. Tajmar, F. Plesescu, B. Seifert, R. Schnitzer, and I. Vasiljevich, "Search for Frame-Dragging in the Vicinity of Spinning Superconductors", to appear in the proceedings of the GRG18 conference (2007), arXiv: 0707.3806

[27] C. Beck, M. C. Mackey, *Phys. Lett. B* **605**, 295 (2005).

[28] C. Beck, *J.Phys.Conf.Ser.* **31**, 123 (2006).

[29] C. Beck, M. C. Mackey, *Fluct. Noise Lett.* **7**, C27 (2007) [astro-ph/0603067]

[30] J. Tate, B. Cabrera, S. B. Felch, J. T. Anderson, *Phys. Rev. Lett.* **62**, 845 (1989).

[31] H. Mazaki, S. Kakiuchi, S. Shimizu, *Z. Physik B* **29**, 285 (1978)

[32] A. Olin, K. T. Bainbridge, *Phys. Rev.* **179**, 2 (1969)

[33] C. J. de Matos, "Electromagnetic dark energy and gravitoelectrodynamics in superconductors", arXiv 0704.2499

[34] C. J. de Matos, M. Tajmar, *Physica C* **432**, 167 (2005).

[35] M. Tajmar, C. J. de Matos, *Physica C* **385**, 551 (2003).

[36] M. Tajmar, C. J. de Matos, *Physica C* **420**, 56 (2005).

[37] B. Mashhoon, *Phys. Lett. A* **173**, 347 (1993)

[38] J. Bicak, B. G. Schmidt, *Astrophys. J.* **521**, 708 (1999) [gr-qc/9903021]

[39] R.H. Koch, D. van Harlingen, and J. Clarke, *Phys. Rev. Lett.* **47**, 1216 (1981)

[40] R.D. Graham et al., "Experiment to detect frame-dragging in a lead superconductor", Preprint University of Canterbury (NZ) (2007)

[41] E. Kim, W. Chan, *Science* **305**, 1941 (2004)

[42] D.J. Kapner et al., *Phys. Rev. Lett.* **98**, 021101 (2007)

[43] E.G. Adelberger et al., "Particle physics implications of a recent test of the gravitational inverse-square law", hep-ph/0611223

[44] R. Y. Chiao, "Millikan oil drops as quantum transducers between electromagnetic and gravitational radiation", gr-qc/0702100

[45] U.R. Fischer et al., *Phys. Rev. B* **64**, 214509 (2001)

In: Dark Energy: Theories, Developments and Implications   ISBN 978-1-61668-271-2
Editors: K. Lefebvre and R. Garcia, pp. 49-88   © 2010 Nova Science Publishers, Inc.

Chapter 4

CROSSING THE PHANTOM DIVIDE

*Hongsheng Zhang**
Shanghai United Center for Astrophysics (SUCA),
Shanghai Normal University, 100 Guilin Road,
Shanghai 200234,China
Korea Astronomy and Space Science Institute,
Daejeon 305-348, Korea
Department of Astronomy, Beijing Normal University,
Beijing 100875, China

Abstract

The cosmic acceleration is one of the most significant cosmological discoveries over the last century. Following the more accurate data a more dramatic result appears: the recent analysis of the observation data (especially from SNe Ia) indicate that the time varying dark energy gives a better fit than a cosmological constant, and in particular, the equation of state parameter w (defined as the ratio of pressure to energy density) crosses -1 at some low redshift region. This crossing behavior is a serious challenge to fundamental physics. In this article, we review a number of approaches which try to explain this remarkable crossing behavior. First we show the key observations which imply the crossing behavior. And then we concentrate on the theoretical progresses on the dark energy models which can realize the crossing -1 phenomenon. We discuss three kinds of dark energy models: 1. two-field models (quintom-like), 2. interacting models (dark energy interacts with dark matter), and 3. the models in frame of modified gravity theory (concentrating on brane world).

PACS numbers: 95.36.+x 04.50.+h

1. The Universe is Accelerating

Cosmology is an old and young branch of science. Every nation had his own creative idea about this subject. However, till the 1920s about the unique observation which had cosmological significance was a dark sky at night. On the other hand, we did not prepare

---

*E-mail address: hongsheng@kasi.re.kr

a proper theoretical foundation till the construction of general relativity. Einstein's 1917 paper is the starting point of modern cosmology [1]. The next mile stone was the discovery of cosmic expansion, that was the recession of galaxies and the recession velocity was proportional to the distance to us.

Except some rare cases, our researches are always based on the cosmological principle, which says that the universe is homogeneous and isotropic. In the early time, this is only a supposition to simplify the discussions. Now we have enough evidences that the universe is homogeneous and isotropic at the scale larger than 100 Mpc. The cosmological principle requires that the metric of the universe is FRW metric,

$$ds^2 = -dt^2 + a^2(t)(dr^2 + r^2 d\Omega_2^2); \tag{1}$$
$$ds^2 = -dt^2 + a^2(t)(dr^2 + \sin(r)^2 d\Omega_2^2); \tag{2}$$
$$ds^2 = -dt^2 + a^2(t)(dr^2 + \sinh(r)^2 d\Omega_2^2), \tag{3}$$

depending on the spatial curvature, which can be Euclidean, spherical or pseudo-spherical. Here, t is the cosmic time, a denotes the scale factor, r represents the comoving radial coordinate of the maximal symmetric 3-space, and $d\Omega_2^2$ stands for a 2-sphere. Which geometry serves our space is decided by observations. FRW metric describes the kinetic evolution of the universe. To describe the dynamical evolution of the universe, that is, the function of $a(t)$, we need the gravity theory which ascribes the space geometry to matter. The present standard gravity theory is general relativity. In 1922 and 1924, Friedmann found that there was no static cosmological solution in general relativity, that is to say, the universe is either expanding or contracting [2]. To get a static universe, Einstein introduce the cosmological constant. However, even in the Einstein universe, where the contraction of the dust is exactly counteracted by the repulsion of the cosmological constant, the equilibrium is only tentative since it is a non-stationary equilibrium. Any small perturbation will cause it to contract or expand. Hence, in some sense we can say that general relativity predicts an expanding (or contracting) universe, which should be regarded as one of the most important prediction of relativity.

In almost 70 years since the discovery of the cosmic expansion in 1929 [3], people generally believe that the universe is expanding but the velocity is slowing down. People try to understand via observation that the universe will expand forever or become contracting at some stage. A striking result appeared in 1998, which demonstrated that the universe is accelerating rather than decelerating. Now we show how to conclude that our universe is accelerating. We introduce the standard general relativity,

$$G_{\mu\nu} + \Lambda g_{\mu\nu} = 8\pi G t_{\mu\nu}, \tag{4}$$

where $G_{\mu\nu}$ is Einstein tensor, G is (4-dimensional) Newton constant, $t_{\mu\nu}$ denotes the energy-momentum tensor, $g_{\mu\nu}$ stands for the metric of a spacetime, μ, ν run from 0 to 3. Throughout this article, we take a convention that $c = \hbar = 1$ without special notation. Define a new energy momentum $T_{\mu\nu}$,

$$8\pi G T_{\mu\nu} = 8\pi G t_{\mu\nu} - \Lambda g_{\mu\nu}. \tag{5}$$

$T_{\mu\nu}$ has included the contribution of the cosmological constant, whose effect can not be distinguished with vacuum if we only consider gravity.

The 00 component of Einstein equation (4) is called Friedmann equation,

$$H^2 + \frac{k}{a^2} = \frac{8\pi G}{3}\rho, \qquad (6)$$

where $H = \frac{\dot{a}}{a}$ is the Hubble parameter, an overdot stands for the derivative with respect to the cosmic time, k is the spatial curvature of the FRW metric, for (1), $k = 0$; for (2), $k = 1$; for (3), $k = -1$, $\rho = -T_0^0$. Throughout this article, we take the signature $(-, +, ..., +)$. The spatial component of Einstein equation can be replaced by the continuity equation, which is much more convenient,

$$\dot{\rho} + 3H(\rho + p) = 0, \qquad (7)$$

where $p = T_1^1 = T_2^2 = T_3^3$. Here we use a supposition that the source of the universe T_μ^ν is in perfect fluid form. Using (6) and (7), we derive the condition for acceleration,

$$\frac{\ddot{a}}{a} = -\frac{4\pi G}{3}(\rho + 3p). \qquad (8)$$

We see that the universe is accelerating when $\rho + 3p < 0$. We know that the galaxies and dark matter inhabit in the universe long ago. They are dust matters with zero pressure. Thus, if the universe is accelerating, there must exist an exotic matter with negative pressure or we should modify general relativity. The cosmological constant is a far simple candidate for this exotic matter, or dubbed dark energy. For convenience we separate the contribution of the cosmological constant (dark energy) from other sectors in the energy momentum, the Friedmann equation becomes,

$$\frac{H^2}{H_0^2} = \Omega_{\Lambda 0} + \Omega_{m0}(1 + z)^3 + \Omega_{k0}(1 + z)^2, \qquad (9)$$

where z is the redshift, the subscript 0 denotes the present value of a quantity,

$$\Omega_{\Lambda 0} = \frac{\Lambda}{8\pi H_0^2 G}, \qquad (10)$$

$$\Omega_{m0} = \frac{\rho_{m0}}{8\pi H_0^2 G}, \qquad (11)$$

$$\Omega_{k0} = -\frac{k}{H_0^2 a_0^2}. \qquad (12)$$

The type Ia supernova is a most powerful tool to probe the expanding rate of the universe. In short, type Ia supernova is a supernova which just reaches the Chandrasekhar limit (1.4 solar mass) and then explodes. Hence they have the same local luminosity since they have roughly the same mass and the same exploding process. They are the standard candles in the unverse. We can get the distance of a type Ia supernova through its apparent magnitude. A sample of type Ia supernovae will generates a diagram of Hubble parameter versus distance, through which we get the information of the expanding velocity in the history of the universe. In 1998, two independent groups found that the universe is accelerating using the observation data of supernovae [4]. After that the data accumulate fairly quickly.

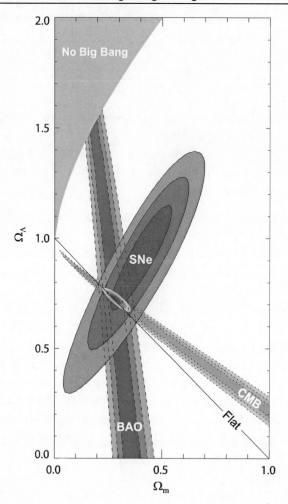

Figure 1. The figure displays the counter constrained by SNe Ia (in blue), by BAO (in green), by CMB (in yellow), and the joint constraint by all the three kinds of observation (in black and white). Ω_Λ is $\Omega_{\Lambda0}$ in (10), Ω_m is Ω_{m0} in (11). This figure is borrowed from [10].

The famous sample includes Gold04 [5], Gold06 [6], SNLS [7], ESSENCE [8], Davis07 [9], Union [10], Constitution [11]. Here, we show some results of one of the most recent sample, Union [10], which is plotted by χ^2 statistics.

From fig 1, we see that: 1. the universe is almost spatially flat, that is the curvature term Ω_{k0} is very small. 2. the present universe is dominated by cosmological constant (dark energy), whose partition is approximately 70%, and the partition of dust is 30%. We introduce a dimensionless parameter, the deceleration parameter q,

$$q = -\frac{\ddot{a}a}{\dot{a}^2}. \tag{13}$$

A negative deceleration parameter denotes acceleration. In a universe with dust and cosmological constant (which is called ΛCDM model), by definition

$$q = \frac{1}{2}\Omega_m - \Omega_\Lambda, \qquad (14)$$

whose present value $q_0 = -0.55$. Hence the present universe is accelerating.

The previous result depends on a special cosmological model, ΛCDM model in frame of general relativity. How about the conclusion if we only consider the kinetics of the universe?

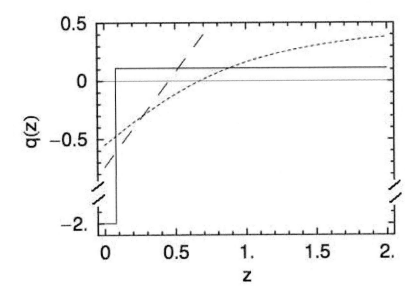

Figure 2. Kinetic universe vs dynamic universe. The fitting results of sudden transition model, linear expansion model, and ΛCDM model. The solid line is the best-fit of the sudden transition model (the deceleration parameter jumps at some redshift); the long-dashed line denotes the best-fit of linear expansion model ($q = q_0 + q_1 z$, q_1 is a constant) [5]; the short-dashed line represents the best-fit of ΛCDM model. From [12].

The simplest kinetic model is a sudden transition model, in which the deceleration parameter is a constant in some high redshift region and jumps to another constant at a critical redshift. The other simple choice is that the deceleration parameter is a linear function of z. We show the two kinetic models with the dynamic model ΛCDM in fig 2. It is clear that the universe accelerates in the present epoch in all the three models. A more rigorous analyze shows that the evidence for an accelerating universe is fairly strong (more than 5 σ) [12]. So we should investigate it seriously.

ΛCDM is the most simple model for the acceleration, which is a concordance model of several observations. As we shown in fig 1, the counters of CMB, BAO, and SNe Ia have cross section, which almost laps over result of the joint fittings. However, ΛCDM has its own theoretical problems. Furthermore, it is found that a dynamical dark energy model fits the observation data better. Especially, there are some evidences that the equation of state (EOS) of dark energy may cross -1, which is a serious challenge to the foundation of theoretical physics.

In the next section we shall study some problems of ΛCDM model and display that a dynamical dark energy model is favored by observations. We'll focus on the crossing behavior implied by the observation. In section III, we study 3 kinds of models with a crossing phantom divide dark energy. In section IV, we present the conclusion and more references of this topic.

2. A Dark Energy with Crossing -1 EOS is Slightly Favored by Observations

2.1. The Problems of ΛCDM

ΛCDM has two famous theoretical problems.

The first is the finetune problem. The effect of the vacuum energy can not be distinguished from the cosmological constant in gravity theory. We can calculate the vacuum energy by a well-constructed theory, quantum field theory (QFT), which says that the vacuum energy should be larger than the observed value by 122 orders of magnitude, if QFT works well up to the Planck scale. In supersymmetric (SUSY) theory, the vacuum energy of the Bosons exactly counteracts the vacuum energy of Fermions, such that we obtain a zero vacuum energy. However, SUSY must break at the electro-weak scale. At that scale, the vacuum energy is still large than the observed value by 60 orders of magnitude. So for getting a vacuum energy we observed, we should introduce a bare cosmological constant Λ_{bare}. The effective vacuum energy ρ_{effect} then becomes,

$$\rho_{\text{effect}} = \frac{1}{8\pi G}\Lambda_{\text{bare}} + \rho_{\text{vacuum}}. \tag{15}$$

$\frac{1}{8\pi G}\Lambda_{\text{bare}}$ and ρ_{vacuum} have to almost counteract each other but do not exactly counteract each other, leaving a tiny tail which is smaller than the ρ_{vacuum} by 60 orders of magnitude. Which mechanism can realize such a miraculous counteraction?

The second problem is coincidence problem, which says that the cosmological constant keeps a constant while the density of the dust evolves as $(1 + z)^3$ in the history of the universe, then why do they approximately equal each other at "our era"? Different from the first problem, the second problem says the present ratio of dark energy and dark matter is sensitively depends on the initial conditions. Essentially, the coincidence problem is the problem of an unnatural initial condition. The densities of different species in the universe redshift with different rate in the evolution of the universe, so if their densities coincidence in *our era*, their density ratio must be a specific, tiny number in the *early universe*. It is also a finetune problem, but a finetune problem of the initial condition.

Except the above theoretical problems, ΛCDM also suffers from observation problem, especially when faced to the fine structure of the universe, including galaxies, clusters and voids. Some specific observations differ from the predictions of ΛCDM (with standard partitions of dust and cosmological constant) at a level of 2σ or higher. Six observations are summarized in [13]: 1. scale velocity flows is much larger than the prediction of ΛCDM, 2. Type Ia Supernovae (Sne Ia) at High Redshift are brighter than what ΛCDM indicates, 3. the void seems more empty than what ΛCDM predicts, 4. the cluster haloes look denser than what ΛCDM says, 5. the density function of galaxy haloes is smooth, while ΛCDM indicates a cusp in the core, 6. there are too much disk galaxies than the prediction of ΛCDM. We do not fully understand the dynamics and galaxies and galaxy clusters, that is, the gravitational perturbation theory at the small scale. The agreement may approve when we advance our perturbation theory with a cosmological constant and the simulation methods at the small scale. However, in the cosmological scale, there are also some evidences that the dark energy is dynamical, including no. 2 of the previous 6 problems.

2.2.   Crossing -1

With data accumulation, observations which favor dynamical dark energy become more and accurate. Now we loose the condition that $p = -\rho$ for the exotic matter (dark energy) which accelerates the universe. We go beyond the ΛCDM model. We permit that the EOS of dark energy is not exactly equal to -1, but still a constant. The fitting results by different samples of SNe Ia are displayed in fig 3. We see that although a cosmological constant is permitted, the dark energy whose EOS < -1 is favored by SNe Ia. The essence whose EOS is less than -1 is called phantom, which can be realized by a scalar field with negative kinetic term. The action for phantom ψ is

$$S_{\text{ph}} = \int d^4x \sqrt{-g} \left(\frac{1}{2} \partial_\mu \psi \partial^\mu \psi - U(\psi) \right), \tag{16}$$

where $\sqrt{-g}$ is the determinate of the metric, the lowercase Greeks run from 0 to 3, $U(\psi)$ denotes the potential of the phantom. In an FRW universe, the density and pressure of the phantom (16) reduce to

$$\rho = -\frac{1}{2}\dot{\psi}^2 + U, \tag{17}$$

$$p = -\frac{1}{2}\dot{\psi}^2 - U, \tag{18}$$

respectively. Now the EOS of the phantom w is given by,

$$w = \frac{p}{\rho} = \frac{-\frac{1}{2}\dot{\psi}^2 - U}{-\frac{1}{2}\dot{\psi}^2 + U}, \tag{19}$$

which is always less than -1 for a positive U. It seems that phantom is proper candidate for the dark energy whose EOS less than -1 [14]. It is very famous that a phantom field is unstable when quantized since the energy has no lower bound. It will transit to a lower and lower energy state. In this article we have no time to discuss this important topic for phantom dark energy. We would point out the basic idea for this issue, which requires the life time of the phantom is much longer than the age of the universe such that we still have no chance to observe the decay of the phantom, though it is fundamentally unstable. For references, see [15]. A completely regular quantum stress with $w < -1$ is suggested in [16].

If the dark energy really behaves as phantom at some low redshift region, it is an unusual discovery. But the dark energy may be more fantastic. In some model-independent fittings, the EOS of dark energy crosses -1, which is a really remarkable property and a serious challenge to our present theory of fundamental physics.

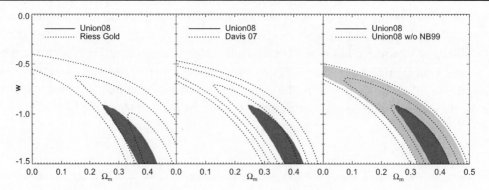

Figure 3. The EOS of dark energy fitted by SNe Ia in a spatially flat universe, the contours display 68.3 %, 95.4 % and 99.7% confidence level on w and Ω_{m0} (Ω_m in the figure). The results of the Union set are shown as filled contours. The empty contours, from left to right, show the results of the Gold sample, Davis 07, and the Union without SCP nearby data. From [10].

Pioneer results of the crossing -1 of EOS of dark energy appeared in [17, 18]. Fig 4 illuminates that the EOS of dark energy may cross -1 in some low redshift. In fig 4, the Gold04 data are applied, a uniform prior of $0.22 \leq \Omega_{m0} \leq 0.38$ is assumed, and a spatially flat universe is the working frame.

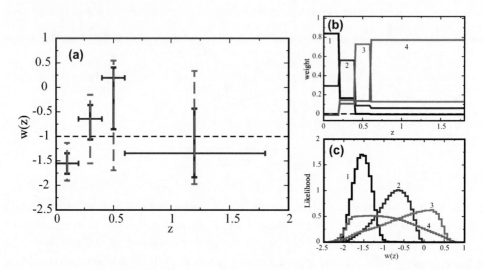

Figure 4. Panel (a): Uncorrelated band-power estimates of the EOS $w(z)$ of dark energy by SNe Ia (Gold set [5]). Vertical error bars show the 1 and 2-σ error bars (in blue and green, respectively). The horizontal error bars denote the data bins used in [18]. Panel (b): The window functions for each bin from low redshift to high redshift. Panel (c): the likelihoods of $w(z)$ in the bins from low redshift to high redshift. From [18].

The perturbation of the dark energy will growth if its EOS is not exactly -1 in the evolution history of the universe. Hence to fit a model with dynamical dark energy with observation, the perturbation of the dark matter should be considered in principle. Such a study was presented in [19], in which a parametrization of the EOS of the dark energy with two constant w_0, w_1 was applied,

$$w = w_0 + w_1 \frac{z}{1+z}. \tag{20}$$

The result is shown in fig 5. We see from fig 5 that there is a mild tendency that the EOS of

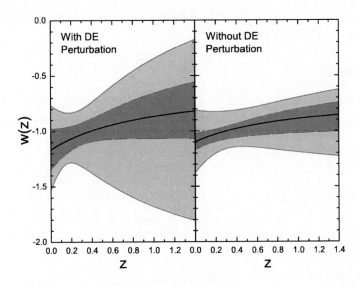

Figure 5. Constraints on the EOS of w(z) by WMAP3 [20] and Gold04 [5]. The light grey region denotes $2\,\sigma$ constraint, while the dark grey for 1σ constraint. The left panel shows the constraint with dark energy perturbation, while the right displays the result without dark energy perturbation. From [19].

the dark energy cross -1. For a more general parametrization of EOS for dark energy, see [21].

   With more and accurate data, the possibility of crossing -1 (phantom divide) seems a little more specific, see for example [22]. This crossing behavior is a significant challenge for theoretical physics. It was proved that the EOS of dark energy can not cross the phantom divide if 1. a dark energy component with an arbitrary scalar-field Lagrangian, which has a general dependence on the field itself and its first derivatives, 2. general relativity holds and 3. the spatially flat Friedmann universe [23]. Thus realizing such a crossing is not a trivial work. In the next section we investigate the theoretical progresses for this extraordinary phenomenon.

3.   Three Roads to Cross the Phantom Divide

To cross the phantom divide, we must break at least one of the conditions in [23]. Now that the dark energy behaves as quintessence at some stage , while evolves as phantom at the other stage, a natural suggestion is that we should consider a 2-field model, a quintessence and a phantom. The potential is carefully chosen such that the quintessence dominates the universe at some stage while the phantom dominates the universe at the other stage. It was invented a name for such 2-field model, "quintom" . There are also some varieties of quintom, such as hessence. We introduce these 2-field models in the first subsection. The next road is to consider an interacting model, in which the dark energy interacts with dark matter. The interaction can realize the crossing behavior which is difficult for independent dark energy. We shall study the interacting models in subsection B. The other possibility is that general relativity fails at the cosmological scale. The ordinary dark energy candidates, such as quintessence or phantom, can cross the phantom divide in a modified gravity theory. We investigate this approach in subsection C.

3.1.   2-Field Model

A typical 2-field model is the quintom model, which was proposed in [24], and was widely investigated later [25]. Generally , the action of a universe with quintom dark energy S is

$$S = \int d^4x \sqrt{-g} \left(\frac{R}{16\pi G} + \mathcal{L}_{\text{stuff}} \right), \tag{21}$$

where R is the Ricci scalar, $\mathcal{L}_{\text{stuff}}$ encloses all kinds of the stuff in the universe, for instance the dust matter, radiation, and quintom. At the late universe, the radiation can be negligible. So, often we only consider the dark energy, here quintom $\mathcal{L}_{\text{quintom}}$, and dust matter $\mathcal{L}_{\text{dm}}$,

$$\mathcal{L}_{\text{quintom}} = -\frac{1}{2}\partial_\mu\phi\partial^\mu\phi + \frac{1}{2}\partial_\mu\psi\partial^\mu\psi - W(\phi, \psi). \tag{22}$$

In (22) the first term is the kinetic term of an ordinary scalar, the second term is the kinetic term of a phantom, and $W(\phi, \psi)$ is an arbitrary function of ϕ and ψ. In an FRW universe, the density and pressure of the quintom are

$$\rho = \frac{1}{2}\dot{\phi}^2 - \frac{1}{2}\dot{\psi}^2 + W, \tag{23}$$

$$p = \frac{1}{2}\dot{\phi}^2 - \frac{1}{2}\dot{\psi}^2 - W. \tag{24}$$

Hence, the EOS of the quintom w is

$$w = \frac{\frac{1}{2}\dot{\phi}^2 - \frac{1}{2}\dot{\psi}^2 - W}{\frac{1}{2}\dot{\phi}^2 - \frac{1}{2}\dot{\psi}^2 + W}. \tag{25}$$

$w = -1$ requires

$$\dot{\phi}^2 = \dot{\psi}^2. \tag{26}$$

We see that in a quintom model, we do not require a static field (a field with zero kinetic term or a field at ground state) to get a cosmological constant. We only need that ψ and ϕ evolves in the same step. $w < -1$ implies,

$$\dot{\phi}^2 - \dot{\psi}^2 < 0, \tag{27}$$

if

$$\frac{1}{2}\dot{\phi}^2 - \frac{1}{2}\dot{\psi}^2 + W > 0; \qquad (28)$$

and

$$\dot{\phi}^2 - \dot{\psi}^2 > 0, \qquad (29)$$

if

$$\frac{1}{2}\dot{\phi}^2 - \frac{1}{2}\dot{\psi}^2 + W < 0. \qquad (30)$$

(30) yields an unnatural physical result,that is, the density of dark energy is negative. However, this is not as serious as the first glance, since we have little knowledge of the dark energy besides its effect of gravitation. Several evidences imply that we should go beyond the standard model of the particle physics when we describe dark energy. There are a few dark energy models permit density of the dark energy, or a component of it is negative (at the same time keep the total density positive), for example, see [26, 27]. But, for a model with only two components, a dust and a quintom, it is difficult to set a negative density dark energy. In that case we need too much dust than we observed or a big curvature term. In the following text of this section, we only consider a dark energy with positive density. So $w > -1$ implies,

$$\dot{\phi}^2 - \dot{\psi}^2 > 0. \qquad (31)$$

In summary, if the kinetic term of the quintessence dominates that of phantom, the quintom behaves as quintessence; else it behaves as phantom. We should select a proper potential to make quintessence and phantom dominate alternatively such that we can realize the crossing behavior.

A simple choice of the potential is that the quintessence and the phantom do not interact with each other, which requires, $W(\phi, \psi) = V(\phi) + U(\psi)$. The exponential potential is an important example which can be solved exactly in the quintessence model (a toy universe only composed by quintessence). In addition, we know that such exponential potentials of scalar fields occur naturally in some fundamental theories such as string/M theories. We introduce a model with such potentials in [28], in which the potential $V(\phi, \psi)$ is given by

$$W(\phi, \psi) = V(\phi) + U(\psi) = A_\phi e^{-\lambda_\phi \kappa \phi} + A_\psi e^{-\lambda_\psi \kappa \psi}, \qquad (32)$$

where A_ϕ and A_ψ are the amplitude of the potentials, $\kappa^2 = 8\pi G$, λ_ϕ and λ_ψ are two constants. Since there is no direct couple between the quintessence and the phantom, the equations of motion of the quintessence and the phantom are two independent equations,

$$\ddot{\phi} + 3H\dot{\phi} + \frac{dV}{d\phi} = 0, \qquad (33)$$

$$\ddot{\psi} + 3H\dot{\psi} - \frac{dU}{d\psi} = 0. \qquad (34)$$

The continuity equation of the dust reads,

$$\dot{\rho}_{\text{dust}} + 3H\rho_{\text{dust}} = 0, \qquad (35)$$

where ρ_{dust} denotes the density of the dust. The method of dynamical system has been widely used in cosmology. This method can offer a clear history of the cosmic evolution, especially the final states of the university. For applying this method, first we define the following dimensionless variables,

$$x_\phi \equiv \frac{\kappa \dot{\phi}}{\sqrt{6}H} \quad , \quad y_\phi \equiv \frac{\kappa \sqrt{V_\phi}}{\sqrt{3}H},$$

$$x_\psi \equiv \frac{\kappa \dot{\psi}_i}{\sqrt{6}H} \quad , \quad y_\psi \equiv \frac{\kappa \sqrt{V_\psi}}{\sqrt{3}H}, \tag{36}$$

$$z \equiv \frac{\kappa \sqrt{\rho_{\rm dust}}}{\sqrt{3}H} \quad ,$$

the evolution equations (33)-(35) become,

$$x'_\phi = -3x_\phi \left(1 + x_\phi^2 - x_\psi^2 - \frac{1}{2}z^2\right) + \lambda_\phi \frac{\sqrt{6}}{2} y_\phi^2, \tag{37}$$

$$y'_\phi = 3y_\phi \left(-x_\phi^2 + x_\psi^2 + \frac{1}{2}z^2 - \lambda_\phi \frac{\sqrt{6}}{6} x_\phi\right), \tag{38}$$

$$x'_\psi = -3x_\psi \left(1 + x_\phi^2 - x_\psi^2 - \frac{1}{2}z^2\right) - \lambda_\psi \frac{\sqrt{6}}{2} y_\psi^2, \tag{39}$$

$$y'_\psi = 3y_\psi \left(-x_\phi^2 + x_\psi^2 + \frac{1}{2}z^2 - \lambda_\psi \frac{\sqrt{6}}{6} x_\psi\right), \tag{40}$$

$$z' = 3z \left(-x_\phi^2 + x_\psi^2 + \frac{1}{2}z^2 - \frac{1}{2}\right), \tag{41}$$

in which a prime denotes derivative with respect to $\ln a$. Generally, z in the above set will not be confused with redshift. The five equations in this system are not independent. They are constrained by Fridemann equation,

$$H^2 = \frac{\kappa^2}{3} \left(\frac{1}{2}\dot{\phi}^2 + V - \frac{1}{2}\dot{\psi}^2 + U + \rho_{\rm dust}\right), \tag{42}$$

which becomes

$$x_\phi^2 + y_\phi^2 - x_\psi^2 + y_\psi^2 + z^2 = 1. \tag{43}$$

with the dimensionless variables defined before. The critical points dwell at $x'_\phi = y'_\phi = x'_\psi = y'_\psi = z' = 0$. We present the result in table 1. For detailed discussion of the critical points, see [28]. We would like to show a numerical example in which the EOS of the quintom crosses the phantom divide. Fig 6 illuminates that the EOS crosses -1.

The previous quintom model includes two fields, which are completely independent and rather arbitrary. We can impose some symmetry in the quintom model. An interesting model with an internal symmetry between the two fields which work as dark energy is hessence [29]. Rather than two uncorrelated fields, we consider one complex scalar field with internal symmetry between the real and the imaginary parts,

$$\Phi = \phi_1 + i\phi_2, \tag{44}$$

Table 1. The critical points, from [28]

Label	x_ψ	y_ψ	x_ϕ	y_ϕ	z	Stability
K	$-x_\psi^2 + x_\phi^2 = 1$	0		0	0	unstable
P	$-\frac{\lambda_\psi}{\sqrt{6}}$	$\sqrt{(1 + \frac{\lambda_\psi^2}{6})}$	0	0	0	stable
S	0	0	$\frac{\lambda_\phi}{\sqrt{6}}$	$\sqrt{(1 - \frac{\lambda_\phi^2}{6})}$	0	unstable
F	0	0	0	0	1	unstable
T	0	0	$\frac{3}{\sqrt{6}\lambda_\phi}$	$\frac{\sqrt{3}}{\lambda_\phi}$	$\sqrt{1 - \frac{3}{\lambda_\phi^2}}$	unstable

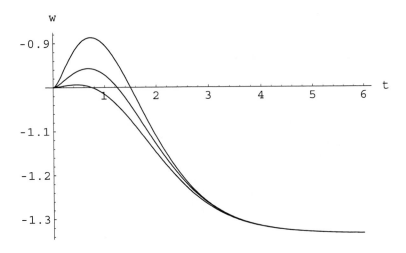

Figure 6. The evolution of the effective equation of state of the phantom and normal scalar fields with $W(\phi, \sigma)$ for the case $\lambda_\phi = 1$. From [28].

with a Lagrangian density

$$\mathcal{L}_{\text{hess}} = -\frac{1}{4} \left[(\partial_\mu \Phi)^2 + (\partial_\mu \Phi^*)^2 \right] - V(\xi, \Phi^*) = -\frac{1}{2} \left[(\partial_\mu \xi)^2 - \xi^2 (\partial_\mu \theta)^2 \right] - V(\xi), \quad (45)$$

which is invariant under the transformation,

$$\phi_1 \rightarrow \phi_1 \cos\alpha - i\phi_2 \sin\alpha, \quad (46)$$
$$\phi_2 \rightarrow -i\phi_1 \sin\alpha + \phi_2 \cos\alpha, \quad (47)$$

if the potential is only a function of $\Phi^2 + (\Phi^*)^2$. For convenience, in (45) we have introduced two new variables (ξ, θ),

$$\phi_1 = \xi \cosh\theta, \qquad \phi_2 = \xi \sinh\theta, \quad (48)$$

which are defined by

$$\xi^2 = \phi_1^2 - \phi_2^2, \qquad \coth\theta = \frac{\phi_1}{\phi_2}. \quad (49)$$

The equations of motion of ξ and θ are

$$\ddot{\xi} + 3H\dot{\xi} + \xi\dot{\theta}^2 + \frac{dV}{d\xi} = 0, \tag{50}$$

$$\xi^2\ddot{\theta} + (2\xi\dot{\xi} + 3H\xi^2)\dot{\theta} = 0. \tag{51}$$

Clearly, ξ and θ couple to each other. The pressure and density of the hessence read,

$$p_{\text{hess}} = \frac{1}{2}\left(\dot{\xi}^2 - \xi^2\dot{\theta}^2\right) - V(\xi), \tag{52}$$

$$\rho_{\text{hess}} = \frac{1}{2}\left(\dot{\xi}^2 - \xi^2\dot{\theta}^2\right) + V(\xi), \tag{53}$$

respectively. The EOS of hessence, playing as dark energy,

$$w = \frac{\frac{1}{2}\left(\dot{\xi}^2 - \xi^2\dot{\theta}^2\right) - V(\xi)}{\frac{1}{2}\left(\dot{\xi}^2 - \xi^2\dot{\theta}^2\right) + V(\xi)}. \tag{54}$$

Qualitatively, hessence evolves as quintessence when $\dot{\xi}^2 \geq \xi^2\dot{\theta}^2$, while as phantom when $\dot{\xi}^2 < \xi^2\dot{\theta}^2$. The Lagrangian (45) does not include θ, hence the canonical momentum π_θ^μ corresponding to the cyclic coordinate θ are conserved quantities,

$$\pi_\theta^\mu = \frac{\partial(\mathcal{L}_{\text{hess}}\sqrt{-g})}{\partial(\partial_\mu\theta)}. \tag{55}$$

In an FRW universe, only π_θ^0 exists. We define a conserved quantity Q which is proportional to π_θ^0,

$$Q = a^3\xi^2\dot{\theta}. \tag{56}$$

With this conserved quantity, the EOS becomes,

$$w = \frac{\frac{1}{2}\dot{\xi}^2 - \frac{Q^2}{2a^6\xi^2} - V(\xi)}{\frac{1}{2}\dot{\xi}^2 - \frac{Q^2}{2a^6\xi^2} + V(\xi)}, \tag{57}$$

which is only a function of ξ. The Friedmann equations read as

$$H^2 = \frac{8\pi G}{3}\left[\rho_{\text{dust}} + \frac{1}{2}\left(\dot{\xi}^2 - \xi^2\dot{\theta}^2\right) + V(\xi)\right], \tag{58}$$

where ρ_{dust} is the energy density of dust. The continuity equation of dust is (35). The continuity equations of hessence are identical to the equations of motion (50) and (51). Then the system is closed and we present a numerical example in fig 7. Evidently, the EOS of hessence, playing the role of dark energy, crosses -1 at about $a = 0.95$ ($z = 0.06$).

After the presentation of hessence model, several aspects of this model have been investigated, including to avoid the big rip [30], attractor solutions for general hessence [31], reconstruction of hessence by recent observations [32], dynamics of hessence in frame of loop quantum cosmology [33], and holographic hessence model[34].

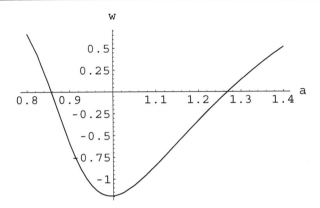

Figure 7. The EOS of hessence w as a function of scale factor with the potential $V(\xi) = \lambda\xi^4$. The parameters for this plot are as follows: $\Omega_{m0} = \rho_{m0}/(3H_0^2) = 0.3$, $\lambda = 5.0$, $Q = 1.0$, $a_0 = 1$ and the unit $8\pi G = 1$. From [29].

3.2. Interacting Model

Two-field model is a natural and obvious construction to realize the crossing -1 behavior of dark energy. However, there are two many parameters in the set-up, though we can impose some symmetries to reduce the parameters to a smaller region. One symmetry decrease one parameter, but we have little clue to impose the symmetries since we have no evidence in the ground labs.

Interaction is a universal phenomenon in the physics world. An interaction term is helpful to cross the phantom divide. To illuminate this point, we first carefully analyze the previous observations which imply the crossing. Both the results of [18] and [19], which are shown in fig 4 and fig 5 respectively, are derived with a presupposition, that is, the dark energy evolves freely. In fact, what we observed is the effective EOS of the dark energy in the sense of gravity at the cosmological scale. When we suppose it evolves freely, we find that its EOS may cross the phantom divide. We can demonstrate for an essence with (local) EOS< -1, the cosmological effective EOS can cross -1 by aids of an interacting term. For the case with interaction, the continuity equation for dark energy becomes,

$$\dot{\rho}_{de} + 3H(\rho_{de} + p_{de}) = -\Gamma, \tag{59}$$

or

$$\dot{\rho}_{de} + 3H(\rho_{de} + p_{de} + \frac{\Gamma}{3H}) = 0. \tag{60}$$

Here ρ_{de} is the density of dark energy, p_{de} denotes the local pressure measured in the lab (if we can measure), Γ stands for the interaction term, and $p_{eff} = p_{de} + \frac{\Gamma}{3H}$ is the effective pressure in the cosmological sense. In a universe without expanding or contracting, $H = 0$, the interaction does no effect on the continuity equation, or energy conservation law, and thus does not yield surplus pressure [1]. Two special cases are interesting: 1. $\frac{\Gamma}{3H}$ is a constant,

---

[1]One may think that Γ/H is meaningless when $H = 0$. But in fact, in most realistic cases, we always assume that Γ is proportional to H.

under which the interaction term contributes a constant pressure throughout the history of the universe. 2. $\frac{\Gamma}{3H\rho_{de}}$ is a constant, under which the interaction term contributes a constant EOS in the history of the universe. In frame of a quintessence or phantom dark energy, the interaction term $\frac{\Gamma}{3H\rho_{de}}$ only shifts the EOS up or down by a constant distance in the $w - z$ plane, without changing the profile of the curve of w. While the term $\frac{\Gamma}{3H}$ shifts the pressure, which can change the EOS significantly since the density ρ_{de} is a variable in the history of the universe.

If the dark energy can couple to some stuff of the universe, the dark matter is the best candidate. Although non-minimal coupling between the dark energy and ordinary matter fluids is strongly restricted by the experimental tests in the solar system [36], due to the unknown nature of the dark matter as part of the background, it is possible to have non-gravitational interactions between the dark energy and the dark matter components, without conflict with the experimental data. The continuity equation for dust-like dark matter reads,

$$\dot{\rho}_{dm} + 3H\rho_{dm} = \Gamma. \tag{61}$$

Based on the previous discussion, we assume a most simple case

$$\Gamma = H\delta\rho_{dm}, \tag{62}$$

where δ is constant [37, 38, 39]. This interaction term shifts a constant to the EOS of *dark matter*, that is, it is no longer evolving as $(1 + z)^3$. We uniformly deal with quintessence and phantom, which are often labeled by X, with a constant EOS w_X. So, the continuity equation of dark energy can be written as,

$$\dot{\rho}_X + 3H(\rho_X + w_X\rho_X) = -H\delta\rho_{dm}. \tag{63}$$

Integrating (61), we derive

$$\rho_{dm} = \rho_{dm0}a^{-3+\delta} = \rho_{dm0}(1 + z)^{3-\delta}. \tag{64}$$

Substituting to (63), we reach

$$\rho_X = \rho_{X0}(1 + z)^{3(1+w_X)} + \rho_{dm0}\frac{\delta}{\delta + 3w_X}\left[(1 + z)^{3(1+w_X)} - (1 + z)^{3-\delta}\right]. \tag{65}$$

Only from the above equation, we can extract the effective EOS of the dark energy. To see this point, we make a short discussion. In a dynamical universe with interaction, the effective EOS of dark energy reads,

$$w_{de} = \frac{p_{eff}}{\rho_{de}} = \frac{p_{de} + \Gamma/3H}{\rho_{de}} = -1 + \frac{1}{3}\frac{d\ln\rho_{de}}{d\ln(1+z)}. \tag{66}$$

Clearly, if $\frac{d\ln\rho_{de}}{d\ln(1+z)}$ is greater than 0, dark energy evolves as quintessence; if $\frac{d\ln\rho_{de}}{d\ln(1+z)}$ is less than 0, it evolves as phantom; if $\frac{d\ln\rho_{de}}{d\ln(1+z)}$ equals 0, it is just cosmological constant. In a more intuitionistic way, if ρ_{de} decreases and then increases with respect to redshift (or time), or increases and then decreases, which implies that EOS of dark energy crosses phantom divide. So, some time we directly use the evolution of density of dark energy to describe the EOS of it. There is a more important motivation to use the density directly: the density is more closely related to observables, hence is more tightly constrained for the same number of redshift bins used [40].

The derivative of ρ_X with respect to $(1+z)$ reads,

$$\frac{d\rho_X}{d(1+z)} = 3(1+w_X)\rho_{X0}(1+z)^{2+3w_X} + \rho_{dm0}\frac{\delta}{\delta+3w_X}$$
$$\left[3(1+w_X)(1+z)^{2+3w_X} - (3-\delta)(1+z)^{2-\delta}\right]. \quad (67)$$

If $\frac{d\rho_X}{d(1+z)} = 0$ at some redshift $z = z_c$, the effective EOS crosses -1. The result is illuminated by fig 8, in which we set $z_c = 0.3$ as an example. This figure displays the corresponding w_X when one fixes a δ, or vice versa if we require the EOS crosses -1 at $z_c = 0.3$. This is an original figure plotted for this review article.

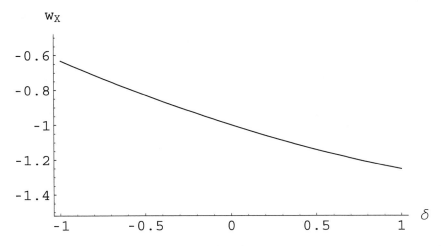

Figure 8. w_X vs δ under the condition $\frac{d\rho_X}{d(1+z)} = 0$.

Then the Friedmann equation reads,

$$\frac{H^2}{H_0^2} = \Omega_{X0}(1+z)^{3(1+w_X)} + \frac{1-\Omega_{X0}}{\delta+3w_X}\left[\delta(1+z)^{3(1+w_X)} + 3w_X(1+z)^{3-\delta}\right], \quad (68)$$

where $\Omega_{X0} = \kappa^2\rho_{X0}/(3H_0^2)$, and we have used $\Omega_{dm0} + \Omega_{X0} = 1$. Thus we need to constrain the three parameters δ, Ω_{X0}, w_X. The constraint result by SNLS data is shown in fig 9. $\delta = 0$ and $w_X = -1$ are indicated by the horizontal and vertical dashed lines, which represent the non-interacting XCDM model and interacting ΛCDM model, respectively.

From fig 8 and 9, we see that the observations leave enough space for the parameters (δ, w_X) to cross the phantom divide.

In the previous interacting model, we consider a phenomenological interaction, which is put in "by hand". We should find a more sound physical foundation for the interactions. We will deduce an interaction term from the low energy limit of string/M theory in the scenario of the interacting Chaplygin gas model [26].

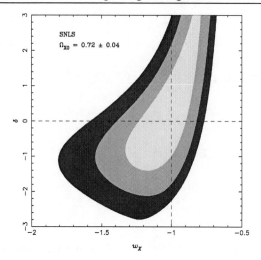

Figure 9. Constraints of (w_X, δ) by SNLS data at 68.3%, 95.4% and 99.7% confidence levels marginalized over Ω_{X0} with priors $\Omega_{X0} = 0.72 \pm 0.04$ and $\delta < 3$. From [37]

The Chaplygin gas model was suggested as a candidate of a unified model of dark energy and dark matter [41]. The Chaplygin gas is characterized by an exotic equation of state

$$p_{ch} = -A/\rho_{ch}, \tag{69}$$

where A is a positive constant. The above equation of state leads to a density evolution in the form

$$\rho_{ch} = \sqrt{A + \frac{B}{a^6}}, \tag{70}$$

where B is an integration constant. The attractive feature of the model is that it naturally unifies both dark energy and dark matter. The reason is that, from (70), the Chaplygin gas behaves as dust-like matter at early stage and as a cosmological constant at later stage.

Though Chaplygin gas has such a nice property, it is a serious flaw when one studies the fluctuation growth in Chaplygin gas model. It is found that Chaplygin gas produces oscillations or exponential blowup of the matter power spectrum, which is inconsistent with observations [35]. So we turn to a model that the Chaplygin gas only plays the role of dark energy. To cross the phantom divide we consider a model in which the Chaplygin gas couples to dark mater.

Although non-minimal coupling between the dark energy and ordinary matter fluids is strongly restricted by the experimental tests in the solar system [36], due to the unknown nature of the dark matter as part of the background, it is possible to have non-gravitational interactions between the dark energy and the dark matter components, without conflict with the experimental data. Thus, the observation constrain the only proper candidate to be coupled to Chaplygin gas is dark matter.

We consider the original Chaplygin gas, whose pressure and energy density satisfy the relation, $p_{ch} = -A/\rho_{ch}$. By assuming the cosmological principle the continuity equations are written as

$$\dot{\rho}_{ch} + 3H\gamma_{ch}\rho_{ch} = -\Gamma, \tag{71}$$

and

$$\dot{\rho}_{dm} + 3H\gamma_{dm}\rho_{dm} = \Gamma, \tag{72}$$

where the subscript dm denotes dark matter, and γ is defined as

$$\gamma = 1 + \frac{p}{\rho} = 1 + w, \tag{73}$$

in which w is the parameter of the state of equation, and $\gamma_{dm} = 1$ throughout the evolution of the universe, whereas γ_{ch} is a variable.

Γ is the interaction term between Chaplygin gas and dark matter. Since there does not exist any microphysical hint on the possible nature of a coupling between dark matter and Chaplygin gas (as dark energy), the interaction terms between dark energy and dark matter are rather arbitrary in literatures [42]. Here we try to present a possible origin from fundamental field theory for Γ.

Whereas we are still lack of a complete formulation of unified theory of all interactions (including gravity, electroweak and strong), there at present is at least one hopeful candidate, string/M theory. However, the theory is far away from mature such that it is still not known in a way that would enable us to ask the questions about space-time in a general manner, say nothing of the properties of realistic particles. Instead, we have to either resort to the effective action approach which takes into account stringy phenomena in perturbation theory, or we could study some special classes of string solutions which can be formulated in the non-perturbative regime. But the latter approach is available only for some special solutions, most notably the BPS states or nearly BPS states in the string spectrum: They seems to have no relation to our realistic Universe. Especially, there still does not exist a non-perturbative formulation of generic cosmological solutions in string theory. Hence nearly all the investigations of realistic string cosmologies have been carried out essentially in the effective action range. Note that the departure of string-theoretic solutions away from general relativity is induced by the presence of additional degrees of freedom which emerge in the massless string spectrum. These fields, including the scalar dilaton field, the torsion tensor field, and others, couple to each other and to gravity non-minimally, and can influence the dynamics significantly. Thus such an effective low energy string theory deserve research to solve the dark energy problem. There a special class of scalar-tensor theories of gravity is considered to avoid singularities in cosmologies in [43]. The action is written below,

$$S_{st} = \int d^4x \sqrt{-g} \left[\frac{1}{16\pi G}R - \frac{1}{2}\partial_\mu\phi\partial^\mu\phi + \frac{1}{q(\phi)^2}L_{\mathrm{dm}}(\xi, \partial\xi, q^{-1}g_{\mu\nu}) \right], \tag{74}$$

where G is the Newton gravitational constant, ϕ is a scalar field, L_{dm} denotes Lagrangian of matter , ξ represents different matter degrees of matter fields, q guarantees the coupling strength between the matter fields and the dilaton. With action (74), the interaction term can be written as follow [43],

$$\Gamma = H\rho_{\mathrm{dm}}\frac{d\ln q'}{d\ln a}. \tag{75}$$

Here we introduce new variable $q(a)' \triangleq q(a)^{(3w_n-1)/2}$, where a is the scale factor in standard FRW metric. By assuming

$$q'(a) = q_0 e^{3\int c(\rho_{\mathrm{dm}}+\rho_\xi)/\rho_{\mathrm{dm}} d\ln a}, \tag{76}$$

where ρ_{dm} and ρ_ξ are the densities of dark matter and the scalar field respectively, one arrive at the interaction term,

$$\Gamma = 3Hc(\rho_{\mathrm{dm}} + \rho_\phi). \tag{77}$$

With this interaction form we study the equation set (71) and (72). Set $s = -\ln(1 + z)$, $\Gamma = 3Hc(\rho_{ch} + \rho_{dm})$, $u = (3H_0^2)^{-1}(3\mu^2)^{-1}\rho_{dm}$, $v = (3H_0^2)^{-1}(3\mu^2)^{-1}\rho_{ch}$, $A' = A(3H_0^2)^{-2}(3\mu^2)^{-2}$, where c is a constant without dimension. Using these variables, (71) and (72) reduce to

$$\frac{du}{ds} = -3u + 3c(u + v), \tag{78}$$

$$\frac{dv}{ds} = -3(v - A'/v) - 3c(u + v). \tag{79}$$

We note that the variable time does not appear in the dynamical system (78) and (79) because time has been completely replaced by redshift $s = -ln(1 + z)$. The critical points of dynamical system (78) and (79) are given by

$$\frac{du}{ds} = \frac{dv}{ds} = 0. \tag{80}$$

The solution of the above equation is

$$u_c = \frac{c}{1 - c} v_c, \tag{81}$$

$$v_c^2 = (1 - c)A'. \tag{82}$$

We see the final state of the model contains both Chaplygin gas and dark matter of constant densities if the singularity is stationary. The final state satisfies perfect cosmological principle: the universe is homogeneous and isotropic in space, as well as constant in time. Physically Γ in (72) plays the role of matter creation term C in the theory of steady state universe at the future time-like infinity. Recall that c is the coupling constant, may be positive or negative, corresponds the energy to transfer from Chaplygin gas to dark matter or reversely. A' must be a positive constant, which denotes the final energy density if c is fixed. Also we can derive an interesting and simple relation between the static energy density ratio

$$c = \frac{r_s}{1 + r_s}, \tag{83}$$

where

$$r_s = \lim_{z \to -1} \frac{\rho_{dm}}{\rho_{ch}}. \tag{84}$$

To investigate the properties of the dynamical system in the neighbourhood of the singularities, impose a perturbation to the critical points,

$$\frac{d(\delta u)}{ds} = -3\delta u + 3c(\delta u + \delta v), \tag{85}$$

$$\frac{d(\delta v)}{ds} = -3(\delta v + \frac{A'}{v_c^2}\delta v) - 3c(\delta u + \delta v). \tag{86}$$

The eigen equation of the above linear dynamical system $(\delta u, \delta v)$ reads

$$(\lambda/3)^2 + (2 + \frac{1}{1-c})\lambda/3 + 2 - 2c^2 = 0, \tag{87}$$

whose discriminant is

$$\Delta = [(1-c)^4 + (3/2 - c)^2]/(1-c)^2 \geq 0. \tag{88}$$

Therefore both of the two roots of eigen equation (87) are real, consequently centre and focus singularities can not appear. Furthermore only $r_s \in (0, \infty)$, such that $c \in (0, 1)$, makes physical sense. Under this condition it is easy to show that both the two roots of (87) are negative. Hence the two singularities are stationary. However it is only the property of the linearized system (85) and (86), or the property of orbits of the neighbourhoods of the singularities, while global Poincare-Hopf theorem requires that the total index of the singularities equals the Euler number of the phase space for the non-linear system (78) and (79). So there exists other singularity except for the two nodes. In fact it is a non-stationary saddle point at $u = 0$, $v = 0$ with index -1. This singularity has been omitted in solving equations (78) and (79). The total index of the three singularities is 1, which equals the Euler number of the phase space of this plane dynamical system. Hence there is no other singularities in this system. From these discussions we conclude that the global outline of the orbits of this non-linear dynamical system (78) and (79) is similar to the electric fluxlines of two negative point charges. Here we plot figs 10 and 11 to show the properties of evolution of the universe controlled by the dynamical system (78) and (79). As an example we set $c = 0.2$, $A' = 0.9$ in figs 10-12.

Further, to compare with observation data we need the explicit forms of $u(x)$ and $v(x)$, especially $v(x)$. We need the properties of γ_{ch} in our model, which is contained in $v(x)$, to compare with observations. Eliminate $u(x)$ by using (78) and (79) we derive

$$\frac{1}{3c}\frac{d^2v}{ds^2} + [1 + (1 + A'/v^2)/c]\frac{dv}{ds} + 3cv + 3(1-c)$$
$$\left\{ v + \left[\frac{dv}{ds} + 3(v - A'/v) \right]/(3c) \right\} = 0, \tag{89}$$

which has no analytic solution. We show some numerical solutions in figure 12. We find that for proper region of parameter spaces, the effective equation of state of Chaplygin gas crosses the phantom divide successfully.

Up to now all of our results do not depend on Einstein field equation. They only depend on the most sound principle in physics, that is, the continuity principle, or the energy conservation law. Different gravity theories correspond to different constraints imposed on our previous discussions. Our improvements show how far we can reach without information of dynamical evolution of the universe.

(78) illuminates that the dark matter in this interacting model does not behaves as dust. Qualitatively, the dark matter gets energy from dark energy for a positive c, and becomes soft, ie, its energy density decreases slower than $(1 + z)^3$ in an expanding universe. The parameter which carries the total effects of cosmic fluids is the deceleration parameter q. From now on we introduce the Friedmann equation of the standard general relativity. As a simple case we study the evolution of q in a spatially flat universe. So q reads

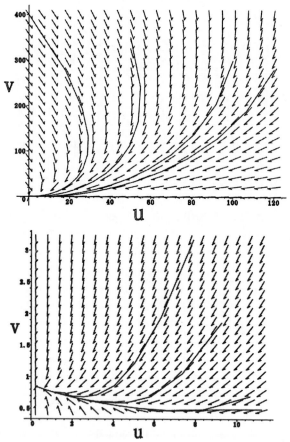

Figure 10. The plane v versus u. **(a)** left panel: We consider the evolution of the universe from redshift $z = e^2 - 1$. The initial condition is taken as $u = 0$, $v = 400$; $u = 50$, $v = 350$; $u = 100$, $v = 300$; $u = 120$, $v = 280$ on the four orbits, from the left to the right, respectively. It is clear that there is a stationary node, which attracts most orbits in the first quadrant. At the same time the orbits around the neibourhood of the singularity is not shown clearly. **(b)** right panel: Orbit distributions around the node $u_c = v_c c/(1 - c)$, $v_c = \sqrt{(1 - c)A'}$. From [26]

$$q = -\frac{\ddot{a}a}{\dot{a}^2} = \frac{1}{2}\left(\frac{u + v - 3A'/v^2}{u + v}\right), \tag{90}$$

and density of Chaplygin gas u and density of dark matter v should satisfy

$$u(0) + v(0) = 1. \tag{91}$$

And then Friedmann equation ensures the spatial flatness in the whole history of the universe. Before analyzing the evolution of q with redshift, we first study its asymptotic behaviors. When $z \to \infty$, q must go to $1/2$ because both Chaplygin gas and dark matter behave like dust , while when $z \to -1$ q is determined by

$$\lim_{z \to -1} q = \frac{1}{2}\left(\frac{u_c + v_c - 3A'/v_c^2}{u_c + v_c}\right). \tag{92}$$

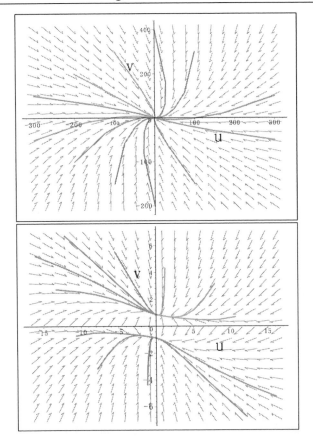

Figure 11. The plane v versus u. **(a)** left panel: To show the global properties of dynamical system (78) and (79) we have to include some "unphysical " initial conditions, such as $u = -100$, $v = -300$, except for physical initial conditions which have been shown in figure 10. **(b)** right panel: Orbits distributions around the nodes. The two nodes $u_c = v_c c/(1 - c)$, $v_c = \sqrt{(1 - c)A'}$ and $u_c = v_c c/(1 - c)$, $v_c = -\sqrt{(1 - c)A'}$ keep reflection symmetry about the original point. Just as we have analyzed, we see that the orbits of this dynamical system are similar to the electric fluxlines of two negative point charges. From [26]

One can finds the parameters $c = 0.2$, $A' = 0.9$ are difficult to content the previous constraint Friedmann constraint (91). Here we carefully choose a new set of parameter which satisfies Friedmann constraint (91), say, $A' = 0.4$, $c = 0.06$. Therefore we obtain

$$\lim_{z \to -1} q = -1.95, \tag{93}$$

by using (81) and (82). Then we plot figure 13 to clearly display the evolution of q. One can check $u(0) = 0.25$, $v(0) = 0.75$; $u(0) = 0.28$, $v(0) = 0.72$; $u(0) = 0.3$, $v(0) = 0.7$, respectively on the curves $v(-2) = 273$; $v(-2) = 250$; $v(-2) = 233$. One may find an interesting property of the deceleration parameter displayed in fig 13: the bigger the proportion of the dark energy, the smaller the absolute value of the deceleration parameter. The reason roots in the extraordinary state of Chaplygin gas (69), in which the pressure p_{ch} is inversely proportional to the energy density ρ_{ch}.

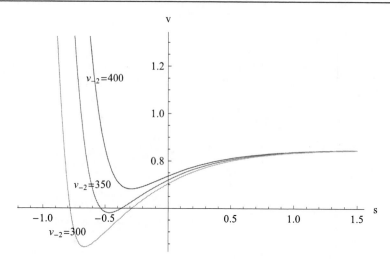

Figure 12. v versus s. The evolution of v with different initial conditions $u(-2) = 0$, $v(-2) = 400$; $u(-2) = 50$, $v(-2) = 350$; $u(-2) = 100$, $v(-2) = 300$ reside on the blue, red, and yellow curves, respectively. Obviously the energy density of Chaplygin gas rolls down and then climbs up in some low redshift region. So the Chaplygin gas dark energy can cross the phantom divide $w = -1$ in a fitting where the dark energy is treated as an independent component to dark matter. From [26], this figure has been re-plotted.

Also we note that maybe an FRW universe with non-zero spatial curvature fits deceleration parameter better than spatially flat FRW universe. This point deservers to research further.

After the presentation of the original interacting Chaplygin gas model, there are several generalizations. For details of these generalizations, see [44].

3.3. Model in Frame of Modified Gravity

The judgement that there exists an exotic component with negative pressure, or dark energy, which accelerates the universe, is derived in frame of general relativity. The validity of general relativity has been well tested from the scale of millimeter to the scale of the solar system. Beyond this scale, the evidences are not so sound. So we should not be surprised if general relativity fails at the scale of the Hubble radius. Surely, any new gravity theory must reduce to general relativity at the scale between millimeter to the solar system. In frame of the new gravity theories, the cosmic acceleration may be a natural result even we only have dust in the universe.

There are various suggestions on how to modify general relativity. In this brief review we concentrates on the brane world theory. Inspired by the developments of string/M theory, the idea that our universe is a 3-brane embedded in a higher dimensional spacetime has received a great deal of attention in recent years. In this brane world scenario, the standard model particles are confined on the 3-brane, while the gravitation can propagate in the whole space. In this picture, the gravity field equation gets modified at the left hand side (LHS) in (4), while the dark energy is a stuff put at the right hand side (RHS) in (4). In the modified gravity model, the surplus geometric terms respective to the Einstein tensor play

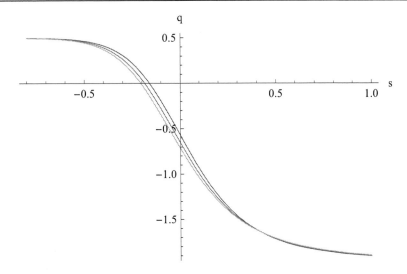

Figure 13. q versus s. The evolution of q with different initial conditions $u(-2) = 0$, $v(-2) = 273$; $u(-2) = 15$, $v(-2) = 250$; $u(-2) = 25$, $v(-2) = 233$, reside on the blue, red, and yellow curves, respectively. Evidently the deceleration parameter q of Chaplygin gas rolls down and crosses $q = 0$ in some low redshift region. The transition from deceleration phase to acceleration phase occurs at $z = 0.18$; $z = 0.21$; $z = 0.23$ to the curves $u(-2) = 0$, $v(-2) = 273$; $u(-2) = 15$, $v(-2) = 250$; $u(-2) = 25$, $v(-2) = 233$, respectively. One finds $-q \approx 0.5 \sim 0.6$ at $z = 0$, which is well consistent with observations. From [26], this figure has been re-plotted.

the role of the dark energy in general relativity.

We consider a 3-brane imbedded in a 5-dimensional bulk. The action includes the action of the bulk and the action of the brane,

$$S = S_{\text{bulk}} + S_{\text{brane}}. \tag{94}$$

Here

$$S_{\text{bulk}} = \int_{\mathcal{M}} d^5 X \sqrt{-g_5} \mathcal{L}_{\text{bulk}}, \tag{95}$$

where $X = (t, z, x^1, x^2, x^3)$ is the bulk coordinate, x^1, x^2, x^3 are the coordinates of the maximally symmetric space. $\mathcal{M}$ denotes the bulk manifold. The bulk Lagrangian can be

$$\mathcal{L}_{\text{bulk}} = \frac{1}{2\kappa_5^2} [R_5 + \alpha F(R_5)] + \mathcal{L}_{\text{m}} + \Lambda_5, \tag{96}$$

where g_5, κ_5, R_5, $\mathcal{L}_{\text{m}}$, denote the bulk manifold, the determinant of the bulk metric, the 5-dimensional Newton constant, the 5-dimensional Ricci scalar, and the bulk matter Lagrangian, respectively. $F(R_5)$ denotes the higher order term of scalar curvature R_5, the Ricci curvature R_{5AB}, the Riemann curvature R_{5ABCD}.

There are too much possibilities and rather arbitrary to choose the higher order terms. Generally the resulting equations of motion of such a term give more than second derivatives of metric and the resulting theory is plagued by ghosts. However there exists a combination of quadratic terms, called Gauss-Bonnet term, which generates equation of motion without the terms more than second derivatives of metric and the theory is free of ghosts [45]. Another important property of Gauss-Bonnet term is that, just like Hilbert Lagrangian is a pure divergence in 2 dimensions and Einstein tensor identifies zero in 1 and 2 dimensions, we have that in 4 or less dimension the Gauss-Bonnet Lagrangian is a pure divergence. We see the dilemma of quadratic term in 4 dimensional theory: if we include it with non pure divergence we shall confront ghosts; if we want to remove ghosts we get a pure divergence term. So only in theories in more than 4 dimensional Gauss-Bonnet combination provides physical effects. Moreover the Gauss-Bonnet term also appears in both low energy effective action of Bosonic string theory [46] and low energy effective action of Bosonic modes of heterotic and type II super string theory [47]. An investigation into the effects of a Gauss-Bonnet term in the 5 dimensional bulk of brane world models is therefore well motivated. The Gauss-Bonnet term in 5 dimension reads,

$$F(R_5) = R_5^2 - 4R_{5\mathrm{AB}}R_5^{\mathrm{AB}} + R_{5\mathrm{ABCD}}R_5^{\mathrm{ABCD}}. \tag{97}$$

The action of the brane can be written as,

$$S_{\mathrm{brane}} = \int_M d^4x\sqrt{-g}\left(\kappa_5^{-2}K + L_{\mathrm{brane}}\right), \tag{98}$$

where M indicates the brane manifold, g denotes the determinant of the brane metric, L_{brane} stands for the Lagrangian confined to the brane, and K marks the trace of the second fundamental form of the brane. $x = (\tau, x^1, x^2, x^3)$ is the brane coordinate. Note that τ is not identified with t if the the brane is not fixed at a position in the extra dimension $z =$ constant. We will investigate the cosmology of a moving brane along the extra dimension z in the bulk, and such that τ is different from t.

We set the Lagrangian confined to the brane as follows,

$$L_{\mathrm{brane}} = \frac{1}{16\pi G}R - \lambda + L_{\mathrm{m}}, \tag{99}$$

where λ is the brane tension and L_{m} denotes the ordinary matter, such as dust and radiation, located at the brane. R denotes the 4 dimensional scalar curvature term on the brane, which is an important one except a Gauss-Bonnet term in the bulk. This induced gravity correction arises because the localized matter fields on the brane, which couple to bulk gravitons, can generate via quantum loops a localized four-dimensional world-volume kinetic term for gravitons [48].

Assuming there is a mirror symmetry in the bulk, we have the Friedmann equation on the brane [49], see also [50],

$$\frac{4}{r_c^2}\left[1 + \frac{8}{3}\alpha\left(H^2 + \frac{k}{a^2} + \frac{U}{2}\right)\right]^2\left(H^2 + \frac{k}{a^2} - U\right) = \left(H^2 + \frac{k}{a^2} - \frac{8\pi G}{3}(\rho + \lambda)\right)^2 \tag{100}$$

where

$$U = -\frac{1}{4\alpha} \pm \frac{1}{4\alpha}\sqrt{1 + 4\alpha\left(\frac{\Lambda_5}{6} + \frac{M\kappa_5^2}{4\pi a^4}\right)}, \tag{101}$$

$$r_c = \kappa_5^2\mu^2. \tag{102}$$

Here M is a constant, standing for the mass of bulk black hole. For various limits of (100), see [51].

For convenience, we introduce the following new variables and parameters,

$$x \equiv \frac{H^2}{H_0^2} + \frac{k}{a^2 H_0^2} = \frac{H^2}{H_0^2} - \Omega_{k0}(1+z)^2,$$

$$u \equiv \frac{8\pi\, {}^{(4)}G}{3H_0^2}(\rho + \lambda) = \Omega_{m0}(1+z)^3 + \Omega_\lambda,$$

$$m \equiv \frac{8}{3}\alpha H_0^2,$$

$$n \equiv \frac{1}{H_0^2 r_c^2},$$

$$y \equiv \frac{1}{2}U H_0^{-2} = \frac{1}{3m}\left(-1 + \sqrt{1 + \frac{4\alpha\Lambda_5}{6} + \frac{8\alpha M^{(5)}G}{a^4}}\right)$$

$$= \frac{1}{3m}\left(-1 + \sqrt{1 + m\Omega_{\Lambda_5} + m\Omega_{M0}(1+z)^4}\right), \tag{103}$$

and we have assumed that there is only pressureless dust in the universe. As before, we have used the following notations

$$\Omega_{k0} = -\frac{k}{a_0^2 H_0^2}, \quad \Omega_{m0} = \frac{8\pi G}{3}\frac{\rho_{m0}}{H_0^2}, \quad \Omega_\lambda = \frac{8\pi G}{3}\frac{\lambda}{H_0^2}, \quad \Omega_{\Lambda_5} = \frac{3\Lambda_5}{8H_0^2}, \quad \Omega_{M0} = \frac{3M\kappa_5^2}{8\pi a_0^4 H_0^2}. \tag{104}$$

With these new variables and parameters, (100) can be rewritten as

$$4n(x - 2y)[1 + m(x + y)]^2 = (x - u)^2. \tag{105}$$

This is a cubic equation of the variable x. According to algebraic theory it has 3 roots. One can explicitly write down three roots. But they are too lengthy and complicated to present here. Instead we only express those three roots formally in the order given in *Mathematica*

$$x_1 = x_1(y, u | m, n),$$
$$x_2 = x_2(y, u | m, n),$$
$$x_3 = x_3(y, u | m, n), \tag{106}$$

where y and u are two variables, m and n stand for two parameters. The root on x of the equation (105) gives us the modified Friedmann equation on the Gauss-Bonnet brane world with induced gravity. From the solutions given in (106), this model seems to have three branches. In addition, note that all parameters introduced in (103) and (104) are not independent of each other. According to the Friedmann equation (106), when all variables are taken current values, for example, $z = 0$, the Friedmann equation will give us a constraint on those parameters,

$$1 = f(\Omega_{k0}, \Omega_{m0}, \Omega_{M0}, \Omega_{\Lambda_5}, \Omega_\lambda, m, n). \tag{107}$$

To compare with observation, we introduce the concept "equivalent dark energy" or "virtual dark energy" in the modified gravity models, since almost all the properties of dark energy are deduced in the frame of general relativity with a dark energy.

The Friedmann equation in the four dimensional general relativity can be written as

$$H^2 + \frac{k}{a^2} = \frac{8\pi G}{3}(\rho + \rho_{de}), \tag{108}$$

where the first term of RHS of the above equation represents the dust matter and the second term stands for the dark energy. Generally speaking the Bianchi identity requires,

$$\frac{d\rho_{de}}{dt} + 3H(\rho_{de} + p_{de}) = 0, \tag{109}$$

we can then express the equation of state for the dark energy as

$$w_{de} = \frac{p_{de}}{\rho_{de}} = -1 - \frac{1}{3}\frac{d\ln\rho_{de}}{dlna}. \tag{110}$$

Note that we can rewrite the Friedmann equation (106) in the form of (108) as

$$xH_0^2 = \frac{8\pi G}{3}\rho + \left(H_0^2 x(y, u|m, n) - \frac{8\pi G}{3}\rho\right) = \frac{8\pi G}{3}(\rho + Q), \tag{111}$$

where ρ is the energy density of dust matter on the brane and the term

$$Q \equiv \frac{3H_0^2}{8\pi G}x(y, u|m, n) - \rho \tag{112}$$

corresponds to ρ_{de} in (108).

In fig 14 we show the equation of state for the virtual dark energy when we take $m = 1.036$ and $n = 0.04917$. In this case, from the constraint equation (128), one has $\Omega_{M0} = 2.08$. From the figure we see that $w_{eff} < -1$ at $z = 0$ and

$$\left.\frac{dw_{eff}}{dz}\right|_{z=0} < 0.$$

Therefore the equation of state for the virtual dark energy can indeed cross the phantom divide $w = -1$ near $z \sim 0$.

Fig 14 illuminates that the behavior of the virtual dark energy seems rather strange. However we should remember that it is only virtual dark energy, not actual stuff. The whole evolution of the universe is described by the Hubble parameter. We plot the Hubble parameter H corresponding to fig 14 in fig 15.

Fig 15 displays that the universe will eventually becomes a de Sitter one. For more figures with different parameters, see [51]. The constraint of this brane model with induced scalar term on the brane and Gauss-Bonnet term in the bulk has been investigated in [52].

The above is an example of "pure geometric" dark energy. We can also consider some mixed dark energy model, ie, the cosmic acceleration is driven by an exotic matter and some geometric effect in part. Why such an apparently complicated suggestion? There are many interesting models are proposed to explain the cosmic acceleration, including dark energy and modified gravity models. However, several influential and hopeful models, such as quintessence and DGP model, fundamentally can not account for the crossing -1 behavior of dark energy. By contrast, some hybrid model of the dark energy and modified gravity

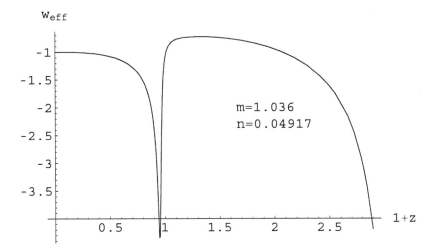

Figure 14. The equation of state w_{eff} with respect to the red shift $1 + z$, with $\Omega_{m0} = 0.28$ and $\Omega_A = 2.08$. From [51].

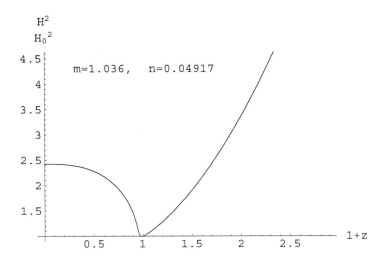

Figure 15. H^2/H_0^2 versus $1 + z$, with $\Omega_{m0} = 0.28$ and $\Omega_A = 2.08$. From [51].

may realize such a crossing. As an example we study the quintessence and phantom in frame of DGP [53].

Our starting point is still action (94). In a DGP model with a scalar, (96) becomes a pure Einstein-Hilbert action,

$$\mathcal{L}_{\text{bulk}} = \frac{1}{2\kappa_5^2} R_5, \tag{113}$$

and we add a scalar term in (99),

$$\mathcal{L}_{\text{brane}} = \frac{1}{16\pi G} R + L_{\text{m}} + L_{\text{scalar}}. \tag{114}$$

Here the scalar term can be ordinary scalar (quintessence) or phantom (scalar with negative kinetic term). The Lagrangian of a quintessence reads,

$$L_\phi = -\frac{1}{2}\partial_\mu\phi\partial^\mu\phi - V(\phi), \tag{115}$$

and for phantom,

$$L_\psi = \frac{1}{2}\partial_\mu\psi\partial^\mu\psi - U(\psi). \tag{116}$$

In an FRW universe we have

$$\rho_\phi = \frac{1}{2}\dot\phi^2 + V(\phi), \tag{117}$$

$$p_\phi = \frac{1}{2}\dot\phi^2 - V(\phi). \tag{118}$$

The exponential potential is an important example which can be solved exactly in the standard model. Also it has been shown that the inflation driven by a scalar with exponential potential can exit naturally in the warped DGP model [54]. It is therefore quite interesting to investigate a scalar with such a potential in late time universe on a DGP brane. Here we set

$$V = V_0 e^{-\lambda_1 \frac{\phi}{\mu}}. \tag{119}$$

Here λ_1 is a constant and V_0 denotes the initial value of the potential.

The Friedmann equation (100) becomes

$$H^2 + \frac{k}{a^2} = \frac{1}{3\mu^2}\left[\rho + \rho_0 + \theta\rho_0(1 + \frac{2\rho}{\rho_0})^{1/2}\right], \tag{120}$$

where

$$\rho_0 = \frac{6\mu^2}{r_c^2}. \tag{121}$$

Similar to the previous case, we derive the virtual dark energy by comparing (120) and (108),

$$\rho_{de} = \rho_\phi + \rho_0 + \theta\rho_0\left[\rho + \rho_0 + \theta\rho_0(1 + \frac{2\rho}{\rho_0})^{1/2}\right]. \tag{122}$$

From (110), we calculate the derivation of effective density of dark energy with respective to $\ln(1 + z)$ for a ordinary scalar,

$$\frac{d\rho_{de}}{d\ln(1 + z)} = 3[\dot\phi^2 + \theta(1 + \frac{\dot\phi^2 + 2V + 2\rho_{dm}}{\rho_0})^{-1/2}(\dot\phi^2 + \rho_{dm})]. \tag{123}$$

If $\theta = 1$, both terms of RHS are positive, hence it never goes to zero at finite time. But if $\theta = -1$, the two terms of RHS carry opposite sign, therefore it is possible that the EOS of dark energy crosses phantom divide. In a scalar-driven DGP, we only consider the case of $\theta = -1$.

For convenience, we define some dimensionless variables,

$$y_1 \triangleq \frac{\dot{\phi}}{\sqrt{6}\mu H}, \tag{124}$$

$$y_2 \triangleq \frac{\sqrt{V}}{\sqrt{3}\mu H}, \tag{125}$$

$$y_3 \triangleq \frac{\sqrt{\rho_m}}{\sqrt{3}\mu H}, \tag{126}$$

$$y_4 \triangleq \frac{\sqrt{\rho_0}}{\sqrt{3}\mu H}. \tag{127}$$

The Friedmann equation (120) becomes

$$y_1^2 + y_2^2 + y_3^2 + y_4^2 - y_4^2 \left(1 + 2\frac{y_1^2 + y_2^2 + y_3^2}{y_4^2}\right)^{1/2} = 1. \tag{128}$$

The stagnation point, that is, $d\rho_{de}/d\ln(1+z) = 0$ dwells at

$$\frac{y_4}{\sqrt{2}+y_4}\left(2 + \frac{y_3^2}{y_1^2}\right) = 2, \tag{129}$$

which can be derived from (123) and (128).

One concludes from the above equation that a smaller r_c, a smaller Ω_{m0} (Recall that it is defined as the present value of the energy density of dust matter over the critical density), or a larger Ω_{ki} (which is defined as the present value of the kinetic energy density of the scalar over the critical density) is helpful to shift the stagnation point to lower redshift region. We show a concrete numerical example of this crossing behaviours in fig 16. For convenience we introduce the dimensionless density and rate of change with respect to redshift of dark energy as below,

$$\beta = \frac{\rho_{de}}{\rho_c} = \frac{\Omega_{r_c}}{b^2}\left[y_1^2 + y_2^2 + y_4^2 - y_4^2(1 + 2\frac{y_1^2 + y_2^2 + y_3^2}{y_4^2})^{1/2}\right], \tag{130}$$

where ρ_c denotes the present critical density of the universe, and

$$\begin{aligned}
\gamma &= \frac{1}{\rho_c}\frac{y_4^2}{\Omega_{r_c}}\frac{d\rho_{de}}{ds} \\
&= 3\left[(1 + 2\frac{y_1^2 + y_2^2 + y_3^2}{y_4^2})^{-1/2}(2y_1^2 + y_3^2) - 2y_1^2\right].
\end{aligned} \tag{131}$$

A significant parameters from the viewpoint of observations is the deceleration parameter q, which carries the total effects of cosmic fluids. We plot q in these figures for corresponding density curve of dark energy. In the fig 16 we set $\Omega_m = 0.3$. Ω_{r_c} is defined as the present value of the energy density of ρ_0 over the critical density $\Omega_{r_c} = \rho_0/\rho_c$.

Now, we turn to the evolution of a universe with a phantom (116) in DGP. In an FRW universe the density and pressure of a phantom can be written as (17), (18).

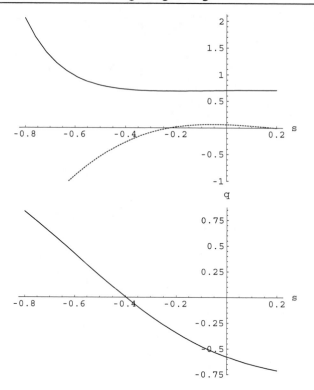

Figure 16. For this figure, $\Omega_{ki} = 0.01$, $\Omega_{r_c} = 0.01$, $\lambda_1 = 0.5$. **(a)** The left panel: β and γ as functions of s, in which β resides on the solid line, while γ dwells at the dotted line. The EOS of dark energy crosses -1 at about $s = -0.22$, or $z = 0.25$. **(b)** The right panel: the corresponding deceleration parameter, which crosses 0 at about $s = -0.40$, or $z = 0.49$. From [53].

To compare with the results of the ordinary scalar, here we set a same potential as before,

$$U = U_0 e^{-\lambda_2 \frac{\psi}{\mu}}. \tag{132}$$

The ratio of change of density of virtual dark energy with respective to $\ln(1 + z)$ becomes,

$$\frac{d\rho_{de}}{d\ln(1+z)} = 3[-\dot{\psi}^2 + \theta(1 + \frac{-\dot{\psi}^2 + 2U + 2\rho_{dm}}{\rho_0})^{-1/2}(-\dot{\psi}^2 + \rho_{dm})]. \tag{133}$$

To study the behaviour of the EOS of dark energy, we first take a look at the signs of the terms of RHS of the above equation. $(-\dot{\psi}^2 + \rho_{dm})$ represents the total energy density of the cosmic fluids, which should be positive. The term $(1 + \frac{-\dot{\psi}^2 + 2U + 2\rho_{dm}}{\rho_0})^{-1/2}$ should also be positive. Hence if $\theta = -1$, both terms of RHS are negative: it never goes to zero at finite time. Contrarily, if $\theta = 1$, the two terms of RHS carry opposite sign: the EOS of dark energy is able to cross phantom divide. In the following of the present subsection we consider the branch of $\theta = 1$.

Now the Friedmann constraint becomes

$$-y_1^2 + y_2^2 + y_3^2 + y_4^2 + y_4^2 \left(1 + 2\frac{-y_1^2 + y_2^2 + y_3^2}{y_4^2}\right)^{1/2} = 1. \tag{134}$$

Again, one will see that in reasonable regions of parameters, the EOS of dark energy crosses -1, but from below -1 to above -1.

The stagnation point of ρ_{de} inhabits at

$$\frac{y_4}{\sqrt{2} - y_4}\left(-2 + \frac{y_3^2}{y_1^2}\right) = 2, \tag{135}$$

which can be derived from (133) and (134). One concludes from the above equation that a smaller r_c, a smaller Ω_m, or a larger Ω_{ki} is helpful to shift the stagnation point to lower redshift region, which is the same as the case of an ordinary scalar. Then we show a concrete numerical example of the crossing behaviour of this case in fig 17. The dimensionless density and rate of change with respect to redshift of dark energy become,

$$\beta = \frac{\rho_{de}}{\rho_c} = \frac{\Omega_{r_c}}{b^2}\left[-y_1^2 + y_2^2 + y_4^2 + y_4^2(1 + 2\frac{-y_1^2 + y_2^2 + y_3^2}{y_4^2})^{1/2}\right], \tag{136}$$

and

$$\gamma = 3\left[-(1 + 2\frac{-y_1^2 + y_2^2 + y_3^2}{y_4^2})^{-1/2}(-2y_1^2 + y_3^2) + 2y_1^2\right]. \tag{137}$$

Similarly, the deceleration parameter is plotted in the figure 17 for corresponding density curve of dark energy. In this figures we also set $\Omega_m = 0.3$.

Fig 17 explicitly illuminates that the EOS of virtual dark energy crosses -1, as expected. At the same time the deceleration parameter is consistent with observations.

4. Summary

The recent observations imply that the EOS of dark energy may cross -1. This is a remarkable phenomenon and attracts much theoretical attention.

We review three typical models for the crossing behavior. They are two-field model, interacting model, and modified gravity model.

There are several other interesting suggestions in or beyond the three categories mentioned above. We try to list them here for the future researches. We are apologized for this incomplete reference list on this topic.

Almost in all dark energy models the dark energy is suggested as scalar. However, 3 orthogonal vectors can also play this role. For the interacting vector dark energy and phantom divide crossing, see [55]. For the suggestion of crossing the phantom divide with a spinor, see [56]. Multiple k-essence sources are helpful to fulfil the condition for phantom divide crossing [57]. Phantom divide crossing can be realized by non-minimal coupling and Lorentz invariance violation [58]. An exact solution of a two-field model for this crossing has been found in [59].

For previous interacting X (quintessence or phantom) models with crossing -1, see [60]. The cosmology of interacting X in loop gravity has been studied in [61].

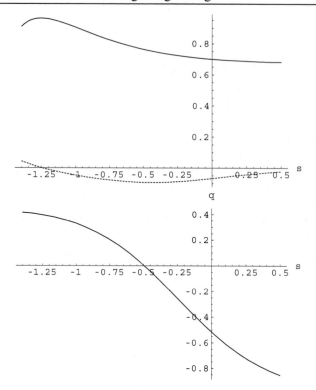

Figure 17. For this figure, $\Omega_{ki} = 0.01$, $\Omega_{r_c} = 0.01$, $\lambda = 0.01$. **(a)** The left panel: β and γ as functions of s, in which β resides on the solid line, while γ dwells at the dotted line. The EOS of dark energy crosses -1 at about $s = -1.25$, or $z = 1.49$. **(b)** The right panel: The corresponding deceleration parameter, which crosses 0 at about $s = -0.50$, or $z = 0.65$. From [53].

Interacting holographic dark energy is a possible mechanism for the phantom divide crossing [62]. And the thermodynamics of interacting holographic dark energy with phantom divide crossing is investigated in [63]. An explicit model of $F(R)$ gravity in which the dark energy crosses the phantom divide is reconstructed in [64]. The phantom-like effects in a DGP-inspired $F(R, \phi)$ gravity model is investigated in [65]. Based on the recent progress in studies of source of Taub space [66], a new braneworld in the sourced-Taub background is proposed [67], while the previous brane world models are imbedded in AdS (RS) or Minkowski (DGP). In this model the EOS for the virtual dark energy of a dust brane in the source region can cross the phantom divide. For other suggestions in brane world model, see [68].

Similar to the coincidence problem of dark energy, we can ask why the EOS crosses -1 recently? This problem is studied in [69].

On the observational side, the present data only mildly favor the crossing behavior. We need more data to confirm or exclude it.

Theoretically, we should find more natural model which has less parameters. We must go beyond the standard model of particle physics. The problem cosmic acceleration is a pivotal problem to access new physics. To study the problem of crossing -1 EOS will impel the investigation to the new Laws of nature.

References

[1] A. Einstein, Sitzungsber. K. *Akad.,* **6**, 142(1917). Amazingly, this paper offered a now standard explanation of the present acceleration of the universe, though Einstein did not intend to do so.

[2] A. A. Friedmann, *Z. Phys.* **10**, 377 (1922); 1924, *Z. Phys.* **21**, 326 (1924).

[3] E.P. Hubble, 1929, *Proc. U.S. Nat. Acad. Sci.* **15**, 168.

[4] A. G. Riess et al. , *Astron. J.* **116**, 1009 (1998), astro-ph/9805201; S. Perlmutter et al., *Astrophys. J.* **517**, 565 (1999), astro-ph/9812133.

[5] A. G. Riess et al., *Astrophys. J.* **607**, 665 (2004) [arXiv:astro-ph/0402512].

[6] A. G. Riess et al., *Astrophys. J.* **659**, 98 (2007) [arXiv:astro-ph/0611572].

[7] P. Astier et al. [SNLS Collaboration], *Astron. Astrophys.* **447**, 31 (2006) [arXiv:astro-ph/0510447].

[8] W. M.Wood-Vasey et al. [ESSENCE Collaboration], Astrophys. J. 666, 694 (2007) [arXiv:astro-ph/0701041].

[9] T. M. Davis et al., *Astrophys. J.* **666**, 716 (2007) [astro-ph/0701510].

[10] M. Kowalski et al., *Astrophys. J.* **686**, 749 (2008) [arXiv:0804.4142]. The numerical data of the full sample are also available at http://supernova.lbl.gov/Union

[11] M. Hicken et al., arXiv:0901.4804 [astro-ph.CO]; M. Hicken et al., arXiv:0901.4787 [astro-ph.CO].

[12] C. Shapiro and M. S. Turner, *Astrophys. J.* **649**, 563 (2006) [arXiv:astro-ph/0512586].

[13] A. Tikhonov and A. Klypin, arXiv:0807.0924 [astro-ph]; L. Perivolaropoulos, arXiv:0811.4684 [astro-ph]; L. Perivolaropoulos and A. Shafieloo, *Phys. Rev. D* **79**, 123502 (2009) [arXiv:0811.2802 [astro-ph]].

[14] R.R. Caldwell, Phys.Lett. B545 (2002) 23, astro-ph/9908168; P. Singh, M. Sami and N. Dadhich, *Phys. Rev.* **D68** (2003) 023522, hep-th/0305110.

[15] S. M. Carroll, M. Hoffman and M. Trodden, *Phys. Rev. D* **68**, 023509 (2003) [arXiv:astro-ph/0301273]; E. O. Kahya and V. K. Onemli, *Phys. Rev. D* **76**, 043512 (2007) [arXiv:gr-qc/0612026]; Hongsheng Zhang and Zong-Hong Zhu, JCAP03(2008)007;

[16] E. O. Kahya, V. K. Onemli and R. P. Woodard, arXiv:0904.4811 [gr-qc].

[17] U. Alam, V. Sahni and A. A. Starobinsky, *JCAP* **0406**, 008 (2004) [arXiv:astro-ph/0403687].

[18] D. Huterer and A. Cooray, *Phys. Rev. D* **71**, 023506 (2005) [arXiv:astro-ph/0404062].

[19] J. Q. Xia, G. B. Zhao, B. Feng, H. Li and X. Zhang, *Phys. Rev. D* **73**, 063521 (2006) [arXiv:astro-ph/0511625].

[20] C. L. Bennett et al. (WMAP Collaboration), *Astrophys. J. Suppl.* **148**, 1 (2003).

[21] E. M. . Barboza, J. S. Alcaniz, Z. H. Zhu and R. Silva, arXiv:0905.4052 [astro-ph.CO].

[22] J. Q. Xia, H. Li, G. B. Zhao and X. Zhang, *Phys. Rev. D* **78**, 083524 (2008) [arXiv:0807.3878 [astro-ph]];G. B. Zhao, D. Huterer and X. Zhang, *Phys. Rev. D* **77**, 121302 (2008) [arXiv:0712.2277 [astro-ph]]; H. Li *et al.*, arXiv:0812.1672 [astro-ph];S. Nesseris and L. Perivolaropoulos, *JCAP* **0701** (2007) 018 [arXiv:astro-ph/0610092].

[23] A. Vikman, *Phys. Rev. D* **71**, 023515 (2005) [arXiv:astro-ph/0407107]

[24] B. Feng, X. L. Wang and X. M. Zhang, *Phys. Lett. B* **607**, 35 (2005) [arXiv:astro-ph/0404224]; B. Feng, M. Li, Y. S. Piao and X. Zhang, *Phys. Lett. B* **634**, 101 (2006) [arXiv:astro-ph/0407432].

[25] H. H. Xiong, T. Qiu, Y. F. Cai and X. Zhang, *Mod. Phys. Lett. A* **24** (2009) 1237; M. R. Setare and A. Rozas-Fernandez, arXiv:0906.1936 [hep-th]; J. Sadeghi, M. R. Setare and A. Banijamali, arXiv:0903.4073 [hep-th];J. Wang and S. Yang, arXiv:0901.1441 [gr-qc];J. Wang, S. Cui and S. Yang, arXiv:0901.1439 [gr-qc]; K. Nozari, M. R. Setare, T. Azizi and S. Akhshabi, arXiv:0901.0090 [hep-th]; G. Leon, R. Cardenas and J. L. Morales, arXiv:0812.0830 [gr-qc];L. P. Chimento, M. Forte, R. Lazkoz and M. G. Richarte, *Phys. Rev. D* **79**, 043502 (2009) [arXiv:0811.3643 [astro-ph]];M. R. Setare, J. Sadeghi and A. R. Amani, arXiv:0811.3343 [hep-th];M. R. Setare and E. N. Saridakis, *Phys. Rev. D* **79**, 043005 (2009) [arXiv:0810.4775 [astro-ph]];C. J. Feng, *Phys. Lett. B* **672**, 94 (2009) [arXiv:0810.2594 [hep-th]];arXiv:0810.1427 [hep-th];M. R. Setare and E. N. Saridakis, *Phys. Lett. B* **671**, 331 (2009) [arXiv:0810.0645 [hep-th]];M. R. Setare and E. N. Saridakis, *JCAP* **0809**, 026 (2008) [arXiv:0809.0114 [hep-th]]; Y. F. Cai and X. Zhang, *JCAP* **0906**, 003 (2009) [arXiv:0808.2551 [astro-ph]];M. R. Setare and E. N. Saridakis, *Int. J. Mod. Phys. D* **18**, 549 (2009) [arXiv:0807.3807 [hep-th]]; M. R. Setare, J. Sadeghi and A. Banijamali, *Phys. Lett. B* **669**, 9 (2008) [arXiv:0807.0077 [hep-th]]; S. Zhang and B. Chen, *Phys. Lett. B* **669**, 4 (2008) [arXiv:0806.4435 [hep-ph]];Y. F. Cai and J. Wang, *Class. Quant. Grav.* **25**, 165014 (2008) [arXiv:0806.3890 [hep-th]]; M. R. Setare and J. Sadeghi, *Int. J. Theor. Phys.* **47**, 3219 (2008) [arXiv:0805.1117 [gr-qc]];H. H. Xiong, Y. F. Cai, T. Qiu, Y. S. Piao and X. Zhang, *Phys. Lett. B* **666**, 212 (2008) [arXiv:0805.0413 [astro-ph]];J. Sadeghi, M. R. Setare, A. Banijamali and F. Milani, *Phys. Lett. B* **662**, 92 (2008) [arXiv:0804.0553 [hep-th]];M. R. Setare and E. N. Saridakis, *Phys. Lett. B* **668**, 177 (2008) [arXiv:0802.2595 [hep-th]];M. R. Setare, J. Sadeghi and A. R. Amani, *Phys. Lett. B* **660**, 299 (2008) [arXiv:0712.1873 [hep-th]];H. H. Xiong, T. Qiu,

Y. F. Cai and X. Zhang, arXiv:0711.4469 [hep-th];Y. F. Cai, T. Qiu, R. Branden-berger, Y. S. Piao and X. Zhang, *JCAP* **0803**, 013 (2008) [arXiv:0711.2187 [hep-th]]; M. Alimohammadi, *Gen. Rel. Grav.* **40**, 107 (2008) [arXiv:0706.1360 [gr-qc]];H. Wei and S. N. Zhang, *Phys. Rev. D* **76**, 063005 (2007) [arXiv:0705.4002 [gr-qc]];Y. F. Cai, T. Qiu, Y. S. Piao, M. Li and X. Zhang, *JHEP* **0710**, 071 (2007) [arXiv:0704.1090 [gr-qc]];R. Lazkoz, G. Leon and I. Quiros, *Phys. Lett. B* **649**, 103 (2007) [arXiv:astro-ph/0701353];Y. f. Cai, M. z. Li, J. X. Lu, Y. S. Piao, T. t. Qiu and X. m. Zhang, *Phys. Lett. B* **651**, 1 (2007) [arXiv:hep-th/0701016];M. R. Setare, *Phys. Lett. B* **641**, 130 (2006) [arXiv:hep-th/0611165];X. Zhang, *Phys. Rev. D* **74**, 103505 (2006) [arXiv:astro-ph/0609699]; Y. f. Cai, H. Li, Y. S. Piao and X. m. Zhang, *Phys. Lett. B* **646**, 141 (2007) [arXiv:gr-qc/0609039];Z. K. Guo, Y. S. Piao, X. Zhang and Y. Z. Zhang, *Phys. Rev. D* **74**, 127304 (2006) [arXiv:astro-ph/0608165]; M. Alimohammadi and H. M. Sadjadi, *Phys. Lett. B* **648**, 113 (2007) [arXiv:gr-qc/0608016];H. Mohseni Sadjadi and M. Alimohammadi, *Phys. Rev. D* **74**, 043506 (2006) [arXiv:gr-qc/0605143];W. Wang, Y. X. Gui and Y. Shao, *Chin. Phys. Lett.* **23**, 762 (2006);W. Zhao, *Phys. Rev. D* **73**, 123509 (2006) [arXiv:astro-ph/0604460]; X. F. Zhang and T. Qiu, *Phys. Lett. B* **642**, 187 (2006) [arXiv:astro-ph/0603824];R. Lazkoz and G. Leon, *Phys. Lett. B* **638**, 303 (2006) [arXiv:astro-ph/0602590];B. Feng, arXiv:astro-ph/0602156;X. Zhang, *Commun. Theor. Phys.* **44** (2005) 762;P. x. Wu and H. w. Yu, *Int. J. Mod. Phys. D* **14**, 1873 (2005) [arXiv:gr-qc/0509036];G. B. Zhao, J. Q. Xia, M. Li, B. Feng and X. Zhang, *Phys. Rev. D* **72**, 123515 (2005) [arXiv:astro-ph/0507482];J. Q. Xia, B. Feng and X. M. Zhang, *Mod. Phys. Lett. A* **20**, 2409 (2005) [arXiv:astro-ph/0411501].

[26] H. S. Zhang and Z. H. Zhu, *Phys. Rev. D* **73**, 043518 (2006).

[27] Hongsheng Zhang and Zong-Hong Zhu, *Modern Physics Letters A,* Vol. 24, No. 7 (2009) 541, arXiv:0704.3121.

[28] Z. K. Guo, Y. S. Piao, X. M. Zhang and Y. Z. Zhang, *Phys. Lett. B* **608**, 177 (2005) [arXiv:astro-ph/0410654].

[29] H. Wei, R. G. Cai and D. F. Zeng, *Class. Quant. Grav.* **22**, 3189 (2005) [arXiv:hep-th/0501160].

[30] H. Wei and R. G. Cai, *Phys. Rev. D* **72**, 123507 (2005) [arXiv:astro-ph/0509328].

[31] M. Alimohammadi and H. Mohseni Sadjadi, *Phys. Rev. D* **73**, 083527 (2006) [arXiv:hep-th/0602268].

[32] H. Wei, N. N. Tang and S. N. Zhang, *Phys. Rev. D* **75**, 043009 (2007) [arXiv:astro-ph/0612746].

[33] H. Wei and S. N. Zhang, *Phys. Rev. D* **76**, 063005 (2007) [arXiv:0705.4002 [gr-qc]].

[34] W. Zhao, *Phys. Lett. B* **655**, 97 (2007) [arXiv:0706.2211 [astro-ph]].

[35] H. Sandvik, M. Tegmark, M. Zaldarriaga, I. Waga, *Phys.Rev.* **D69** (2004) 123524, astro-ph/0212114.

[36] C. M. Will, *Living Rev. Rel.* **4**, 4 (2001) ,gr-qc/0103036.

[37] Z. K. Guo, N. Ohta and S. Tsujikawa, *Phys. Rev. D* **76**, 023508 (2007) [arXiv:astro-ph/0702015].

[38] H. Wei and S. N. Zhang, *Phys. Lett. B* **644**, 7 (2007).

[39] L. Amendola, G. Camargo Campos and R. Rosenfeld, *Phys. Rev. D* **75**, 083506 (2007).

[40] Y. Wang, and P. Garnavich, ApJ, 552, 445 (2001); M. Tegmark, *Phys. Rev.* **D66**, 103507 (2002); Y. Wang, and K. Freese, *Phys.Lett.* **B632**, 449 (2006); astro-ph/0402208.

[41] A. Kamenshchik, U. Moschella and V. Pasquier, *Phys. Lett.* **B511** (2001) 265.

[42] N. Bartolo and M. Pietroni, Phys. Rev. D61, 023518 (2000); T. Damour, G. W. Gibbons and C. Gundlach , *Phys. Rev. Lett.* **64**, 123 (1990).

[43] R. Curbelo, T. Gonzalez and I. Quiros, astro-ph/0502141; N. Kaloper and K. A. Olive, *Phys.Rev.D* **57**, 811 (1998).

[44] M. Jamil, arXiv:0906.3913 [gr-qc];M. Jamil and M. A. Rashid, *Eur. Phys. J. C* **58**, 111 (2008) [arXiv:0802.1146 [astro-ph]];H. Garcia-Compean, G. Garcia-Jimenez, O. Obregon and C. Ramirez, *JCAP* **0807**, 016 (2008) [arXiv:0710.4283 [hep-th]];S. Li, Y. Ma and Y. Chen, arXiv:0809.0617 [gr-qc].

[45] C. Lanczos, *Z. Phys.*, **73** (1932) 147; D. Lovelock, *J. Math. Phys.*, **12** (1971) 498

[46] R. R. Metsaev and A. A. Tseytlin, *Phys. Lett. B* **191**, 354 1987.

[47] D. J. Gross and J. H. Sloan, *Nucl. Phys.* **B291**, 41 1987.

[48] G. Dvali, G. Gabadadze and M. Porrati, hep-th/0005016; G. Dvali and G. Gabadadze, *Phys. Rev. D* **63**, 065007 (2001).

[49] G. Kofinas, R. Maartens and E. Papantonopoulos, *JHEP* **0310** (2003) 066.

[50] K. Maeda and T. Torii, *Phys. Rev. D* **69**, 024002 (2004).

[51] R. G. Cai, H. S. Zhang and A. Wang, *Commun. Theor. Phys.* **44**, 948 (2005) [arXiv:hep-th/0505186].

[52] J. H. He, B. Wang and E. Papantonopoulos, *Phys. Lett. B* **654**, 133 (2007) [arXiv:0707.1180 [gr-qc]].

[53] H. S. Zhang and Z. H. Zhu, *Phys. Rev. D* **75**, 023510 (2007) [arXiv:astro-ph/0611834].

[54] H. s. Zhang and R. G. Cai, *JCAP* **0408**, 017 (2004) [arXiv:hep-th/0403234].

[55] H. Wei and R. G. Cai, *Phys. Rev. D* **73**, 083002 (2006) [arXiv:astro-ph/0603052].

[56] M. Cataldo and L. P. Chimento, arXiv:0710.4306 [astro-ph].

[57] L. P. Chimento and R. Lazkoz, Phys. Lett. B **639**, 591 (2006) [arXiv:astro-ph/0604090].

[58] K. Nozari and S. D. Sadatian, *Eur. Phys. J. C* **58**, 499 (2008) [arXiv:0809.4744 [gr-qc]].

[59] S. Y. Vernov, *Teor. Mat. Fiz.* **155**, 47 (2008) [*Theor. Math. Phys.* **155**, 544 (2008)] [arXiv:astro-ph/0612487].

[60] T. Gonzalez and I. Quiros, *Class. Quant. Grav.* **25**, 175019 (2008) [arXiv:0707.2089 [gr-qc]];

H. Wei and S. N. Zhang, *Phys. Lett. B* **654**, 139 (2007) [arXiv:0704.3330 [astro-ph]];

H. M. Sadjadi, arXiv:0904.1349 [gr-qc].

[61] P. Wu and S. N. Zhang, *JCAP* **0806**, 007 (2008) [arXiv:0805.2255 [astro-ph]].

[62] B. Wang, Y. g. Gong and E. Abdalla, *Phys. Lett. B* **624**, 141 (2005) [arXiv:hep-th/0506069]; H. M. Sadjadi, *JCAP* **0702**, 026 (2007) [arXiv:gr-qc/0701074].

[63] H. M. Sadjadi and M. Honardoost, *Phys. Lett. B* **647**, 231 (2007) [arXiv:gr-qc/0609076].

[64] K. Bamba, C. Q. Geng, S. Nojiri and S. D. Odintsov, *Phys. Rev. D* **79**, 083014 (2009) [arXiv:0810.4296 [hep-th]].

[65] K. Nozari and F. Kiani, *JCAP* **0907**, 010 (2009) [arXiv:0906.3806 [gr-qc]]; K. Nozari, N. Behrouz and B. Fazlpour, arXiv:0808.0318 [gr-qc]; M. Bouhmadi-Lopez and R. Lazkoz, *Phys. Lett. B* **654**, 51 (2007) [arXiv:0706.3896 [astro-ph]].

[66] H. s. Zhang, H. Noh and Z. H. Zhu, *Phys. lett.* **B663**,291(2008) [arXiv:0804.2931 [gr-qc]]; H. s. Zhang, H. Noh, *Phys. Lett.* **B670**, 271(2009),arXiv:0904.0063; H. s. Zhang, H. Noh, *Phys. Lett.* **B671**,428 (2009), arXiv:0904.0065.

[67] H. s. Zhang and H. Noh, *Phys. Lett. B* **679**, 81 (2009) [arXiv:0904.0067 [gr-qc]].

[68] M. Bouhmadi-Lopez and A. Ferrera, *JCAP* **0810**, 011 (2008) [arXiv:0807.4678 [hep-th]]; I. Quiros, R. Garcia-Salcedo, T. Matos and C. Moreno, *Phys. Lett. B* **670**, 259 (2009) [arXiv:0802.3362 [gr-qc]]; X. Wu, R. G. Cai and Z. H. Zhu, *Phys. Rev. D* **77**, 043502 (2008) [arXiv:0712.3604 [astro-ph]];I. Y. Aref'eva, A. S. Koshelev and S. Y. Vernov, *Phys. Rev. D* **72**, 064017 (2005) [arXiv:astro-ph/0507067]; G. Kofinas, G. Panotopoulos and T. N. Tomaras, *JHEP* **0601**, 107 (2006) [arXiv:hep-th/0510207];M. Bouhmadi-Lopez, *Nucl. Phys. B* **797**, 78 (2008) [arXiv:astro-ph/0512124];P. S. Apostolopoulos and N. Tetradis, *Phys. Rev. D* **74**, 064021 (2006) [arXiv:hep-th/0604014];P. S. Apostolopoulos, N. Brouzakis, N. Tetradis and E. Tzavara, *Phys. Rev. D* **76**, 084029 (2007) [arXiv:0708.0469 [hep-th]];S. Yin, B. Wang, E. Abdalla and C. Y. Lin, *Phys. Rev. D* **76**, 124026 (2007) [arXiv:0708.0992

[hep-th]]; E. N. Saridakis, *Phys. Lett. B* **661**, 335 (2008) [arXiv:0712.3806 [gr-qc]];K. Nozari and B. Fazlpour, *JCAP* **0806**, 032 (2008) [arXiv:0805.1537 [hep-th]];K. Nozari and N. Rashidi, arXiv:0906.4263 [gr-qc].

[69] H. Wei and R. G. Cai, *Phys. Lett. B* **634**, 9 (2006) [arXiv:astro-ph/0512018].

In: Dark Energy: Theories, Developments and Implications   ISBN 978-1-61668-271-2
Editors: K. Lefebvre and R. Garcia, pp. 89-109   © 2010 Nova Science Publishers, Inc.

Chapter 5

QUANTUM YANG-MILLS CONDENSATE DARK ENERGY MODELS

Wen Zhao[1], Yang Zhang[2] and Minglei Tong[3]
[1]School of Physics and Astronomy, Cardiff University,
Cardiff, CF24 3AA, United Kingdom
[2]Center for Astrophysics,
University of Science and Technology of China,
Hefei, 230026, People's Republic of China
[3]Center for Astrophysics,
University of Science and Technology of China,
Hefei, 230026, People's Republic of China

Abstract

We review the quantum Yang-Mills condensate (YMC) dark energy models. As
the effective Yang-Mills Lagrangian is completely determined by the quantum field
theory, there is no adjustable parameter in the model except the energy scale. In this
model, the equation-of-state (EOS) of the YMC dark energy, $w_y > -1$ and $w_y < -1$,
can both be naturally realized. By studying the evolution of various components in the
model, we find that, in the early stage of the universe, dark energy tracked the evolution
of the radiation, i.e. $w_y \rightarrow 1/3$. However, in the late stage, w_y naturally runs to the
critical state with $w_y = -1$, and the universe transits from matter-dominated into dark
energy dominated stage only at recently $z \sim 0.3$. These characters are independent
of the choice of the initial condition, and the cosmic coincidence problem is avoided
in the models. We also find that, if the possible interaction between YMC and dust
matter is considered, the late time attractor solution may exist. In this case, the EOS
of YMC must evolve from $w_y > 0$ into $w_y < -1$, which is slightly suggested by the
observations. At the same time, the total EOS in the attractor solution is $w_{tot} = -1$,
the universe being the de Sitter expansion in the late stage, and the cosmic big rip is
naturally avoided. These features are all independent of the interacting forms.

PACS number: 98.80.-k, 98.80.Es, 04.30.-w, 04.62.+v

1. Introduction

The observations of Type Ia Supernova (SNIa) [1], together with the cosmic microwave background radiation (CMB)[2] and the larger scale structure [3], suggest that the present universe is accelerating expansion, which needs a kind of mysterious dark energy with negative equation-of-state (EOS). The simplest model is by introducing the cosmological constant term Λ, which has a constant effective EOS $w = -1$, and drive the acceleration of the universe. If assuming the effective energy of the Λ term occupies $\sim 73\%$ of the total energy, together with $\sim 23\%$ dark matter, $\sim 4\%$ baryon matter and $\sim 10^{-5}$ radiation component, constitute the so-called ΛCDM model.

Although this simple model satisfies nearly all the cosmological observations, it still remains a phenomenological model. Since the major components, Λ and dark matter, still keep unclear for us. For the Λ as the candidate of dark energy, also suffers from the following dilemmas (see for instant [4]): First, the effective energy density is quite tiny, $\rho_\Lambda \sim 5.8h^2 \times 10^{-11} \text{eV}^4$. If we consider it as the vacuum energy of the particle physics, this energy density is 120 order smaller than the Planck energy scale! This is the so-called 'fine-tunning' problem. Second, the density scale of Λ keeps constant in all the stage of the universe. The observations show that the present value of the matter component (including dark matter and baryon components) is some one third of ρ_Λ, but it varies with time as $\rho_m \propto a^{-3}$. So, for example, at an earlier time of radiation-matter equality with redshift $z \sim 3454$, ρ_Λ should be a very fine tuned value $\sim 6.3 \times 10^{-11} \rho_m$. Otherwise, a slightly variant initial value of ρ_Λ would lead to a value of the ratio ρ_Λ / ρ_m drastically different from the observed one. This is the so-called 'coincidence' problem. In addition, the ΛCDM model also faces some observational challenges: there are mild evidences show that, the EOS of dark energy might evolve from $w > -1$ in the early stage to $w < -1$ in the current stage [5], which is expected to be confirmed by the future sensitive observations [6].

So, it is necessary to look for other candidates as the dark energy, especially the dynamical models. One possibility is proposing a canonical scalar field ϕ with the langrangian $\mathcal{L}_\phi = \dot{\phi}^2/2 - V(\phi)$, which is dubbed as the 'quintessence' models (see the review [7]). Similar to the inflationary field, when the potential term $V(\phi)$ is dominant, the EOS of the quintessence field approaches to -1, i.e. Λ-like, and accounts for the observations. The most interesting is that, in [8], the authors introduced a kind of potential forms, such as $V(\phi) = M^{4+\alpha}\phi^{-\alpha}$ or $V(\phi) = M^4[\exp(M_{\rm pl}/\phi) - 1]$, which have the tracker solutions, i.e. the field ϕ tracked the evolution of the background components in the early universe. So they address the coincidence problem, i.e. removing the need to tune initial conditions in order for the matter and dark energy densities to nearly coincide today. Although, this kind of models have been excluded by the cosmological observations, it provides the excellent possibility to naturally avoided the coincidence problems.

However, the quintessence models also suffer from some dilemmas. The EOS of the quintessence models are in the range $-1 < w < 1$. In order to obtain a EOS with $w < -1$, one always has to introduce the so-called 'phantom' field, which includes the non-canonical negative kinetic terms [9][10]. However, the 'phantom' field lacks the strong physical motivation, and also leads to the problem of quantum instabilities [11]. In [12], the authors also pointed that, if $w < -1$ keeps, the universe shall face the 'big-rip' problem.

In addition to proposing the dynamical dark energy models, some efforts have been paid to modify the general relativity (GR) [13], which can also speed-up the universe in the recent stage. However, it should be pointed that, any revised GR should prepare to go through the strict solar system test [14], as well as to explain the various cosmological observations, such as the CMB temperature and polarization anisotropies power spectra. In addition, to our view, the current observations have not provided the strong reasons to answer: Why we should modify GR and how to modify it.

In this paper, we shall propose the Yang-Mills condensate (YMC) dark energy models, where, instead of the scalar field, the quantum Yang-Mills field is considered as the candidate of dark energy component. Recently, the similar models are also discussed by a number of authors [15][16]. Different from the scalar field models, the Yang-Mills fields are the indispensable cornerstone to particle physics, gauge bosons have been observed. There is no room for adjusting the form of effective Yang-Mills Lagrangian as it is predicted by quantum corrections according to field theory. In this review, we shall firstly introduce physical motivations of the YMC dark energy models in Section II. In Section III, we simply introduce and discuss the Lagrangian of effective quantum Yang-Mills field.

As the main part of this paper, in Section IV, we apply the YMC into the cosmology as the candidate of dark energy, and investigate the cosmic evolution of the various components, especially the evolution of dark energy. We find the excellent characters of this kind of models. Different from the quintessence models, both the EOS $w_y > -1$ and $w_y < -1$ can be naturally realized. In the free YMC dark energy models, $w_y \to 1/3$ in the early stage and tracked the evolution of radiation component. Only in the recent stage, w_y rolls to the critical state with $w_y = -1$, i.e. Λ-like, and accounts for the observations. This feature is independent of the choice of the initial condition, so the coincidence problem is naturally avoided. We also find that, if the possible interaction between YMC and dust components is considered, the late time attractor solution can exist, where the EOS of YMC naturally runs from $w_y \to 1/3$ to the the phantom-like attractor state, i.e. $w_y < -1$. However, the total effective EOS is $w_{tot} = -1$ in the attractor solution, and the so-called 'big-rip' problem is also avoided. We should pointed that, these new features are all independent of the interaction forms. In Section V, we calculate the statefinder and Om diagnostics for the YMC dark energy models, which are helpful to differentiate the YMC models from the other dark energy models from the observations.

Section VI is contributed as a summary of this paper.

Throughout this paper, we will work with unit in which $c = \hbar = k_B = 1$.

2. Physical Motivation

The introduction of the quantum effective YMC into cosmology [17] has been motivated by the fact that the $SU(3)$ YMC has given a phenomenological description of the vacuum within hadrons confining quarks, and yet at the same time all the important properties of a proper quantum field are kept, such as the Lorentz invariance, the gauge symmetry, and the correct trace anomaly [18] [19]. Quarks inside a hadron would experience the existence of the Bag constant, B, which is equivalent to an energy density B and a pressure $-B$. So quarks would feel an energy-momentum tensor of the vacuum as $T_{\mu\nu} = B\text{diag}(1, -1, -1, -1)$. This non-trivial vacuum has been formed mainly by the contributions from the quantum effective YMC, and from the possible interactions with quarks. Our thinking has been that if the vacuum inside a hadron is filled with the quantum effective YMC, what if the vacuum of the universe as a whole is also filled with some kind of YMC.

Gauge fields play a very important role in, and are the indispensable cornerstone to, particle physics. All known fundamental interactions between particles are mediated through gauge bosons. Generally speaking, as a gauge field, the YMC under consideration may have interactions with other species of particles in the universe. However, unlike those well known interactions in QED, QCD, and the electro-weak unification, here at the moment we do not yet have a model for the details of the microscopic interactions between the YMC and other particle. Therefore, in this paper on the dark energy model based on the quantum effective YMC, we will adopt a simple description of the possible interactions between the YMC and other cosmic particles, in addition to the simplest model with free YMC component. We will investigate in these models the cosmic evolution of the universe from the radiation-dominated era up to the present.

3.   Yang-Mills Field Model

In the effective YMC dark energy model, the effective Yang-Mills field Lagrangian is given by [18][20, 21]:

$$\mathcal{L}_{\text{eff}} = \frac{1}{2}bF\left(\ln\left|F/\kappa^2\right| - 1\right),\tag{1}$$

where κ is the renormalization scale of dimension of squared mass, $F \equiv -(1/2)F^a_{\mu\nu}F^{a\mu\nu}$ plays the role of the order parameter of the YMC. The Callan-Symanzik coefficient $b = (11N - 2N_f)/24\pi^2$ for $SU(N)$ with N_f being the number of quark flavors. For the gauge group $SU(2)$ considered here, one has $b = 11/12\pi^2$ when the fermion's contribution is neglected. For the case of $SU(3)$ the effective Lagrangian in (1) leads to a phenomenological description of the asymptotic freedom for quarks inside hadrons [18]. It should be noted that the $SU(2)$ Yang-Mills field is introduced here as a model for the cosmic dark energy, in may not be directly identified as the QCD fields, nor the weak-electromagnetic unification gauge fields.

An explanation can be given for the form in (1) as an effective Lagrangian up to 1-loop quantum correct [18, 22, 19]. A classical $SU(N)$ Yang-Mills field Lagrangian is $\mathcal{L} = F/2g_0^2$, where g_0 is the bare coupling constant. As is known. when 1-loop quantum corrections are included, the bare coupling g_0 will be replaced by a running one g as the following [23], $g_0^2 \to g^2 = 2/(b\ln(k^2/k_0^2))$, where k is the momentum transfer and k_0 is the energy scale. To build up an effective theory [18, 22, 19], one may just replace the momentum transfer k^2 by the field strength F in the following manner: $\ln(k^2/k_0^2) \to 2\ln|F/e\kappa^2|$, yielding equation (1). We would like to point out that the renormalization scale κ is the only parameter of this effective Yang-Mills model, and its value should be determined by comparing the observations. In contrast to the scalar field dark energy models, the YMC Lagrangian is completely fixed by quantum corrections up to 1-loops, and there is no room for adjusting its functional form. This is an attractive feature of the effective YMC dark energy model. We should mention that, the YMC dark energy models including 2-loop and

3-loop quantum corrects are also discussed in the recent papers [24][25]. It was found that, these more complicated models have the exactly same characters with the 1-loop models. So in this paper, we mainly focus on the 1-loop models with the Lagrangian in (1).

4. YMC as Dark Energy

Let us work in the flat Friedmann-Lemaitre-Robertson-Walker (FLRW) universe, which is described by

$$ds^2 = dt^2 - a^2(t)\delta_{ij}dx^i dx^j = a^2(\tau)[d\tau^2 - \delta_{ij}dx^i dx^j],$$

where t time and τ time are related by $cdt = ad\tau$. Considering the simplest case with only the YMC in the universe, which minimally coupled to the gravity, the effective action is,

$$S = \int \sqrt{-g} \left[-\frac{R}{16\pi G} + \mathcal{L}_{\text{eff}} \right] d^4x. \tag{2}$$

Here, g is the determinant of the metric $g_{\mu\nu}$. $\mathcal{R}$ is the scalar Ricci curvature, and $\mathcal{L}_{\text{eff}}$ is the effective Lagrangian of YMC, described by Eq. (1). By variation of S with respect to the metric $g^{\mu\nu}$, one obtains the Einstein equation $G_{\mu\nu} = 8\pi G T_{\mu\nu}$, where the energy-momentum tensor of YMC is given by

$$T_{\mu\nu} = \sum_{a=1}^{3} \frac{g_{\mu\nu}}{4g^2} F^a_{\sigma\delta} F^{a\sigma\delta} + \epsilon F^a_{\mu\sigma} F^{a\sigma}_\nu. \tag{3}$$

The dielectric constant is defined by $\epsilon \equiv 2\partial\mathcal{L}_{\text{eff}}/\partial F$. In the one-loop order, it is given by

$$\epsilon = b \ln \left| F/\kappa^2 \right|. \tag{4}$$

This energy-momentum tensor (3) is the sum of three different energy-momentum tensors of the vectors, $T_{\mu\nu} = \sum_a {}^{(a)}T_{\mu\nu}$, neither of which is of prefect-fluid form. Here we assume the gauge fields are only the functions of time t, and $A_\mu = \frac{1}{2}\sigma_a A^a_\mu(t)$ (here σ_a are the Pauli's matrices) are given by $A_0 = 0$ and $A^a_i = \delta^a_i A(t)$. Thus, we will find that, the total energy-momentum tensor $T_{\mu\nu}$ is homogeneous and isotropic.

Define the Yang-Mills field tensor as usual:

$$F^a_{\mu\nu} = \partial_\mu A^a_\nu - \partial_\nu A^a_\mu + f^{abc} A^b_\mu A^c_\nu, \tag{5}$$

where f^{abc} is the structure constant of gauge group and $f^{abc} = \epsilon^{abc}$ for the $SU(2)$ case. This tensor can be written in the form with the electric and magnetic field as

$$F^{a\mu}_\nu = \begin{pmatrix} 0 & E_1 & E_2 & E_3 \\ -E_1 & 0 & B_3 & -B_2 \\ -E_2 & -B_3 & 0 & B_1 \\ -E_3 & B_2 & -B_1 & 0 \end{pmatrix}. \tag{6}$$

It can be easily found that $E_1^2 = E_2^2 = E_3^2$, and $B_1^2 = B_2^2 = B_3^2$. Thus F has a simple form with $F = E^2 - B^2$, where $E^2 = \sum_{i=1}^3 E_i^2$ and $B^2 = \sum_{i=1}^3 B_i^2$. In this case, each component of the energy-momentum tensor is

$$^{(a)}T_\mu^0 = \frac{1}{6}(\epsilon - b)(B^2 - E^2)\delta_\mu^0 + \frac{\epsilon}{3}E^2\delta_\mu^0, \tag{7}$$

$$^{(a)}T_j^i = \frac{1}{6}(\epsilon - b)(B^2 - E^2)\delta_j^i + \frac{\epsilon}{3}E^2\delta_j^i\delta_j^a - \frac{\epsilon}{3}B^2\delta_j^i(1 - \delta_j^a). \tag{8}$$

Although this tensor is not isotropic, its value along the $j = a$ direction is different from the ones along the directions perpendicular to it. However, the total energy-momentum tensor $T_{\mu\nu} = \sum_{a=1}^3 {}^{(a)}T_{\mu\nu}$ has isotropic stresses, and the corresponding energy density and pressure are given by

$$\rho_y = \frac{1}{2}\epsilon(E^2 + B^2) + \frac{1}{2}b(E^2 - B^2), \quad p_y = \frac{1}{6}\epsilon(E^2 + B^2) - \frac{1}{2}b(E^2 - B^2). \tag{9}$$

In this paper, for simplicity, we only discuss the pure 'electric' case, $F = E^2$. This is a typical consideration, since in the expanding universe, a given 'magnetic' component of Yang-Mills field decreases quite rapidly, and the Yang-Mills field becomes the 'electric' type [26]. The energy density and pressure of YMC are reduced to

$$\rho_y = \frac{E^2}{2}(\epsilon + b), \quad p_y = \frac{E^2}{2}\left(\frac{\epsilon}{3} - b\right).$$

It is convenient to introduce a dimensionless quantity $y \equiv \epsilon/b = \ln|F/\kappa^2|$. So the quantities ρ_y and p_y can be rewritten as

$$\rho_y = \frac{1}{2}b\kappa^2(y + 1)e^y, \quad p_y = \frac{1}{6}b\kappa^2(y - 3)e^y. \tag{10}$$

One sees that, to ensure that energy density ρ_y be positive in any physically viable model, the allowance for the quantity y should be $y > -1$, i.e. $F > \kappa^2/e \simeq 0.368\kappa^2$. The EOS of the YMC is

$$w_y \equiv \frac{p_y}{\rho_y} = \frac{y - 3}{3y + 3}. \tag{11}$$

Before setting up a cosmological model, the EOS w_y itself as a function of F is interesting. At the state with $F = \kappa^2$, which is called critical point, one has $y = 0$ and the YMC has an EOS of the cosmological constant with $w_y = -1$. Around this critical point, $F < \kappa^2$ gives $y < 0$ and $w < -1$, and $F > \kappa^2$ gives $y > 0$ and $w > -1$. So in the YMC dark energy models, EOS with $w > -1$ and $w < -1$ all can be naturally realized. On the other hand, in the high energy scale with $F \gg \kappa^2$, one finds that the YMC exhibits an EOS of radiation with $w_y = 1/3$. In the follows, we will detailed show that the YMC was evolving from the state with $w_y = 1/3$ to $w_y = -1$ (or even to $w_y < -1$) in the expanding universe. This is the main context in this paper. In addition, the characteristic statefinder parameters and the perturbations of the YMC dark energy is also discussed in [27, 28][29], which are helpful to distinguish the YMC model from other dark energy models.

As is known, an effective theory is a simple representation for an interacting quantum system of many degrees of freedom at and around its respective low energies. Commonly, it applies only in low energies. However, it is interesting to note that the YMC model as an effective theory intrinsically incorporates the appropriate states for both high and low temperature. As has been shown, the same expression in (11) simultaneously gives $p_y \sim -\rho_y$ at low energies, and $p_y \to \rho_y/3$ at high energies. Therefore, the model of effective YMC can be used even at higher energies than the renormalization scale κ.

4.1. Free YMC Models

Let us discuss the cosmological model, which filled with three kinds of major energy components, the dark energy, the matter, including both baryons and dark matter, and the radiation. Here, the dark energy component is represented by YMC, and the matter component is simply described by a non-relativistic dust with negligible pressure, and the radiation component consists of the photons and possibly other particles, such as the neutrino, if they are massless.

Since the universe is assumed to be flat, the sum of the fraction densities is $\Omega_y + \Omega_m + \Omega_r = 1$, where the fractional energy densities are $\Omega_y \equiv \rho_y/\rho_{tot}$, $\Omega_m \equiv \rho_m/\rho_{tot}$, $\Omega_r \equiv \rho_r/\rho_{tot}$, and the total energy density is $\rho_{tot} \equiv \rho_y + \rho_m + \rho_r$. The overall expansion of the universe is determined by the Friedmann equations:

$$\left(\frac{\dot{a}}{a}\right)^2 = \frac{8\pi G}{3}(\rho_y + \rho_m + \rho_r), \tag{12}$$

$$\frac{\ddot{a}}{a} = -\frac{4\pi G}{3}(\rho_y + 3p_y + \rho_m + \rho_r + 3p_r), \tag{13}$$

where and what follows, the superscript dot denote the d/dt. These three components of energy contribute to the source on the right-hand side of the equations. We should notice that, $p_r = \rho_r/3$. In this subsection, we shall assume there is no interaction between these three components. The dynamical evolutions of the three components are determined by their equations of motion, which can be written as equations of the conservation of energy,

$$\dot{\rho}_y + 3\frac{\dot{a}}{a}(\rho_y + p_y) = 0, \tag{14}$$

$$\dot{\rho}_m + 3\frac{\dot{a}}{a}\rho_m = 0, \tag{15}$$

$$\dot{\rho}_r + 3\frac{\dot{a}}{a}(\rho_r + p_r) = 0. \tag{16}$$

As is known, Eq. (13) is not independent and can be derived from Eqs. (14), (15), (16) and (12). From Eqs. (15) and (16), we can obtain the evolutions of the matter and radiation components, $\rho_m \propto a^{-3}$ and $\rho_m \propto a^{-4}$.

Now, let us focus on the evolution of YMC. Inserting (10) into (14), we can obtain the following simple relation,

$$ye^{y/2} \propto a^{-2}, \tag{17}$$

where the coefficient of proportionality in the above depends on the initial condition. At the early stages, $a \to 0$, Eq. (17) leads to $y \gg 1$, and Eq. (11) gives $w_y \to 1/3$, so the

YMC behaves as the radiation component. With the expansion of the universe, the value of y runs to the critical state of $y = 0$, and the EOS goes to $w_y = -1$. So, in the late stage of the universe, the YMC behaves as the cosmological constant. This is one of the most attractive character of the YMC models. Around the critical point with $y = 0$, Eq.(17) yields $y \propto a^{-2}$, and the EOS of YMC is $w_y + 1 \simeq 4y/3 \propto a^{-2}$. The YMC can achieve the states of $w_y > -1$ and $w_y < -1$, but it can not cross over -1, just like in the scalar models [30].

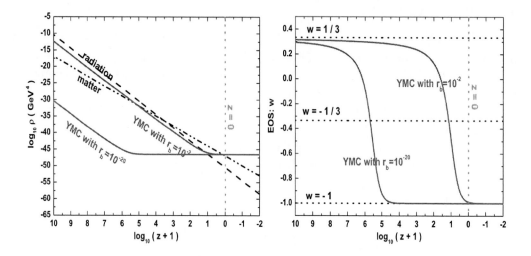

Figure 1. In the free YMC dark energy models, the evolution of the YMC energy density (left panels) and EOS (right panels) for the models with different initial conditions.

We should notice that, the relation in (17) can also be derived by the effective Yang-Mills equations. By variation of the action S with respect to gauge field A_μ^a, we can obtain the Yang-Mills equations, which are

$$\partial_\mu(a^4 \epsilon \, F^{a\mu\nu}) + f^{abc}A_\mu^b(a^4 \epsilon \, F^{c\mu\nu}) = 0. \qquad (18)$$

The $\nu = 0$ component of which is an identity, and the $\nu = 1, 2, 3$ spatial components can be reduced to $\partial_\tau(a^2 \epsilon E) = 0$. At the critical point ($\epsilon = 0$), this equation is an identity. When $\epsilon \neq 0$, this equation has an exact solution as (17) [21].

Now, let us fix the value of κ, the only parameter in the model. At the present time, the YMC energy density is

$$\rho_y = \frac{bE^2}{2}(y + 1) \simeq \frac{b\kappa^2}{2}, \qquad (19)$$

and, as the dark energy, it should be $\Omega_y\rho_{tot}$, where the present total energy density in the universe $\rho_{tot} \approx 8.099 \, h^2 \times 10^{-11} \text{eV}^4$. We choose $\Omega_y = 0.73$ as has been observed, yielding

$$\kappa = 3.57 \, h \times 10^{-5} \text{eV}^2. \qquad (20)$$

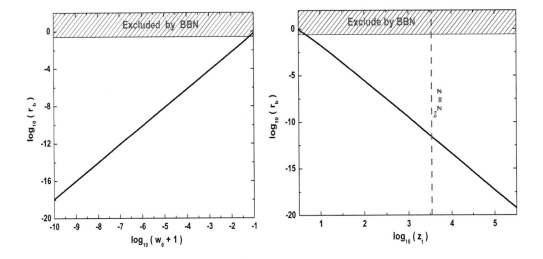

Figure 2. Left panel: The present EOS of the YMC w_0 depends on the initial value r_b. Right panel: The transition redshift z_t depends on the initial value r_b. In both panels, the shadow region has been excluded by the observation of BBN.

This energy scale is low compared to typical energy scales in particle physics, such as the QCD and the weak-electromagnetic unification. This is the reason, why the $SU(2)$ Yang-Millls field introduced in this paper cannot be directly identified as the the QCD gluon fields, nor the weak-electromagnetic unification gauge fields.

To be more specific about how the YMC evolves in the expanding universe, we look at an early stage when the Big Bang nucleosynthesis (BBN) processes occur around a redshift $z \sim 10^{10}$ with an energy scale $\sim 1\text{MeV}$. To see how the evolution of ρ_y depends the the initial condition, we introduce the ratio of energies of the two components

$$r_b \equiv \left. \frac{\rho_y}{\rho_r} \right|_{z=10^{10}}, \qquad (21)$$

where ρ_r is the radiation energy density. We consider $r_b < 1$, i.e. the YMC is subdominant to the radiation component initially. Of course, the YMC evolves differently for different initial values of r_b. Nevertheless, we will see that, as the result of evolution, the present universe is always dominated by the YMC $\Omega_y \sim 0.73$ for a very wide range of initial values r_b.

Now we use the exact solution (17) to plot the evolution of ρ_y as a function of the redshift z in Fig. 1 (left panel). As specific examples, here we take $r_b = 10^{-2}$, and $r_b = 10^{-20}$. In comparison, also plotted are the energy densities of radiation, and of matter. It is seen that, in the early stages, ρ_y decreases as $\rho_y \propto a^{-4}$. So the YMC density is subdominant and tracks the radiation, a scaling solution. The corresponding EOS of Yang-Mills field is $w_y \simeq 1/3$ shown in Fig. 1 (right panel). At late stages, with the expansion of the universe, $a \to \infty$, y decreases to nearly zero, and $w_y \to -1$ asymptotically. Moreover, this asymptotic region is arrived at some redshift z before the present time, and this z has different values for different initial values of r_b. For smaller r_b, the transition redshift is larger (seen in Fig. 1), and the transition happens earlier. Once the asymptotic region is achieved, the density of the YMC levels off and remain a constant forever, like a cosmological constant. We have also checked that the present value $\Omega_y \sim 0.73$ is also the outcome of the cosmic evolution for any value of r_b in the very wide range $(10^{-20}, 10^{-2})$. So the coincidence problem do not exist in the YMC model.

The present EOS of the YMC w_0 is nearly -1. Fig. 2 (left panel) plots the dependence of the present EOS w_0 on the initial condition r_b. The function $\log_{10}(r_b)$ versus $\log_{10}(w_0 + 1)$ is nearly linear: a smaller r_b leads to a smaller w_0. For a value $r_b = 10^{-2}$, one has $w_0 = -0.99$. For a value $r_b = 10^{-20}$, w_0 would be -1 accurately up to one in 10^{11}. Therefore, at present the YMC is very similar to the cosmological constant.

The solution in Eq.(17) can converted into the following form

$$z = \sqrt{\frac{y}{y_0}} \exp\left[\frac{y - y_0}{4}\right] - 1, \tag{22}$$

where y_0 is the value of y at $z = 0$, depending on the initial value r_b. For a fixed y_0, this formula tells a one-one relation between the EOS (through y) and the corresponding redshift z. As is seen from Fig. 1, the transition of ω from $1/3$ to -1 occurs during a finite period of time, instead of instantly. To characterize the time of transition, we use z_t to denote the redshift when $\omega = -1/3$, i.e. $y = 1$, as given by Eq.(11). This is, in fact, the time when the strong energy condition begins to be violated, i.e., $\rho_y + 3p_y \leq 0$. Then

$$z_t = \sqrt{\frac{1}{y_0}} \exp\left[\frac{1 - y_0}{4}\right] - 1. \tag{23}$$

Therefore, this gives a function $z_t = z_t(r_b)$. Fig. 2 (right panel) shows how the transition redshift z_t depends on the ratio r_b. Interestingly, this transition can occur before, or after the radiation-matter equality ($z_{eq} = 3454$). This feature is different from the tracked quintessence models in which transition occurs during the matter dominated era [8]. A larger r_b leads to a smaller z_t. For example, $r_b = 10^{-2}$ leads to $z_t \simeq 12.4 \ll z_{eq}$, and the transition occurs in the matter dominated stage, and $r_b = 10^{-20}$ leads to $z_t \simeq 5.0 \times 10^5 \gg z_{eq}$, and the transition occurs in the radiation dominated stage.

The value of r_b cannot be chosen to arbitrarily large. In fact, there is a constraint from the observation result of the BBN. As is known, the presence of dark energy during nucleosynthesis epoch will speedup the expansion, enhancing the effective species N_ν of neutrinos [31]. The latest analysis gives a constrain on the extra neutrino species $\delta N_\nu \equiv N_\nu - 3 < 1.60$ [31]. Here in our model, the dark energy is played by the YM field. By a similar analysis, the ratio r_b is related to δN_ν through $r_b = \frac{7\delta N_\nu/4}{10.75}$. This leads to an upper limit $r_b < 0.26$, the present EoS $w_0 < -0.94$ by Fig. 2 (left panel), and the transition redshift $z_t > 5.8$ by Fig. 2 (right panel). The range of initial conditions $r_b \in (10^{-20}, 10^{-2})$ that we have taken satisfies this constraint.

4.2. Coupled YMC Models

In this subsection, we shall generalize the original YMC dark energy model to include the interaction between the YMC and dust matter. We should mention that, the possible interaction between YMC and radiation component may also exist, which has been discussed in

[32]. (The similar models on the scalar field dark energy have been discussed by a number of authors (see [33] for instant)). In this section, we assume the YMC dark energy and background matter interact through an interaction term Q. Thus the equations of the conservation of energy in (14) and (15) should be changed into

$$\dot{\rho}_y + 3\frac{\dot{a}}{a}(\rho_y + p_y) = -Q, \tag{24}$$

$$\dot{\rho}_m + 3\frac{\dot{a}}{a}\rho_m = Q, \tag{25}$$

and the equation for radiation in (16) is still held. The sum of Eqs. (24) and (25) guarantees that the total energy of YMC and dust matter is still conserved. It is worth noting that the free Yang-Mills equation in (18) is not satisfied when $Q \neq 0$. In the natural unit, the interaction term Q has the dimension of $[\text{energy}]^5$. The coupling Q is phenomenological, and their specific forms will be addressed later. When $Q > 0$, the YMC transfers energy into the matter, and this could be implemented, for instance, by the processes with the YMC decaying into pairs of matter particles. On the other hand, when $Q < 0$, the matter transfers energy into the YMC. Note that, once Q is introduced as above, it will bring another new parameters in the models, in addition the free parameter κ.

We introduce the following dimensionless variables, rescaled by the critical energy density $b\kappa^2/2$ of the YMC,

$$x \equiv \frac{2\rho_m}{b\kappa^2}, \quad f \equiv \frac{2Q}{b\kappa^2 H}, \tag{26}$$

where f is the function of x and y. By the help of the definition of y, the evolution equations (24) and (25) can be rewritten as a dynamical system, i.e.

$$y' = -\frac{4y}{2+y} - \frac{f(x,y)}{(2+y)e^y}, \tag{27}$$

$$x' = -3x + f(x,y). \tag{28}$$

Here, a prime denotes derivative with respect to the so-called e-folding time $N \equiv \ln a$. The fractional energy densities of dark energy and background matter are given by

$$\Omega_y = \frac{(1+y)e^y}{(1+y)e^y + x}, \quad \text{and} \quad \Omega_m = \frac{x}{(1+y)e^y + x}. \tag{29}$$

Before discussing the specific form of the interaction term Q, let us first investigate the general feature of this dynamical system, described by (27) and (28). We can obtain the critical point (y_c, x_c) of the autonomous system by imposing the conditions $y'_c = x'_c = 0$. From the equations (27) and (28), we obtain that the critical state satisfies the following simple relations

$$3x_c = f(x_c, y_c), \tag{30}$$

$$3x_c = -4y_c e^{y_c}. \tag{31}$$

So we can obtain the critical state (y_c, x_c) by solving these two equations. In order to study the stability of the critical point, we substitute linear perturbations $y \to y_c + \delta y$ and

$x \rightarrow x_c + \delta x$ about the critical point into dynamical system equations (27) and (28) and linearize them. Thus, two independent evolutive equations are derived,

$$\begin{pmatrix} \delta y' \\ \delta x' \end{pmatrix} \equiv M \begin{pmatrix} \delta y \\ \delta x \end{pmatrix} = \begin{pmatrix} G_{,y} + R_{,y} & R_{,x} \\ f_{,y} & f_{,x} - 3 \end{pmatrix} \begin{pmatrix} \delta y \\ \delta x \end{pmatrix},$$

where $R_{,y} \equiv \partial R/\partial y$ at $(y, x) = (y_c, x_c)$. The definitions of $R_{,x}$, $f_{,y}$, $f_{,x}$ and $G_{,y}$ are similar. The functions G and R are defined by

$$G \equiv G(y) = -\frac{4y}{2 + y}, \quad R \equiv R(x, y) = -\frac{f(x, y)}{(2 + y)e^y},$$

which are used for the simplification of the notation. The two eigenvalues of the coefficient matrix M determine the stability of the corresponding critical point. The critical point is an attractor solution, which is stable, only if both the these two eigenvalues are negative (stable node), or real parts of these two eigenvalues are negative and the determinant of the matrix M is negative (stable spiral).

Here we discuss some general features of the attractor solutions, regardless the special form of the interaction term Q. From the expression (31), we find that $x_c = -(4/3)y_c e^{y_c}$. Substitute this into the formula (29), one obtains

$$\Omega_y = \frac{(y_c + 1)e^{y_c}}{(y_c + 1)e^{y_c} + x_c} = \frac{3 + 3y_c}{3 - y_c}. \tag{32}$$

Since $0 \leq \Omega_y \leq 1$, this formula follows a constraint of the critical point

$$-1 \leq y_c \leq 0. \tag{33}$$

From the formulae (11) and (32), we find a simple, but interesting relation,

$$\Omega_y w_y = -1. \tag{34}$$

This relation is held for all attractor solutions, independent of the special form of the interaction. Since the value of Ω_y is smaller than or equal to unity in the attractor solution, we obtain that

$$w_y \leq -1. \tag{35}$$

This means that, the EOS of the YMC dark energy must be smaller than or equal to -1, i.e. phantom-like or Λ-like. Since in the early universe, the value of the order parameter of the YM field F is much larger than that of κ^2, i.e. $y \gg 1$, the YM field is a kind of radiation component. However, in the late attractor solution, the dark energy is phantom-like or Λ-like. So the phantom divide must be crossed in the former case, which is different from the interacting quintessence models.

It is interesting to investigate the total EOS of the YMC and dust matter, which determines the finial fate of the universe, when the radiation becomes negligible. The total EOS is defined by

$$w_{tot} \equiv \frac{p_{tot}}{\rho_{tot}} = \frac{p_y + p_m}{\rho_y + \rho_m} = \Omega_y w_y, \tag{36}$$

where $p_m = 0$ is used. From the relation (34), we obtain that, in the attractor solution,

$$w_{tot} = -1. \tag{37}$$

This result is also independent of the specific form of the interaction. So the universe is an exact de Sitter expansion, and the cosmic big rip is naturally avoided, although the YMC dark energy can be phantom-like.

Now, let us discuss the evolution of the various components in the universe for some specific interaction models. In this paper, we shall focus on the following three phenomeno-logical interaction models: $Q \propto H\rho_y$, $Q \propto H\rho_m$ and $Q \propto H(\rho_y + \rho_m)$, separately. Some other phenomenological models have also been discussed in the papers [32, 34].

4.2.1. $Q \propto H\rho_y$

In this case, we can write the function $f(x, y)$ as the following form, $f(x, y) = \alpha(1+y)e^y$. Of course, when $\alpha = 0$. the system returns to the model with free YMC dark energy. Here, we consider the simplest case with α being a non-zero dimensionless constant. Thus, the dynamical equations in (27) and (28) becomes

$$y' = -\frac{4y}{2+y} - \frac{\alpha(1+y)}{(2+y)}, \tag{38}$$

$$x' = -3x + \alpha(1+y)e^y. \tag{39}$$

Obviously, the evolution of dust matter and YMC are influenced by the interaction by the function $f(x, y)$. We expect, when the fraction density of YMC was sub-dominant in the universe, the effect on the dust is small. Only in the latest stage of the universe, where the YMC dark energy dominates the evolution of the universe, the effect of interaction on the dust becomes important.

The critical point (y_c, x_c) is obtained by imposing the condition $y'_c = x'_c = 0$, which are

$$y_c = -\frac{\alpha}{4+\alpha}, \quad x_c = -\frac{4y_c}{3}e^{y_c}, \tag{40}$$

and the fractional energy density and the EOS of the YMC at this critical point are

$$\Omega_y = -\frac{1}{w_y} = \frac{3}{\alpha + 3}. \tag{41}$$

The constraint $0 \leq \Omega_y \leq 1$ requires that $\alpha > 0$ (note, we have set $\alpha \neq 0$ throughout the discussion). In this condition, we find that $w_y < -1$ is satisfied. In order to keep this critical point being stable, i.e. (y_c, x_c) is the attractor solution, a constraint on α can be derived: $\alpha > -8$, which is auto-satisfied when $\alpha > 0$ is required. So the attractor solution requires that the constraint on the coefficient being

$$\alpha > 0, \tag{42}$$

which follows that the EOS of YMC in this attractor solution must be negative. If we still require that, the fraction density of YMC is larger than the value at the present stage, i.e. $\Omega_y > 0.73$, the constraint on α becomes much tighter $0 < \alpha < 1.11$.

In order to have a much clear picture for this system, let us study the evolution of the various components in the universe by adopting $\alpha = 0.5$. We still choose the initial condition at the BBN stage, i.e. $z = 10^{10}$, where the value of x is chosen as $x_i = 1.8 \times 10^{-29}$ to keep the present value of Ω_y being $\Omega_y = 0.73$. For the YMC, we consider the following two choices as the initial condition: i.e. $y_i = 60$ and $y_i = 20$. In the former case, we have $\Omega_{yi}/\Omega_{mi} = 0.16$, and in the latter case, we have $\Omega_{yi}/\Omega_{mi} = 3 \times 10^{-19}$. Although, the difference between these two Ω_{yi}/Ω_{mi} cases is larger than 17 orders. we will show that, these two models follow the similar present state of universe.

We should also fix the value of κ. In order to keep the present total energy density in the universe being $\rho_{tot} \simeq 8.099h^2 \times 10^{-11}\text{eV}^4$, we find the $\kappa = 4.00h \times 10^{-5}\text{eV}^2$ in both initial condiitons. If the Hubble constant $h = 0.72$ is adopted, we get $\kappa = 2.88 \times 10^{-5}\text{eV}^2$. Again, we find the value of κ is much smaller than the typical energy scale in the particle physics.

In Fig. 3 (left panel), we plot the evolution of various components in the universe. Note that, the evolution line for dust component covered each other in both initial cases, as well as the radiation component. Similar to the free YMC models (shown in the left panel in Fig. 1), in the early stage of the universe, the YMC was tracking the radiation, and it transfered to the attractor solution in the later stage. The smaller y_i induces the earlier transfer. In the right panel of Fig. 3, we plot the evolution of EOS of YMC w_y and the total EOS w_{tot}. Note that, the evolution line w_{tot} covered each other in both initial cases. As expected, in the both initial conditions, in the early stage, $w_y \to 1/3$. In the late stage, w_y runs to the attractor solution with $w_y = -(\alpha+3)/3$ with $\alpha = 0.5$, i.e. $w_y \to -7/6$. In the intermedial stage, the EOS w_y must cross the line with $w_y = -1$. For the total EOS w_{tot}, in the early stage, $w_{tot} \to 0$, where dust component is dominant than YMC. However, in the late stage, $w_{tot} \to -1$, which is consistent with the previous discussions.

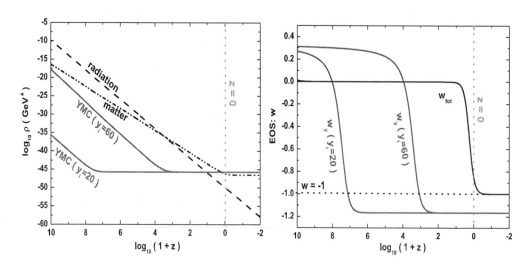

Figure 3. In the coupled YMC dark energy models, the evolution of the YMC energy density (left panels) and EOS (right panels) for the models with different initial conditions.

4.2.2. $Q \propto H\rho_m$

Now, let us turn to the interaction cases with $Q \propto H\rho_m$, which is equivalent to the form $f(x,y) = \beta x$. The dynamical equations in (27) and (28) becomes

$$y' = -\frac{4y}{2+y} - \frac{\beta x}{(2+y)e^y}, \tag{43}$$

$$x' = (\beta - 3)x. \tag{44}$$

If we consider the simplest, where β is a constant, the equation (44) follows that $x \propto a^{\beta-3}$. From the definition of x, we derive that $\rho_m \propto a^{\beta-3}$. When $\beta = 0$, the model returns to the free YMC cases, and $\rho_m \propto a^{-3}$ as usual. However, when $\beta \neq 0$, the evolution of the dust component is changed, which is conflicted with the evolution of dust in the standard hot big-bang model. So it is dangerous to consider this kind of interaction term in the early stage of the universe.

However, it is allowed to consider the form of $Q \propto H\rho_m$ as a kind of phenomenological model in the late stage of the universe. The critical point (y_c, x_c) of (43) and (44) is obtained by imposing the condition $y_c' = x_c' = 0$, we find these is no solution at all. So we conclude that, it is impossible to obtain an attractor solution for this kind of system.

4.2.3. $Q \propto H(\rho_y + \rho_m)$

In the end, let us discuss the another kind of phenomenological model, where $Q \propto H(\rho_y + \rho_m)$ is satisfied. This is equivalent to set the form $f(x,y) = \gamma[(y+1)e^y + x]$. We consider the simplest case, where γ is a non-zero dimensionless constant. In the dark energy dominant stage, this system returns to the first case with $Q \propto H\rho_y$, and in the dust dominant stage, it returns to the second case with $Q \propto H\rho_m$. Similar to the second case, if we directly apply this system to the early universe, the evolution of dust would be changed, which is conflicted with the prediction of the standard hot big-bang models.

In this paper, we also consider this interaction form as a kind of phenomenological model in the late stage of the universe. The dynamical equations in (27) and (28) becomes

$$y' = -\frac{4y}{2+y} - \frac{\gamma[(y+1)e^y + x]}{(2+y)e^y}, \tag{45}$$

$$x' = -3x + \gamma[(y+1)e^y + x]. \tag{46}$$

From the equations (30) and (31), we obtain the critical point

$$y_c = \frac{3\gamma}{\gamma - 12}, \quad x_c = -\frac{4y_c}{3}e^{y_c}. \tag{47}$$

The fractional energy density and the EOS of the YMC at this critical point are

$$\Omega_y = -\frac{1}{w_y} = \frac{3-\gamma}{3}. \tag{48}$$

The constraint of $0 \leq \Omega_y \leq 1$ requires that $0 < \gamma \leq 3$. In order to keep this critical point being stable, i.e. (y_c, x_c) is the attractor solution, another constraint of γ can be derived:

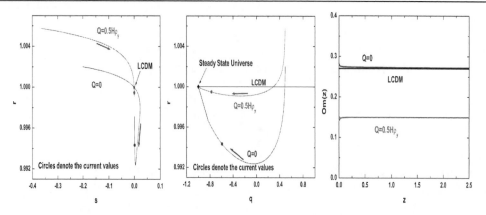

Figure 4. Left panel: The $r - s$ diagram of the YMC models; Middle panel: The $r - q$ diagram of the YMC models; The right panel: $Om(z)$ diagnostic in the YMC models.

$\gamma > -120/31$, which is auto-satisfied when $0 < \gamma \leq 3$ is required. So the attractor solution requires that the constraint on the coefficient being

$$0 < \gamma \leq 3, \tag{49}$$

which follows that, the EOS of YMC in this attractor solution must be negative. If we still require that, the fraction density of YMC is larger than the value at the present stage, i.e. $\Omega_y > 0.73$, the constraint on α becomes much tighter $0 < \gamma < 0.81$.

5.   Statefinder and Om Diagnosis in the YMC Models

In this section, we shall present the way to discriminate between the YMC models and the other dark energy models. In the previous works [35][36], the authors suggested to use the so-called "statefinder" $\{r, s\}$ pair and Om diagnostics. Here, we shall also apply the statefinder diagnosis into the YMC models. The similar discussion can be found in the previous works [27][25][28][37].

The statefinder diagnostic pair $\{r, s\}$ is defined as

$$r \equiv \frac{\dddot{a}}{aH^3}, \quad s \equiv \frac{r - 1}{3(q - 1/2)}, \tag{50}$$

where $q = -\ddot{a}/(aH^2)$ is the deceleration parameters. r forms the next step in the hierarchy of geometrical cosmological parameters beyond H and q, and s is a linear combination of r and q. Apparently, the statefinder parameters depends only on scale factor a and its derivatives, and thus it is a geometrical diagnostic. These parameters can also been expressed in terms of ρ_{tot} and p_{tot} as follows

$$q = \frac{1}{2}\left(1 + \frac{3p_{tot}}{\rho_{tot}}\right), \quad r = 1 + \frac{9(\rho_{tot} + p_{tot})\dot{p}_{tot}}{2\dot{\rho}_{tot}\rho_{tot}}, \quad s = \frac{\rho_{tot} + p_{tot}}{p_{tot}}\frac{\dot{p}_{tot}}{\dot{\rho}_{tot}}. \tag{51}$$

By using Eqs. (24) and (25), we plot the trajectories of (r, s) and of (r, q) in Fig. 4 from the redshift $z = 10$, where we have considered two cases with $Q = 0$ and $Q = 0.5H\rho_y$.

The arrows along the curves indicate the direction of evolution. In both cases, we have chosen the initial condition at the redshift $z = 3454$, where the densities of matter and radiation equate to each other. We choose the initial condition to make the $\Omega_y = 0.01$ at this high redshift.

From this figure, we find that, the model with different interaction terms have the difference in both (r, s) and (r, q) trajectories. However, in the low redshift ($z < 10$), especially when $z \sim 0$, the trajectories in both models become quite close to those of the ΛCDM model. However, we should also mention that, the trajectories of the statefinder shows in this figure are quite different from the other dark energy models, such as the decaying vacuum model, the Quintessence models, the K-essence models (see for instant [37]). In order to break the degeneracy between the YMC models and the ΛCDM models, let us consider the Om diagnostic, which is defined as

$$Om(x) = \frac{h^2(x) - 1}{x^3 - 1}, \tag{52}$$

where $x \equiv (1+z)$, and $h(x) \equiv H(x)/H_0$. Thus Om involves only the first derivative of the scale factor through the Hubble parameter and is easier to reconstruct from the observational data. For the ΛCDM model, it is simple, i.e. $Om(x) = \Omega_m$, independent of the redshift. For YMC model, we plot Om as a function of z in the Fig.4 (right panel). It is interesting to find that, for either model, Om is nearly a constant at the low redshift. For the model without interaction, as expected, $Om \simeq 0.27$, very close to that of ΛCDM models. However, for the model with $Q = 0.5H\rho_y$, $Om = 0.15$, which is helpful to differentiate the coupled YMC models from the ΛCDM model.

6.   Conclusion

In order to answer the observed accelerating expansion of the universe in the present stage, we introduce the quantum Yang-Mills condensate dark energy models. Different from the general scalar field models, the quantum Yang-Mills fields are the indispensable cornerstone to particles, and the Lagrangian of the Yang-Mills field is predicted by quantum corrections according to field theory.

In this paper, we review the main characters of the YMC dark energy models, where both free YMC model and possible coupled YMC models are considered. In all these models, the EOS of YMC is close to $w_y = 1/3$ in the early stage, and tracked the evolution of radiation components. In the late stage, the value of w_y runs to the attractor solution with $w_y = -1$ for the free YMC model, or with $w_y < -1$ for the coupled YMC models. This naturally explains the observations. The present state of the universe is independent of the initial state of Yang-Mills field, and the coincidence problem is naturally avoided. In the coupled YMC models, not only the state of $w_y < -1$ can be naturally realized, mildly suggested by observations, but also the 'big-rip' problem is auto-avoided. The most importance we should mention is that, as a dynamic dark energy model motivated by quantum effective YM field Lagrangian, all the YM models based upon 1-loop, 2-loop, and 3-loop quantum corrections, respectively, have similar dynamic behaviors. That is, the main properties of YM model for dark energy remain stable when the number of loops for quantum corrections increases up to 3-loop.

However, in all the models, the fine-tunning problem still exists, i.e. the value of $\kappa \sim 10^{-5} eV^2$, which is too low comparing with the typical scales in particle physics. We should notice that, this problem exists in nearly all the dark energy models, which may suggest a new physics at this low energy scale.

References

[1] Kowalski, M. et al. (2008). Improved Cosmological Constraints from New, Old and Combined Supernova Datasets. *Astrophysical Journal*, **686**, 749.

[2] Komatsu, E. et al. (2009). Five-Year Wilkinson Microwave Anisotropy Probe (WMAP) Observations: Cosmological Interpretation. *Astrophysical Journal Supplement Series*, **180**, 330.

[3] Tegmark, M. et al. (2006). Cosmological Constraints from the SDSS Luminous Red Galaxies. *Physical Review D*, **74**, 123507.

[4] Padmanabhan, T. (2006). Dark Energy: Mystery of the Millennium. *Albert Einstein Century International Conference*, **861**, 179.

[5] Huterer, D. & Cooray, A. (2005). Uncorrelated Estimates of Dark Energy Evolution. *Physical Review D*, **71**, 023506.

[6] Albrecht A. et al. (2006). Report of the Dark Energy Task Force. *arXiv:astro-ph/0609591*.

[7] Copeland, E. J.; Sami, M. & Tsujikawa, S. (2006). Dynamics of Dark Energy. *International Journal of Modern Physics D*, **15**, 1753.

[8] Zlatev, I.; Wang, L. & Steinhardt, P. J. (1999). Quintessence, Cosmic Coincidence, and the Cosmological Constant. *Physical Review Letters*, **82**, 896.

[9] Caldwell, R. R. (2002).A Phantom Menace? Cosmological Consequences of a Dark Energy Component with Super-negative Equation of State. *Physics Letters B*, **545**, 23.

[10] Feng, B.; Wang, X. L. & Zhang, X. M. (2005). Dark Energy Constraints from the Cosmic Age and Supernova. *Physics Letters B*, **607**, 35;

Zhao, W & Zhang, Y. (2006). Quintom models with an equation of state crossing -1. *Physical Review D*, **73**, 123509.

[11] Carroll, S. M.; Hoffman, M. & Trodden, M. (2003). Can the Dark Energy Equation-of-state Parameter w Be Less Than -1? *Physical Review D*, **68**, 023509.

[12] Caldwell, R. R.; Kamionkowski, M. & Weinberg, N. N. (2003). Phantom Energy and Cosmic Doomsday. *Physical Review Letters*, 91, 071301.

[13] Carroll, S. M.; Duvvuri, V.; Trodden M. & Turner, M. S. (2004). Is Cosmic Speed-Up Due to New Gravitational Physics? *Physical Review D*, **70**, 043528;

Carroll, S. M. et al. (2005). The Cosmology of Generalized Modified Gravity Models. *Physical Review D*, **71**, 063513.

[14] Chiba, T.; Smith, T. L. & Erickcek, A. L. (2007). Solar System Constraints to General f(R) Gravity. *Physical Review D*, **75**, 124014.

[15] Armendariz-Picon, C. (2004). Could Dark Energy Be Vector-like? *Journal of Cosmology and Astroparticle Physics*, **0407**, 007;

Wei, H. & Cai, R. G. (2006). Interacting Vector-like Dark Energy, the First and Second Cosmological. *Physical Review D*, **73**, 083002;

Wei H. & Cai, R. G. (2007). Cheng-Weyl Vector Field and its Cosmological Application. *Journal of Cosmology and Astroparticle Physics*, **0709**, 015;

Bamba, K.; Nojiri, S. & Odintsov, S. D. (2008). Inflationary Cosmology and the Late-time Accelerated Expansion of the Universe in Nonminimal Yang-Mills-F(R) Gravity and Nonminimal Vector-F(R) Gravity. *Physical Review D*, **77**, 123532;

Koivisto, T. S. & Mota, D. F. (2008). Vector Field Models of Inflation and Dark Energy. *arXiv:0805.4229*;

Gal'tsov, D. V. (2009). Non-Abelian Condensates As Alternative for Dark Energy. *arXiv:0901.0115*;

De Lorenci, V. A. (2009). Nonsingular and Accelerated Expanding Universe from Effective Yang-Mills Theory. *arXiv:0902.2672*;

Zhang, Y. (2009). The Slow-Roll and Rapid-Roll Conditions in The Space-likeVector Field Scenario. *arXiv:0903.3269*;

Koivisto, T. S. & Nunes, N. J. (2009). Three-form Cosmology. *arXiv:0907.3883*;

Nunes, N. J. & Koivisto, T. S. (2009). Ination and Dark Energy from Three-forms. *arXiv:0908.0920*.

[16] Kiselev, V. V. (2004). Vector Field as a Quintessence Partner. *Classical and Quantum Gravity*, **21**, 3323;

Novello, M.; Bergliaffa, S. E. P. & Salim, J. (2004). Non-linear Electrodynamics and the Acceleration of the Universe. *Physical Review D*, **69**, 127301;

Boehmer, C. G. & Harko, T. (2007). Dark Energy as a Massive Vector Field. *The European Physical Journal C*, **50**, 423;

Jimenez, J. B. & Maroto, A. L. (2008). A Cosmic Vector for Dark Energy. *Physical Review D*, **78**, 063005;

Jimenez, J. B.; Lazkoz, R. & Maroto, A. L. (2009). Cosmic Vector for Dark Energy: Constraints from SN, CMB and BAO. *arXiv:0904.0433*;

Jimenez, J. B. & Maroto, A. L. (2009). Cosmological Evolution in Vector-tensor Theories of Gravity. *arXiv:0905.1245*;

Dimopoulos, K.; Karciauskas, M. & Wagstaff, J. M. (2009). Vector Curvaton with Varying Kinetic Function. *arXiv:0907.1838*.

[17] Zhang, Y. (1994). Inflation with Quantum Yang-Mills Condensate. *Physics Letters B*, 340, 18;

Zhang, Y. (1996). An Exact Solution of a Quark Field Coupled with a Yang - Mills Field in de Sitter Space. *Classical and Quantum Gravity*, **13**, 2145;

Zhang, Y. (2003). Dark Energy Coupled with Relativistic Dark Matter in Accelerating Universe. *Chinese Physics Letters*, **20**, 1899.

[18] Adler, S. L. (1981). Effective-action Approach to Mean-field non-Abelian statics, and a Model for Bag Formation. *Physical Review D*, **23**, 2905;

Adler, S. L. (1983). Short-distance Perturbation Theory for the Leading Logarithm Models. *Nuclear Physics B*, **217**, 381.

[19] Pagels, H. & Tomboulis, E. (1978). Vacuum of the Quantum Yang-Mills Theory and Magnetostatics. *Nuclear Physics B*, **143**, 485.

[20] Zhao, W. & Zhang, Y. (2006). The State Equation of the Yang-Mills Field Dark Energy Models. *Classical and Quantum Gravity*, **23**, 3405.

[21] Zhao, W. & Zhang, Y. (2006). Coincidence Problem in YM Field Dark Energy Model. *Physics Letters B*, **640**, 69.

[22] Matinyan, S. G. & Savvidy, G. K. (1978). Vacuum Polarization Induced by the Intense Gauge Field. *Nuclear Physics B*, **134**, 539.

[23] Gross, D. J. & Wilczez, F. (1973). Ultraviolet Behavior of Non-Abelian Gauge Theories. *Physical Review Letters*, **30**, 1343;

Gross, D. J. (1973). Asymptotically Free Gauge Theories. I. *Physical Review D*, **8**, 3633.

[24] Xia, T. Y. & Zhang, Y. (2007). 2-loop Quantum Yang-Mills Condensate As Dark Energy. *Physics Letters B*, **656**, 19.

[25] Wang, S.; Zhang, Y. & Xia, T. Y. (2008). 3-loop Yang-Mills Condensate Dark Energy Model and its Cosmological Constraints. *Journal of Cosmology and Astroparticle Physics*, **10**, 037.

[26] Zhao, W. & Xu, D. H. (2007). Evolution of Magnetic Component in Yang-Mills Condensate Dark Energy Models. *International Journal of Modern Physics D*, **16**, 1735.

[27] Zhao, W. (2008). Statefinder Diagnostic for Yang-Mills Dark Energy Model. *International Journal of Modern Physics D*, **17**, 1245.

[28] Tong, M. L.; Zhang, Y. & Xia, T. Y. (2009). Statefinder Parameters for Quantum Effective Yang-Mills Condensate Dark Energy Model. *International Journal of Modern Physics D*, **18**, 797.

[29] Zhao, W. (2009). Perturbations of the Yang-Mills Field in the Universe. *Research in Astronomy and Astrophysics*, **9**, 874.

[30] Vikman, A. (2005). Can Dark Energy Evolve to the Phantom? *Physical Review D*, **71**, 023515.

[31] Cyburt, R. H.; Fields, B. D.; Olive, K. A. & Skillman, E. (2005). New BBN Limits on Physics Beyond the Standard Model from He4. *Astroparticle Physics*, **23**, 313.

[32] Zhang, Y.; Xia, T. Y. & Zhao, W. (2007). Yang-Mills Condensate Dark Energy Coupled with Matter and Radiation. *Classical and Quantum Gravity*, **24**, 3309.

[33] Wang, B. et al. (2007). Interacting Dark Energy and Dark Matter: Observational Constraints From Cosmological Parameters. *Nuclear Physics B*, **778**, 69;

He, J. H.; Wang, B. & Zhang, P. (2009). The Imprint of the Interaction Between Dark Sectors in Large Scale Cosmic Microwave Background Anisotropies. *arXiv:0906.0677*.

[34] Zhao, W. (2009). Attractor Solution in Coupled Yang-Mills Field Dark Energy Models. *arXiv:0810.5506*.

[35] Sahni, V,; Saini, T. D.; Starobinsky, A. A. & Alam, U. (2003). Statefinder - A New Geometrical Diagnostic of Dark Energy. *JETP Letters*, **77**, 201.

[36] Sahni, V.; Shafieloo, A. & Starobinsky, A. A. (2008). Two New Diagnostics of Dark Energy. *Physical Review D*, **78**, 103502.

[37] Tong, M. L. & Zhang, Y. (2009). Cosmic Age, Statefinder and Om Diagnostics in the Decaying Vacuum Cosmology. *Physical Review D*, **80**, 023503.

In: Dark Energy: Theories, Developments and Implications   ISBN 978-1-61668-271-2
Editors: K. Lefebvre and R. Garcia, pp. 111-126   © 2010 Nova Science Publishers, Inc.

Chapter 6

CONSTRAINTS ON DARK ENERGY AND DARK MATTER FROM SUPERNOVAE AND GAMMA RAY BURST DATA

Michael L. Smith[1,*], *Bishmer Sekaran*[1], *Ahmet M. Öztaş*[2] *and Jan Paul*[3]
[1]21st Century HealthCare, Inc., Tempe, AZ 85282, USA
[2]Department of Engineering Physics, Hacettepe University,
TR-06800 Ankara, Turkey
[3]Division of Physics, Luleå University of Technology S-971 87 Luleå, Sweden

Abstract

We test models of cosmology for the best estimates of important parameters, that is, for matter, dark energy(DE), spacetime(ST) and cold dark matter(CDM) within variants and special cases of the Friedman-Robertson-Walker(FRW) approximation. We use the largest available collection of supernovae Ia data(SNe Ia), 307 SNe Ia along with 69 data pairs reported from gamma ray bursts(GRB) and our present situation for a total of 377 data pairs extending back perhaps 10 billion light years. Modeling with this large set allows better definition of the limits of DE, CDM and ordinary matter than previously. While the ancient GRB data are quite noisy they do allow better estimates of the fitted curve asymptotes towards singularity than without. The results from models employing the commonly used luminosity distance moduli(log) *versus* redshift data slightly prefer the DE model with H_0 of 70.2 ± 0.6 km/s/Mpc at a normalized matter density, Ω_m, of 0.34 with significant negative spacetime (ST) curvature. The best fit calculation for the model without DE presents a much lower matter density(0.10 ± 0.01) and a very slow Universe expansion rate(41.4 ± 3.2 km/s/Mpc). When tests were made using the actual distances(Mpc) *versus* frequency decline ν_0/ν_e data, which is the preferred method not relying too heavily on the great errors from far distance emissions(some standard deviations are many thousand Mpc), the flat DE model presents the worst fit of all models with H_0 of 75.4 ± 0.7 km/s/Mpc at Ω_m of 0.12 ± 0.02. Most interestingly, the two best fits present Ω_Λ of 1.03 and 0.68 with significant ST curvature. These solutions suggests a Universe consisting primarily of DE with abundant ST. We suggest astronomers cease using the luminosity distance moduli from distant emissions for modeling cosmology and present results from models based

---

*E-mail address: mlsmith55@gmail.com. (Correspondence address)

on distance estimates *vs.* redshift or frequency decline. At least an order magnitude more and better astronomical data are required to unequivocally choose the best FRW model to refine for further work in cosmology. Our results suggesting a Universe with excessive DE should exhibit open or negative curvature might foster some interesting theoretical models in the near future. FRW models with the largest collection of SNe and GRB data tell us the Universe is not flat.

1. Introduction

The concept of Dark Energy(DE) rests completely on one idea and the supernovae emission data collected recently by astronomers. The idea was a suggestion by Einstein during the early 20th century that stellar objects in the Universe were placed in stationary positions by some unknown force, and to allow for this he introduced a new term in his fundamental equation with the cosmic constant[6]. This represents the prior energy required to place the stars at stationary positions in our Milky Way, the entire cosmos at that time, thus avoiding collapse of our world into a disastrous end or beginning from some unimaginable location. The data supporting the DE idea began accumulating in the waning years of that century primarily as observations of very distant type Ia supernovae (SNe Ia). The connection between theory and observation is the cosmic constant, Λ, of Einstein. This is the kernel of the additional term which has recently been shown to aid the mathematical modeling of spacetime expansion to SNe Ia data. That is, inclusion of Λ in the Friedman-Robertson-Walker(FRW) approximation of the Einstein solution models better with than without. Astronomers and physicists are not in agreement if DE operates only "intergalactically" or even sub-atomically. It is now acceptable to use the terms dark energy and cosmic constant interchangeably and we do so[3]. Like Cold Dark Matter(CDM) this energy has been monitored affecting other, distant worlds but not characterized in earthly laboratories up to now.

In the mid-20th century Hubble showed that galaxies are not only far outside the Milky Way but generally moving away from each other[13]. This fact seemed to have dulled Einstein's interest in both a static Universe and the cosmic constant. Over the next several decades spacetime expansion from singularity was proven by observation of the big bang remnant radiation and the necessity for dealing with the cosmic constant waned. Confirmation of the big bang from the discovery of the Cosmic Microwave Background(CMB), with residual radiation reminiscent of a black body, seemed the death knell for the cosmic constant.

A score of years ago astronomers could only guess at the local Hubble constant, H_0, the measure of spacetime expansion. A good estimate at that time would be a value between 50-100 km/s/Mpc with very large uncertainty[35]. Other important values, such as the Universe age and the mean Universe matter density, can be calculated if H_0 is known with some accuracy. Research programs were begun just before this millenia aiming towards a more accurate measure of H_0 using data from distant SNe Ia, because these emissions are thought to be quite uniform in nature, no matter the age of the event[11, 18]. Uniform luminosity - over expansive spacetime and with spectacular intensity - because a type Ia supernova is a thermonuclear explosion of a dwarf star rich in carbon-oxygen at a unique, critical mass[4]. This uniformity has allowed a little better calibration of H_0 but a definitive value still depends on more exact distance calibration.

Some of the first collections of SNe Ia data were acquired from recent emissions from up to about 800 Mpc distant with fairly small errors of luminary magnitude[16]. These observations were used to gauge more distant and difficult emissions, which were required for estimating the Hubble constant as calculated using the FRW approximations. This makes SNe Ia our best distant standard candles, allowing estimates of ancient galactic distances in reference to nearby SNe Ia with well-measured histories[10]. Independent observations were then collected of about 40 really distant SNe Ia, in two independent sets which, being obtained by different groups, were not normalized and so could not be combined into a larger data set[26, 24]. Nevertheless, both groups concluded the FRW model fit the data best when a term for the cosmic constant was included. Both groups simultaneously and strongly suggested their solutions were also evidence for a new form of energy, now termed Dark Energy. The FRW model allows the cosmic constant to be either opposed to or with gravitational pull. The sign of the term extracted from the current *standard model* fit to SNe Ia data strongly suggests an antigravity-like effect, at least recently ($<$ 5 Gyrs lookback time).

Without a doubt SNe Ia are the best available tools for investigating numerical values of cosmology. One drawback which seems difficult to overcome is the emission strength beyond $z \approx 1.5$ being too weak to out shine noise from intergalactic dust, intervening galaxies and other like problems. Another problem is these "standard candles" may not emit so uniformly after all, with early SN($z > 1.5$) displaying a slightly different light-curve than more recent events[12]. A third problem which has been recently reported is that the supernova explosion is not of simple origin, but begins asymmetrically on the surface of the white dwarf, so the SNe Ia signals demand more specialized care during light-curve analysis[14] and can differ significantly in the amounts of radioactive nickel synthesized, and hence light emitted[33]. Several new types of supernovae have recently been reported which may not arise from the Chandrasekhar type mechanism and hence may be misleading with respect to distance if "mistakenly" reported as type Ia[23]. Thankfully, astronomers are continuing to develop more discriminatory mechanisms to either modify/correct the distance data for use and refining standards for inclusion in this important data set[9].

Other related tools, such as Gamma Ray Bursts(GRB) are just now being organized to aid systematic cosmology. These really ancient emissions may allow us to peer almost to the births of the first galaxies. Unlike the SN Ia explosion, the origin and mechanism of these gamma ray signals are still debated. As a rule, determining the exact distances of GRB are more problematic than SNe Ia since data collection seems more demanding and is not as well organized as SNe Ia programs. However, GRB have been observed in numbers enough to be useful tools for investigations in cosmology; 69 normalized observations have been recently published[29]. As we shall see, although both precision and accuracy of some astronomical data are suspect, especially for GRB, a guess into the ancient past is better than ignorance.

Some efforts have been also made to analyze the effects of gravitational lensing to constrain the values of the cosmic constant. Arguments have been published, pro and con, but it seems the effect of DE upon gravitationally distorted light might be too small to allow lensing to become a useful tool[30].

Here we present the efforts of our investigation into the best value for DE using the latest and largest data sets from GRB and SNe Ia observations[17, 29]. We have evaluated the data using both the typical luminosity distance moduli *versus* redshift and the rarely used distance *versus* frequency ratio relationships. For reasons we shall explain, we much prefer analyses using the latter data pairing rather than the more common luminosity distance moduli *versus* redshift set. The GRB data, though extremely noisy, are very useful for both types of evaluations; these data allow much better estimates of the asymptotes towards singularity and emphasize the curved nature of the emission data relationship. We present 3-dimensional figures of the normalized values for matter density, spacetime curvature and the cosmic constant(DE) from multiple curve fittings selecting a fixed value for a single free parameter; for instance, *stepping* from very low to moderate matter density in regular steps. This is done because the FRW generality really presents four parameters including the Hubble constant, and while normalization reduces these to three the data are still not good enough to fit a 3-free parameter model(at least for us) with confidence. This is one reason why the *Classical* model used for fitting SNe Ia data suggesting DE fits best as a flat universe, that is, without spacetime as a free parameter. This special variety of the FRW approximation presents only 3-free parameters, which normalization reduces to two. To keep our problem simple, we chose not to test models allowing w, the equation of state parameter, to freely float but to presume the usual w of -1 for all testing regimes.

We find the most favorable models, as judged by error minimization, to deliver a large Ω_Λ - meaning a universe rich in DE - but also delivering significant spacetime curvature as Ω_k. We also find several varieties of the FRW approximation to present solutions which cannot be judged better than one another on the basis of the usual χ^2 values or cross-over number. These varieties present values of Ω_Λ from 0 to 1, with Ω_k from -0.36 to +0.80. The analyses do deliver values for H_0 between 68 to 75 km/s/Mpc with Ω_m between 0.20 to 0.34, so provide reasonable constraints for H_0 and a fair range for Ω_m. We suggest that much more data are really required before the values for the parameters Ω_k and Ω_Λ are known with any certainty.

2.   Model Regimes

We use the common, normalized solution to the FRW generalization allowing the linear combination of matter density, ST curvature and DE density to be 1

$$\Omega_m + \Omega_k + \Omega_\Lambda = 1. \tag{1}$$

This great approximation allows the combination of the mean universal matter density, cold dark matter (CDM) and all radiation as Ω_m, all forms of DE as Ω_Λ and ST curvature as Ω_k. We presume a positive Ω_k indicating a concave-curved, closed universe and a negative Ω_k indicating a convex-curved, open universe. The value for Ω_m has important meaning since an Ω_m of 1 suggests a universe teetering between expansion and reversal towards the "Big Crunch"; an $\Omega_m > 1$ means the ultimate gathering is unavoidable. Such values would also mean that dark energy, as Ω_Λ, and spacetime curvature would be of little importance. An excellent exposition of the usefulness of this normalization with some graphical examples has been presented by the erudite Sean Carroll[2] and is presently used and referenced by many astronomers and astrophysicists.

We combine Eq. (1) with another general relationship to model the combined SNe Ia and GRB data along with associated redshifts as

$$D_L = \frac{c(1+z)}{H_0\sqrt{|\Omega_k|}} \text{sinn}\left\{\sqrt{|\Omega_k|}\int_0^z \frac{dz}{\sqrt{(1+z)^2(1+\Omega_m z) - z(2+z)\Omega_\Lambda}}\right\} \qquad (2)$$

When Ω_k is positive *sinn* becomes sinh and when negative sin is used and, when flat space-time is presumed the interesting trigonometry disappears and the pre-integral reduces to simply $c(1+z)/H_0$. In either case, the integral of Eq. (2) cannot be solved analytically but can be solved numerically with enough data.

Relationships using the redshift, z, are most often modeled with the astronomical data in terms of luminosity distance moduli rather than distance. The data are usually fit by taking the log of the right side of Eq. (2) and adjusting to match the astronomical observations as the often used approximation

$$\mu = 5Log(D_L) + 25 \qquad (3)$$

where D_L is the luminary distance in megaparsec (Mpc), usually obtained from the energy flux (brightness) of the SN, but sometimes by other related means, μ is the luminosity distance moduli and equal to 0 for an object at a distance of 10 parsec - to a cosmologist this is right next door. More detailed presentations of calculating a luminary distance from the energy flux is given in Peebles (13.51) in[22] with an excellent example of the use of luminosity distance moduli presented in Clocchiatti, *et al.*[5]. The values for μ are therefore interrelated in a logarithmic manner and are not strictly linear.

We shall also use a variant of this general relationship in terms of the ratio of observed emitted frequency over the presumed emitted frequency, replacing $1/(1+z)$ with ξ which recasts the FRW approximation with DE as

$$D_L = \frac{c}{\xi H_0\sqrt{|\Omega_k|}} \text{sinn}\left\{\sqrt{|\Omega_k|}\int_\xi^1 \frac{d\xi}{\sqrt{\frac{\Omega_m}{\xi} + \Omega_\Lambda \xi^2 + \Omega_k}}\right\}. \qquad (4)$$

The advantage of this relationship is that we can add our present situation of zero distance at a frequency ratio of 1, without any error, for one more important data pair.

For evaluation of the FRW model without DE we simply drop the $\Omega_\Lambda \xi^2$ term from the denominator leaving us with an integral which has been solved analytically[20, 22]

$$D_L = \frac{c}{\xi H_0\sqrt{|\Omega_k|}} \text{sinn}[2(\text{arctanh}(\sqrt{|\Omega_k|}) - \text{arctanh}(\frac{\sqrt{|\Omega_k|}}{\sqrt{\frac{\Omega_m}{\xi} + \Omega_k}}))]. \qquad (5)$$

When flat spacetime is presumed (as often) Eq. (2) is simplified to the following for use with redshift-luminosity distance moduli data

$$D_L = \frac{c(1+z)}{H_0}\left\{\int_0^z \frac{dz}{\sqrt{(1+z)^2(1+\Omega_m z) - z(2+z)\Omega_\Lambda}}\right\}. \qquad (6)$$

This is the form of the FRW solution popularized by many astronomers and often termed, the standard model, used to fit the astronomical data, we shall use the term *Classic* for this solution. Since astronomers collect distant emission data in the logarithm form, as luminosity distance moduli, rather than data linear to distance Eq.(4) must be further modified to allowing curve fitting to data as

$$Moduli = 5Log[\frac{c(1+z)}{H_0}\left\{\int_0^z \frac{dz}{\sqrt{(1+z)^2(1+\Omega_m z) - z(2+z)\Omega_\Lambda}}\right\}] + 25 \quad (7)$$

which is the more detailed form of Eq. (3), and this again is the *Classic* model, even though only popular for a decade. Likewise the associated distance errors are reported and applied in logarithm form. For many displays of the fits of the *Classic* model the abscissa is transformed to log z so a nearly straight line is presented[15], and we present one example as Figure 1.

3. Model Application

Most current, computerized curve-fitting routines converge on the best fit by minimizing the sum of the squared errors, that is, the squares of the ordinate values between the fitted curve and the data. Preferred programs inversely weigh the importance of each data pair relative to the assigned standard deviation; the larger the standard deviation the less important that pair is to the placement of the curve and the calculation of the free parameters. To accomplish this, the ordinate is chosen as the portion of the data pair with the larger error. This is the reason why Hubble diagrams from the last century not subjected to computerized analysis, usually presented redshift *vs.* distance (interpreted as radial velocity *vs.* distance) but nowadays one usually reads distance *vs.* redshift. It seems that astronomy will always suffer that problem plaguing mankind from the dawn of time of accurate distance determination.

What astronomers most often observe are the luminosity distance moduli of stars, galaxies and supernovae. Evaluation of "distance" is often performed using these distance moduli, related to galactic distance, with associated relatively small log errors. Because the log error of distance moduli is significantly smaller than the related error of distance, too much dependence of a curve fit can be placed on ancient signals rather than more secure and nearby emissions. This is because the log errors are proportionally too small for distant objects in comparison to the errors of more well-known nearby SNe Ia; hence the most probable values for the free parameters are not necessarily found through computerized curve-fit.

The data sets we have chosen to use are the recent, on-line report of 307 SNe Ia, which are the normalized data from several discovery groups[1, 17, 36] and the collection of 69 GRB data from Schaefer in 2007[29]. By adding our present situation we can evaluate 377 data pair, which may be the largest distance-velocity evaluation yet. We understand that the GRB data has not been adjusted with respect to the larger SNe Ia data set. To avoid extraordinary dependence of model fits on distant and more unreliable data we have also calculated emission distances in Mpc and associated geometric errors for the SNe Ia and GRB data as the ordinate and the associated frequency ratios ν_e/ν_0 as ξ for the abscissa. We present results using both the data *as is* and after conversion to Mpc *vs.* ξ. For all modeling

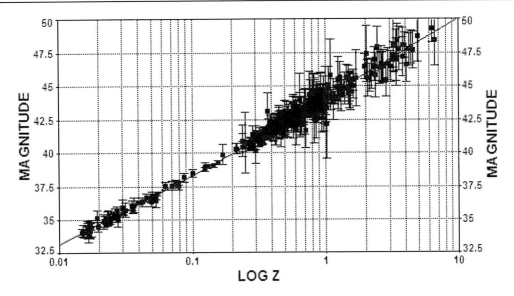

Figure 1. Luminary distance moduli *vs.* LogZ for 376 SNe Ia and GRB data pairs(MR).

we evaluate using the robust routine which well ignores outlying data. As we shall show, there is much to be gained by using data over nearly all time.

We present a "typical" plot of observed luminary distance moduli versus associated log(redshift) in Figure 1, which are the same data and model but with log(z) for presentation. This is a plot of the fit of a flat universe with DE to that data set. One can view this type of display in widespread publications suggesting DE[15, 16]. The one point of exact knowledge, our present time, cannot be used in this analysis and also note the errors increase by a factor of about 10 or so over the entire redshift range, which is much of our time as the Universe.

A plot of distances in Mpc versus ξ is presented in Figure 2, using the function related to that displayed in Figure 1. This figure allows direct comprehension of the errors associated with distance determinations. These errors are seen to increase drastically from nearby SNe Ia(still older than the dinosaurs) to by well over a factor of 50 for those GRB which occurred while our Universe was young. (A few standard deviations seem larger than some estimates of the Universe "diameter".) This presentation also allows one additional data pair without any distance error for our earth-bound observers. (We know exactly where we are and what time it is, relatively.) Also note that Figure 2 presents the best fit with a pronounced curve due to the inclusion of the GRB data; the curve would be much less noticeable using only SNe data.

4.   Results

4.1.   Luminosity Distance Moduli *vs.* Redshift Data

Most of our results are from fits with 2 free parameters(a few with only 1 free parameter) of the combined SNe Ia and GRB data in the largest set of emissions yet examined. This

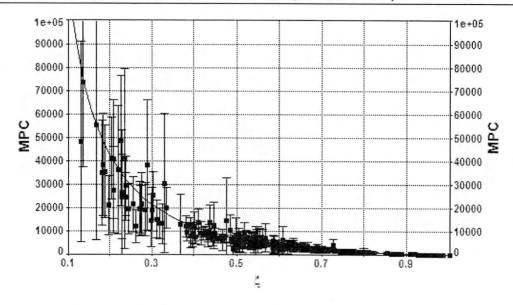

Figure 2. Distance *vs.* Frequency Ratio for 377 SNe Ia, GRB and our earth data pairs(DFR).

allows direct comparison of the goodness of fit between models. We divide the results into two groups; the results from analyzing the combined luminosity distance moduli *vs.* redshift (z) data (**MR**) and the results from analyzing the distance in Mpc *vs.* ξ data (**DFR**). For the first task we simply use the data as presented, which includes log errors and evaluate the goodness of fit in terms of χ^2/N and crossover numbers(X-over), where N is the total number of data pairs used. The lower the χ^2/N the better the fit, the higher the X-over number the better the fit(a rough generalization).

We first model the flat Universe with DE using the MR data allowing Ω_m and H_0 as free parameters. This is the usual, *Classic* solution to the emission data which sometimes yields the best fit. With 376 data pairs from the MR data, this model presents a H_0 of 70 ± 0.6 km/s/Mpc and an Ω_m of 0.30 ± 0.03 and by normalization a Ω_Λ of about 0.70(Table 1). While the value for matter density is slightly high and Ω_Λ slightly lower than previous reports with less data, all of these values are centered within the ranges currently popular. This calculated H_0 is well within the range of recent reports[8, 28, 34].

The best fit for all trials with the MR data, is found by holding Ω_k as a constant, varying

Table 1. Results from luminary distance moduli *vs* redshift (MR)with 376 data pairs

Routine	step/fix	χ^2/N	X-ovr	H_0	Ω_m	Ω_k	Ω_Λ
Classic	Ω_k	1.22	183	70.0 ± 0.6	0.30 ± 0.03	0	0.70
Ω_k-Step	Ω_k	1.22	185	70.2 ± 0.6	0.34 ± 0.03	-0.11	0.77
Ω_m-Step	Ω_m	1.22	183	70.2 ± 0.6	0.38	-0.16 ± 0.05	0.78
Popular	Ω_m	1.25	179	69.6 ± 0.6	0.25	0.19 ± 0.05	0.56
No DE	Ω_Λ	1.31	171	41.5 ± 3.2	0.10 ± 0.01	0.90 ± 0.01	0

H_0 in km/s/Mpc

this in steps of 0.01, and allowing the values for H_0 and Ω_Λ to freely float. This Ω_k-Step routine exhibits a χ^2/N similar to the *Classic*(flat) model and a slightly larger X-over value. It also suggests a precise estimate of H_0 of 70.2 ± 0.6 with a significant negative value for Ω_k hinting at convex ST curvature with an open universe. This results from a large matter density of 0.34 which is quite a bit higher than the popular 0.25. When we allow H_0 and Ω_Λ free reign while stepping through matter density, the Ω_m-Step routine, we are presented with another best fit at Ω_m of a very high 0.38. This also demands a significant negative value for Ω_k which again hints at convex curvature for our Universe. Contrarily, the results of modeling the data with 2-free parameters of H_0 and Ω_Λ while holding matter density at the popular value of 0.25 is not one of the best fits. While H_0 is a very believable 69.6 the Ω_Λ is well low of the remaining 0.75, thus necessitating a significant Ω_k of 0.19 for a closed universe. When we choose the 2-free parameters as H_0 and Ω_m while holding Ω_Λ at 0 we observe a χ^2/N and X-over values far from the best fit, as reported in the last row of Table 1. This is the usual result when the standard, FRW model without Ω_Λ is compared against the *Classic* model with Ω_Λ using MR type data and is not considered a "good fit". The extremely slow H_0 of 41.5 km/s/Mpc allows plenty of time for galaxy and star formation, though.

4.2. Distance *vs.* Frequency Decline Data

For this analyses we choose distances, in terms of Mpc, and associated geometric errors, versus ν_0/ν_e (the **DFR** data) which are all extracted by arithmetic manipulation from the on-line, astronomical data. We point out that for the first time enough data from distant GRB emissions can be used to visualize the full curve in the DFR-type plots, as the GRB data are now much closer to singularity than possible for SNe Ia data alone, which seem to loose value as standard candles at $z > 1.5$.

We first choose to fit the flat FRW *Classic* model with 2 free parameters, with results presented in Table 2. This model is not the best fit to the DFR data by far, and it seems most any other FRW-type model we choose fits these data better. We find the matter density of best fit at 0.12 ± 0.02 with a speedy H_0 of 75.4 ± 0.7 km/s/Mpc(Table 2) for the *Classic*, flat spacetime model. Both the Ω_m and H_0 values are outside the current *Popular* range with matter density about half the 0.25 often reported and the H_0 on the top side of the range now thought probable. Because this model of flat spacetime with Ω_Λ is tested with

Table 2. Results from distance *vs* frequency ratio (DFR) with 377 data pairs

Routine	step/fix	χ^2/N	X-ovr	H_0	Ω_m	Ω_k	Ω_Λ
Classic	Ω_k	1.68	174	75.4 ± 0.7	0.12 ± 0.02	0	0.88
Ω_k-Step	Ω_k	1.51	176	73.2 ± 0.6	0.22 ± 0.02	0.10	0.68
Ω_k-Step	Ω_k	1.52	182	75.4 ± 0.7	0.33 ± 0.02	-0.36	1.03
Ω_m-Step	Ω_m	1.55	172	71.5 ± 0.7	0.20	0.31 ± 0.05	0.49
No DE	Ω_Λ	1.59	163	67.6 ± 0.4	0.20 ± 0.00	0.80 ± 0.00	0
Popular	Ω_m	1.66	152	73.1 ± 0.7	0.25	0.23 ± 0.06	0.52

H_0 in km/s/Mpc

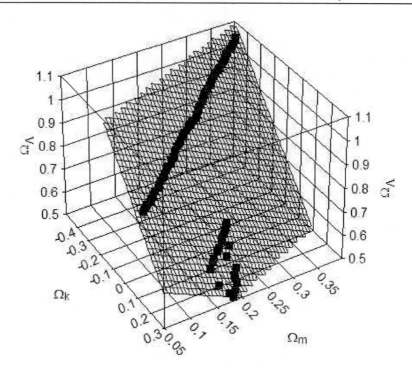

Figure 3. Ω_Λ and Ω_m as functions of Ω_k from the 377 DFR data pairs.

two different, though related data sets, we cannot unambiguously claim the better fit of one analysis over another given that χ^2/N are calculated from different basis. Still, because the latter model is a "flat" version of Eq. 4 and properly weighs the luminary distance errors we prefer these values for Ω_m and the larger for H_0 if the Universe is indeed flat with DE.

The best fitting routines of all are presented when Ω_k is varied in steps of 0.01 from positive to negative curvature territory, the Ω_k-Step routine. Here we have two solutions of nearly equal validity, degeneracy if you wish. In positive curvature territory with $\Omega_k = 0.10$, we are presented with a "typical" matter density of 0.22 and a cosmic constant of 0.68, slightly lower than currently popular. In negative spacetime territory with $\Omega_k = -0.36$ we find a solution, of similar goodness of fit, with more matter density, 0.33 and a Hubble constant a quick 75.4 km/s/Mpc. The real winner in this universe is the cosmic constant which is best fit at over 1. In Figure 3 we present solutions of the two free parameters of Ω_Λ and Ω_m vs. an independent Ω_k. A smooth surface has been added as a visual aid. Though this presents a smooth surface, there is obvious evidence of inconsistencies between values of Ω_k and the two free parameters. Model behavior with low quality data usually leads to a best fit in a broad, shallow area of fit minimization. In this instance, while the best fit occurs at Ω_k of -0.36 and 0.10, the goodness of fit is not much better than that of Ω_k at -0.35 or -0.37(or 0.09 and 0.11). Likewise, it is impossible with these data and analytical techniques to define Ω_Λ with real accuracy.

Another good fit is presented when the matter density is varied stepwise (Ω_m-Step) and where Ω_m becomes 0.20 we find a rather small Ω_Λ of 0.49 with the remainder as Ω_k of 0.31. We present a 3-D figure of Ω_Λ and Ω_k as functions of Ω_m as Figure 4, which

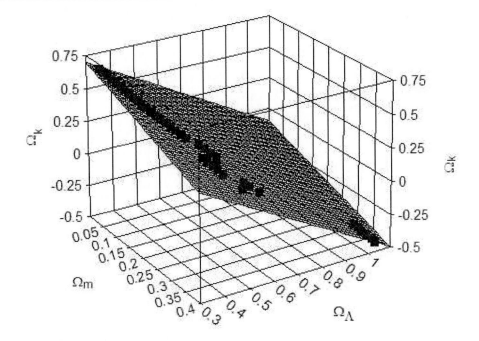

Figure 4. Ω_Λ and Ω_k as functions of Ω_m from the 377 DFR data pairs.

shows the inconsistent results of fitting the FRW with DE to DFR data as sharp deviations from a smooth line over the connecting surface. We have previously warned to expect such results[20] and have reported such findings in a preliminary study[32]. What is worrisome about these results is the appearance of inconsistent solutions throughout the popular range of matter density, > 0.18.

The third good fit is the regime of a universe without any DE, the former standard FRW model, presenting a lower ξ /N and X-over number than the *Classic* model. We again find Ω_m at a reasonable 0.20 with the remainder of existence as ST curvature in a closed universe. Here H_0 is low of the currently popular 69 to 72 km/s/Mpc range, but not unreasonably so. Fitting the FRW approximation with the DE model while staking Ω_m to the popular 0.25 (*Popular*) also did not present a flat universe; with Ω_Λ of 0.52, a significant value for Ω_k and at the cost of poor X-over and χ^2/N values; this fit is only slightly more acceptable than the *Classic* model. At the moment the two best fits to the DFR data are at Ω_m of 0.22 and 0.33 with Ω_k of 0.10 and -0.36, respectively, with a normalized Ω_Λ of 0.68 or greater. The results of using intergalactic distance *vs.* the frequency ratio (DFR) rather than MR data for modeling, does not really recommend any of the common models over another. Rather, the regimes used here with SNe Ia and GRB data suggest we should question the idea of a flat Universe and request more and better data to resolve the several, equally valid, solutions.

4.3. The Hubble Constant from SNe Ia and GRB Data

We present several results suggesting the Hubble constant is narrowly defined by the combination of these two large data sets, somewhat independently of the model, fitting routine

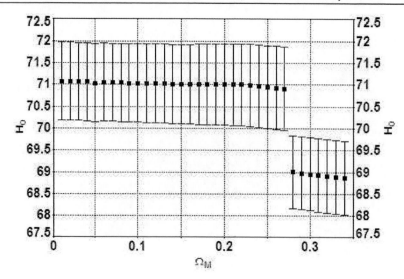

Figure 5. The Hubble constant as a function of matter density with best fit errors.

and version of the data chosen. One can see in Figure 5 that using the DFR data with the former standard FRW model cast as per Eq. (5) with the single free parameter Ω_k, allows significant Ω_m, Ω_k and a Hubble constant between 71 to 69 km/s/Mpc. This almost invariant result occurs over a wide range of matter density (0.01 to 0.35) which satisfies most, if not all, currently popular models for the cosmos. It is also obvious that this model presents an inconsistency between Ω_m of 0.27 and 0.28. This may indicate one of several things, (1) the model is too simple for the data (something like the Ω_Λ term is needed) (2) this large, combination data set needs internal normalization (3) we need more GRB and SNe Ia data of better quality in the search for a more accurate Hubble constant (4) accurate determination of H_0 can only be made by restricting the data to more recent emissions since this is a local constant[21]. If one only desires a Hubble constant to within say $\pm$ 5 km/s/Mpc then we are there.

We find a similar situation when the MR data are used for modeling the 2-parameter fits. Excepting the model without Ω_Λ, H_0 is within the 69 to 71 km/s/Mpc range, exhibiting a similar value spread as solutions using the DFR data set. Again, the values for Ω_Λ and Ω_k vary widely across the range of matter density. It appears the both data sets, while providing an excellent determination of H_0, compared to 20 years ago, are inadequate or too wide in scope to nail down Ω_Λ and Ω_k when these general versions of the FRW-standard model are used.

5.   Conclusion

Data from deep field astronomical surveys are beginning to approach the numbers and accuracy necessary to define H_0 within a narrow range. What is now needed is better calibration of the SNe Ia absolute distances. This is good news for projects which began more than a decade ago as surveys to assay the skies and narrowly refine this important fundamental value. Some of these projects have since widened to include collecting evidence of CDM and DE and so should hopefully be around for some time. The most important of these observations to date are surveys of SNe Ia emissions, which shall continue for some time to come. We see more efforts such as updating instruments on the Hubble telescope, better quality data collection, along with improved raw data analysis and the recent publication of pooled and normalized results of SNe Ia data from several sources. These efforts should help improve the poor signal/noise which plagues our estimates of Ω_m and Ω_Λ.

Here we take for granted that Ω_m is the sum of ordinary matter and radiation at a value between 0.01 and 0.05 with CDM being 3 to 5 times more numerous than this. With this assumption our results indicate the simplest models based on the FRW approximation of the Einstein world still cannot discriminate between a concave or convex universe rich or poor with matter, CDM or DE. Our best fit of this largest data collection yet, indicates the Universe is brimming over with DE and with galaxies surrounded by a wealth of CDM being pushed through spacetime of convex, or maybe concave, curvature? The model which cannot be discarded for reasons of statistics, but which has become unpopular, is a Universe primarily consisting of convex, curved spacetime with no DE and only moderately populated with matter. Still, results from fitting this standard FRW model without DE are pretty good - indicating plenty of convex, curved spacetime - and cannot be easily dismissed. This view is not too different from the Universe envisioned by scientists a few decades ago. Unfortunately, these analyses and results add little to solution of the *cosmological constant problem*, whereby the observed value for this energy differs from expectation by many, many orders of magnitude[3, 25]. Despite great effort and enormous investment in human efforts and money, we have come but only a little way in knowledge from the opinion of Einstein, in his popular paperback *Relativity*, that a universe with uniformly distributed matter shall be spherical or elliptical in shape with nary a hint about dark energy[7].

Since a negative Ω_k pushes the value for DE towards unity(or *vice versa*) the best fit here presents our existence as a universe primarily of dark energy. If this evaluation is correct, that DE is the most abundant entity of all, somebody, somewhere should soon be able to characterize it experimentally. One single experiment delivering a value to the order of magnitude would be a monumental advance, to be appreciated by all. The admittance of significant negative ST associated with a large DE does not necessarily pose a problem since DE may warp ST in a similar manner to matter, except with the opposite geometry; one might think this an obvious property of anti-gravity. That is, a universe with an abundance of DE and a smattering of matter should exhibit a convex, open geometry, and that ST in the neighborhood of DE will suffer negative curvature. This would be an clean change, or shall we write addition, to the Einstein model.

Some astronomers take SNe Ia data as evidence for a "jerk" in the state of the Universe several billion years ago[27]. We see in Figures 3 and 4 surfaces exhibiting inconsistent solutions to fitting regimes, a little evidence for this possibility. However, we judge the "jerks" in this figure as phenomena that can partly be traced to the mathematical deficiencies of the *Classic* model and not necessarily to the data[20]. It is also likely these discontinuities arise because our fitting regimes have assayed data of too broad domain, after all, Ω_m and H_0 at z $\approx$ 5 should have been larger than current. We welcome the day when enough quality emission data has been collected to truly confirm or discredit the notion of the cosmological "jerk" of many epochs past.

It appears to us now that the cosmic constant term introduced by Einstein nearly a century ago was simply a "patch" to correct the problem of a static Universe reported to him in error[7]. His new DE term being introduced to balance the energy between the collapse of the Milky Way and observations at that time, before the man Hubble, of a static "world" - a Milky Way which was not on the path to collapse. One interpretation of Einstein's remark that Λ was his greatest mistake might be that he missed his opportunity to predict our singularity, not really as a disclaimer of Λ. This revised formulation by the great man is now detected with inconsistent solutions to astronomical data and is evidenced by the sharp discontinuities displayed by our results. We suggest that better mathematics, which must conform to the data without inconsistencies when DE is allowed, may be found by investigations discarding the cosmic constant term while looking deeper into the dreaded tensor for solutions to the expanding Universe. Another possible solution is to edit the FRW approximation, with consideration of the local horizon, to better fit the data[19]. Though we have not thoroughly investigated these data here, using a floating state parameter, we think regimes incorporating this parameter will also be found as only a short-term solution to the problem of inconsistency.

In order to narrow the ranges of estimates for $\Omega_m, \Omega_k, \Omega_\Lambda$ and H_0 we suggest an order magnitude more SNe Ia and GRB data are needed[21]. The GRB data especially, suffers huge errors of distance determination and hopefully techniques will evolve to lessen these very uncertain values. In addition, the SNe Ia and GRB data should be merged into a single set with normalization of the GRB data to the more well-calibrated SNe Ia set. There may also come a wonderful day when data from other type supernova explosions can be calibrated and merged with these two sets. When enough data have been collected it may be possible to break the data into epoch subsets. Each subset can then be used to solve for the local H_0 and Ω_m and be used to judge between the many current theories of cosmology. Using better defined data, that is much more data within a smaller redshift range with much, much smaller distance errors, should help avoid presentation of discontinuities which plague current models.

Bishmer, Jan and Michael remember the day when warned during a lecture in freshman physics about the inadvisability of using log data for modeling. We think it time astronomers use luminary distances with realistic errors rather than luminosity distance moduli with depressed error acknowledgment, as a tool for studying cosmology. We also believe astrophysics needs emission data by the hundreds if not thousands to answer the important, the really big questions of our Universe. In the meantime, theoretical physicists should present solutions of the Einstein world which do not exhibit discontinuities when faced with small adjustment to FRW parameters. We think a fruitful pathway might be to discard the Λ term and dig deeper into anti-gravity as applied within the Einstein tensor.

References

[1] Astier, P., *et al. Astron. Astrophys.* 2006, 447, 31-48.

[2] Carroll, S.M., Press, W.H. and Turner, E.L. *Annu. Rev. Astron. Astrophys.* 1992, 30, 499-542.

[3] Carroll, S.M. *The Cosmological Constant, Living Reviews in Relativity* 2001, 4, 1.[Online Article]: http://www.livingreviews.org/Articles/Volume4/2001-1carroll/

[4] Chandrasekhar, S., Nobel Lecture
http://nobelprize.org/nobelprizes/physics/laureates/1983/chandrasekhar-lecture.html
1983.

[5] Clocchiatti, A., *et al. Astrophys. J.* 2006, 642, 1-21.

[6] Einstein, A., *The Principle of Relativity;* Dover Publications: Mineola, NY, 1952.

[7] Einstein, A., *Relativity;* Routledge Classics: NY, NY, 2001.

[8] Freedman, W.L., Madore, B.F., Gibson, B.K., Ferrarese, L., Kelson, D.D., Sakai, S., Mould, J.R., Kennicutt, Jr., R.C., Ford, H.C., Graham, J.A., Huchra, J.P., Hughes, S.M.G., Illingworth, G.D., Macri, L.M., Stetson, P.B. *Astrophys.J.* 2001, 553, 47-72.

[9] Foley, R.J., *et al. Astronom. J.* 2009, 137, 3731-3749.

[10] Hamuy, M., Phillips, M.M., Maza, J., Suntzeff, N.B., Schommer, R.A. and Aviles, R. *Astronom. J.* 1995, 109, 1-13.

[11] Hillebrandt, W. and Niemeyer, J.C. *Ann. Rev. Astron. Astrophys.* 2000, 38, 191-230.

[12] Howell, D.A., Sullivan, M., Conley, A. and Carlberg, R. *Astrophys. J. Lett.* 2007, 667, L37-L40.

[13] Hubble, E. *Proc. Nat. Acad. Sci. U.S.A.* 1929, 15, 168-173.

[14] Kasen, D., Rpke, F.K. and Woosley, S.E. *Nature* 2009, 460, (7257): 869.

[15] Kirshner, R.P. *The Extravagant Universe: exploding stars, dark energy and the accelerating cosmos;* Princeton University Press: Princeton, NJ, 2002.

[16] Kirshner, R.P. *Proc. Nat. Acad. Sci. U.S.A.* 2004, 101, 8-13.

[17] M. Kowalski, *et al. Astrophys. J.* 2008, 686, 749-778.

[18] Leibundgut, B. *Gen. Relativ. Gravit.* 2008, 40, 221-248.

[19] Melai, F. and Abdelqader, M. *Inter. J. Mod. Phys. D* 2009, arXiv:0907.5394v1.

[20] Oztas, A.M. and Smith, M.L. *Inter. J. Theoret. Phys.* 2006, 45, 925-936.

[21] Oztas, A.M., Smith, M.L. and J. Paul, J. *Inter. J. Theoret. Phys.* 2008, 47, 1725-1744.

[22] Pebbles, P.J.E. *Principles of Physical Cosmology;* Princeton Series in Physics; Princeton University Press: Princeton, NJ, 1993.

[23] Perets, H.B., *et al.* arXiv:0906.2003v1 [astro-ph.HE] (2009).

[24] Perlmutter, S., *et al. Astrophys. J.* 1999, 517, 565-586.

[25] Perlmutter, S., *Physics Today* 2003, 56 53-60.

[26] Riess, A.G., *et al. Astron. J.* 1998, 116, 1009-1038.

[27] Riess, A.G., *et al. Astrophys. J.* 2007, 659, 98-121.

[28] Riess, A.G., Lucas Macri, L., Casertano, S., Sosey, M., Lampeitl, H., Ferguson, H.C., Filippenko, A.V., Jha, S.W., Li, W., Chornock, R., Sarkar, D., arXiv:0905.0695v1 [astro-ph.CO] (2009).

[29] Schaefer, B.E. *Astrophys. J.* 2007, 660, 16-46.

[30] Schucker, T. *Gen. Relativ. Gravit.* 2009, 41, 67-75.

[31] Smith, M.L, Oztas, A.M. and Paul, J. *Inter. J. Theoret. Phys.* 2006, 45, 937-952.

[32] Smith, M.L. and Oztas, A.M. *Advan. Stud. Theoret. Phys.* 2008, 2, 1-10.

[33] Timmes, F.X., Brown, E.F. and Truran, J.W. *Astrophys. J.* 2003, 590 L83-L86.

[34] van Leeuwen, F., Feast, M.W., Whitelock, P.A, Laney, C.D., arXiv:0705.1592v1 [astro-ph] (2007).

[35] Webb, S., *Measuring the Universe, the cosmological distance ladder;* Springer-Praxis: UK, 1999.

[36] Wood-Vasey, W.M., *et al. Astrophys. J.* 2007, 666, 694-715.

In: Dark Energy: Theories, Developments and Implications   ISBN 978-1-61668-271-2
Editors: K. Lefebvre and R. Garcia, pp. 127-142   © 2010 Nova Science Publishers, Inc.

Chapter 7

QUANTUM MECHANICAL APPROACH TO OUR EXPANDING UNIVERSE WITH DARK ENERGY: SOLUBLE SECTOR OF QUANTUM GRAVITY

Subodha Mishra
Department of Physics,
Institute of Technical Education and Research,
Jagamohan Nagar, Bhubaneswar-751030, India

Abstract

We study quantum mechanically our expanding universe which is made up of gravitationally interacting particles such as particles of luminous matter, dark matter and dark energy as a self-gravitating system using a well-known many-particle Hamiltonian, but only recently shown as representing a soluble sector of quantum gravity. Describing dark energy by a repulsive harmonic potential among the points in the flat 3-space and incorporating Mach's principle to relativize the problem, we derive a quantum mechanical relation connecting, temperature of the cosmic microwave background radiation, age, and cosmological constant of the universe. When the cosmological constant is zero, we get back Gamow's relation with a much better coefficient. Otherwise, our theory predicts a value of the cosmological constant 2.0×10^{-56} cm^{-2} when the present values of cosmic microwave background temperature of 2.728 K and age of the universe 14 billion years are taken as input. It is interesting to note that in this flat universe, our method dynamically determines the value of the cosmological constant reasonably well compared to General Theory of Relativity where the cosmological is a free parameter.

1. Introduction

Our expanding universe is made up of gravitationally interacting particles which are described by particles of luminous matter, dark matter and dark energy. Representing dark energy by a repulsive harmonic potential among the points in the flat 3-space, we derive a quantum mechanical relation connecting, temperature of the cosmic microwave background radiation, age, and cosmological constant of the universe. When the cosmological constant is zero, we get back Gamow's relation with a much better coefficient. Otherwise, our theory predicts a value of the cosmological constant 2.0×10^{-56} cm^{-2} when the present values of cosmic microwave background temperature of 2.728 K and age of the universe 14 billion years are taken as input. We study[1, 2, 3] the self-gravitating system such as the universe using a well-known many-particle Hamiltonian, which is known from the early days of quantum mechanics, from a condensed matter point of view by using a quantum mechanical variational approach. This can also be viewed as a novel way of looking at the self-gravitating systems and it not only reproduces the results known from Einstein's General Theory of Relativity but also goes beyond by predicting certain relations and specifically the value of the cosmological constant. Instead of looking at the systems through the space-time dynamics, this theory treats the energy of the system directly. The above Hamiltonian has only recently been derived[4] as representing the exactly soluble sector of quantum gravity. Infact from the quantum gravity point of view after quantizing the GTR in the $c \to \infty$ (Newton-Cartan theory with spacial vanishing curvature), we have explicitly the above known Hamiltonian. We then special relativize the problem by using Mach's principle in case of the universe . Figure 1 represents the underling physical theories relating to full-blown quantum theory of gravity (denoted by FQG) which is still an elusive one. But ours is an effective theory of quantum gravity with vanishing spacial curvature ETQGK0 (denoted by a circle with squares inside) of this FQG with special-relativization done on the problem which is described by the quantum many-particle Hamiltonian obtained by quantizing the Newton-Cartan theory which we write as NQG (Fig. 1). Any general-relativistic exotic theory "theory of everything" (string theory, loop quantum gravity etc) would not be physically relevant if it does not reduce to Newton quantum gravity interacting with quantum fields in the Galilean-relativistic limit. Any valid theory of quantum gravity must reduce[4] to NQG (Fig. 1) in the "$c \to \infty$" limit, to GTR in the "$\hbar \to 0$" limit and to QTF in the "$G \to 0$" limit. Since in our theory, we relativise the NQG, the elusive full-blown quantum theory of gravity must be consistent with the results obtained by our approach. The directions of research have been to go from GTR to FQG or from QTF to FQG. But the third possibility of going from NQG/QNG to FQG by special relativising the Newton-Cartan quantum gravity (undoing the $c \to \infty$ limit) opens up an exciting yet unexplored direction in the research of quantum gravity.

An expanding system of self-gravitating particles can be described by the Hamiltonian which is given as

$$H = -\sum_{i=1}^{N} \frac{\hbar^2}{2m}\nabla_i^2 + \frac{1}{2}\sum_{i=1,\,i\neq j,j=1}^{N}\sum^{N} v(|\,\vec{X}_i - \vec{X}_j\,|) - \sum_{i=1}^{N} \Lambda c^2 |\vec{X}_i|^2 \qquad (1)$$

where $v(|\,\vec{X}_i - \vec{X}_j\,|) = -g^2/|\,\vec{X}_i - \vec{X}_j\,|$, with $g^2 = Gm^2$, G being the universal gravitational constant and m the mass of the constituent particles. Incase of our universe, the last term in the above Hamiltonian describes the expansion due to the Dark Energy, which can be taken as the nonrelativistic limit of the Λ term known as Einstein's cosmological constant term in General Theory of Relativity.

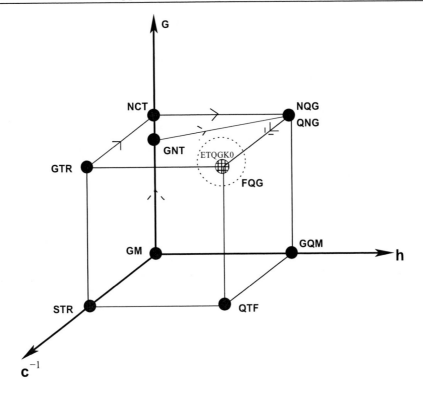

Figure 1. The great dimensional figure of physics indicating the fundamental roles played by G, h and c in various basic physical theories. The theories are GM=Galilean mechanics, STR=special theory of relativity, GTR=general theory of relativity, GQM=Galilean quantum mechanics, QTF=quantum theory of relativistic fields, NCT=Newtonian-Cartan theory, GNT=Galilean Newtonian theory, NQG=Newtonian quantum gravity, QNG=quantum Newtonian gravity. It turns out that when the NCT with (flat 3-space) is quantised the resulting quantum many-particle Hamiltonian is same as one would write the quantum many-particle Hamiltonian for a system of self-gravitating particles that is what we denote as QNG. Here FQG=The elusive full-blown quantum gravity represented by the bigger circle with broken lines. If FQG turns out to require some additional fundamental constants-like the constant $\alpha' \equiv (2\pi T)^{-1}$ where T is the string tension of the string theory then instead of this cube of theories we will have a hypercube. But, in that case NQG would be a limiting theory of FQG with respect to total $\alpha' \to 0$ and $c \to \infty$. The position denoted by a circle with squares inside denotes our effective theory of quantum gravity (ETQGK0) in the limiting case when special relativisation is incorporated but the spacial curvature is zero, ETQGK0$\subset$ FQG. One can note two tracks to this limiting theory one shown by solid-arrow-heads and other by broken-arrow-heads. The cube has been adopted and modified from Ref.[4].

2. Present Universe: Exactly Soluble Sector of Quantum Gravity

In this section we describe how the Hamiltonian given by Eq. 1 can be derived as a soluble sector of quantum gravity. In a seminal work[4], recently a generally covariant but

Galilean-relativistic quantum theory of gravity has been constructed. It is shown that the Galilean-relativistic limit of as-yet unrealized full quantum theory of gravity with matter is exactly soluble and in the very classical ($c = \infty$) domain the problem of time and causality and other related problems do not exist. A space-time reformulation of the Newtonian theory of gravity known as Newton-Cartan theory of gravity is obtained when the nonrelaivistic limit of Einstein's general theory of relativity is considered. The Newton-Cartan[5] theory lies between that of special theory of relativity with specetime completely fixed and general relativity with no background structure. The NC theory has two fixed and degenerate metrics, one temporal and another spatial metric but with a dynamical connection field like GTR metric connection field. Newton-Cartan theory of gravity is in general the true Galilean-relativistic form of GTR and it is a local field theory. It has a mutable space-time unlike Newtonian theory of gravity of immutable spacetime and in Newtonian-Cartan theory the instantaneous gravitational interactions between bodies propagate through the spacetime.

The connection field of Newtonian-Cartan theory depends on the distribution of matter. The quantum mechanically treated matter impels that the NC connection should be treated also quantum mechanically. So here a quantum theory of gravity is constructed at the Galilean-relativistic level such that superposition principle holds for the states of matter and NC connection field. In the seminal paper[4] the following action functional in the inertial frame is constructed

$$\mathfrak{I} = \int dt \int d\vec{x} \left[\frac{\Phi \Delta \Phi}{8\pi G} + \frac{\hbar^2}{2m} \delta^{ab} \partial_a \Psi \partial_b \Psi + \frac{i\hbar}{2} (\Psi \partial_t \bar{\Psi} - \bar{\Psi} \partial_t \Psi) - m\Psi \bar{\Psi} \Phi \right] \quad (2)$$

where $k = 0$ that is a flat space is considered and $\Lambda = 0$ is taken for simplicity.

When we extremize this functional by varying the scalar potential Φ, we get the Newton-Poisson equation

$$\Delta \Phi = \frac{4\pi G}{< \psi | \psi >} m \psi \bar{\psi}, \quad (3)$$

where $\psi := \sqrt{2\pi G} \Psi$. Also the extremization of the action with respect to the matter field Ψ gives the Schrodinger equation

$$i\hbar \frac{\partial}{\partial t} \psi = \left[-\frac{\hbar^2 \Delta}{2m} + m\Phi \right] \psi \quad (4)$$

These two equations describe a single Galilean-relativistic particle gravitationally interacting with its own Newtonian field. Solving the first equation gives

$$\Phi(\vec{x}) = -Gm \int d\vec{x}' \frac{\bar{\psi}(\vec{x}')\psi(\vec{x}')}{|\vec{x}' - \vec{x}|} \quad (5)$$

This when substituted in the second equation we obtain

$$i\hbar \frac{\partial}{\partial t} \psi(\vec{x}, t) = -\frac{\hbar^2 \Delta}{2m} \psi(\vec{x}, t) - Gm^2 \int d\vec{x}' \frac{\bar{\psi}(\vec{x}', t)\psi(\vec{x}', t)}{|\vec{x}' - \vec{x}|} \psi(\vec{x}, t) \quad (6)$$

But when ψ is written in second quantized form $\hat{\psi}$ as a annihilation operator satisfying $[\hat{\psi}(\vec{x}), \hat{\psi}^\dagger(\vec{x}')] = \delta(\vec{x} - \vec{x}')$, it describes a many-particle system in the Heisenberg picture.

The system of particles is now described by a Hamiltonian

$$\hat{\mathcal{H}} = -\int d\vec{x}\,\hat{\psi}^{\dagger}(x)\frac{\hbar^2\Delta}{2m}\hat{\psi}(\vec{x}) - \frac{1}{2}Gm^2\int d\vec{x}d\vec{x}'\frac{\hat{\psi}^{\dagger}(\vec{x}')\hat{\psi}^{\dagger}(\vec{x})\hat{\psi}(\vec{x})\hat{\psi}(\vec{x}')}{|\vec{x}' - \vec{x}|}\psi(\vec{x}, t) \quad (7)$$

The Hamiltonian satisfies the Heisenberg equation of motion

$$i\hbar\frac{\partial}{\partial t}\hat{\psi}(\vec{x}, t) = [\hat{\psi}(\vec{x}, t), \hat{\mathcal{H}}] \quad (8)$$

It can be shown that when the Hamiltonian operator $\hat{\mathcal{H}}$ acts on a many-particle state gives

$$< \vec{x}_1...\vec{x}_N|\hat{\mathcal{H}}|\xi> = \left[-\frac{\hbar^2}{2m}\sum_{i=1}^{N}\nabla_i^2 - \frac{1}{2}Gm^2\sum_{i,j,i\neq j}\frac{1}{|\vec{x}_i - \vec{x}_j|}\right] \times < \vec{x}_1...\vec{x}_N|\xi> \quad (9)$$

This is the quantum mechanical version of the classical manyparticle Hamiltonian with gravitational pair interactions. There is no self interaction since $\hat{\mathcal{H}}|\vec{x}> = 0$ and the number operator $\hat{N} = \int d\vec{x}\hat{\psi}^{\dagger}\hat{\psi}$ commutes with total Hamiltonian and it is conserved.

The Hamiltonian represents a system of N gravitating particles each of mass m interacting through a sum of pair-wise gravitational interaction. We special relativise the problem by using Mach's principle in case of the universe.

We use a variational approach[1, 2, 3] in our theory to study these systems. But the form of the trial densty chosen is so good that the calculated energy obeys the bound earlier known and results are very good. Our theory is provides a frame work to find a relation between time, temperature of the cosmic microwave background radiation and the cosmological constant of the universe. It determines the value of the Cosmological constant dynamically, where as in GTR the Cosmological constant is a free parameter.

3.   Cosmological Constant Λ as the Dark Energy

The most important theory for the origin of the universe is the Big Bang Theory[6] according to which the present universe is considered to have started with a huge explosion from a superhot and a superdense stage. Theoretically one may visualize its starting from a mathematical singularity with infinite density. This also comes from the solutions of the type I and type II form of Einstein's field equations[7]. What follows from all these solutions is that the universe has originated from a point where the scale factor R (to be identified as the radius of the universe) is zero at time $t = 0$, and its derivative with time is taken to be infinite at this time. That is, it is thought that the initial explosion had happened with infinite velocity, although, it is impossible for us to picture the initial moment of the creation of the universe. The accelerated expansion of the universe has been conformed by studying the distances to supernovae of type Ia[8, 9]. For the universe, it is being said that the major constituent of the total mass of the present universe is made of the Dark Energy 70%, Dark Matter about 26% and luminous matter 4%. The Dark energy is responsible for the accelerated expansion of the universe since it has negative pressure and produces repulsive gravity. The cosmological constant[10, 11] of Einstein provides a repulsive force when its value is positive. The cosmological constant is also associated with the vacuum energy density[12] of the space-time.

The vacuum has the lowest energy of any state, but there is no reason in principle for that ground state energy to be zero. There are many different contributions[12] to the ground state energy such as potential energy of scalar fields, vacuum fluctuations as well as of the cosmological constant. The individual contributions can be very large but current observation suggests that the various contributions, large in magnitude but different in sign delicately cancel to yield an extraordinarily small final result. The conventionally defined cosmological constant Λ is proportional to the vacuum energy density ρ_Λ as $\Lambda = (8\pi G/c^2)\rho_\Lambda$. Hence one can guess that $\rho_\Lambda = \Lambda c^2/8\pi G \approx \rho_{Pl} = c^5/G^2\hbar \sim 5 \times 10^{93}$ g cm^{-3}, where ρ_{Pl} is the Plank density. But the recent observations of the luminosities of high redshift supernovae gives the dimensionless density $\Omega_\Lambda = \rho_\Lambda/\rho_{cr} \equiv \Lambda c^2/3H_0^2 \approx 0.7$ where $\rho_{cr} = 3H_0^2/8\pi G \approx 1.9 \times 10^{-29}$ g cm^{-3}, which implies $\rho_\Lambda = \rho_{Pl} \times 10^{-123}$. This shows that the cosmological constant today is 123 orders of magnitude smaller.

This is known as the 'cosmological constant problem'. In the classical big-bang cosmology there is no dynamical theory[13] to relate the cosmological constant to any other physical variable of the universe. There have been some studies[14, 15, 16] regarding the universe to relate the space-time manifold to somekind of condensed matter systems. Here by considering[1, 2, 3] the visible universe made up of self-gravitating particles representing luminous baryons and dark matter such as neutrinos (though only a small fraction) which are fermions and a repulsive potential describing the effect of Dark Energy responsible for the accelerated expansion of the universe, we in this chapter derive quantum mechanically a relation connecting temperature, age and cosmological constant of the universe. When the cosmological constant is zero, we get back Gamow's relation with a much better coefficient. Otherwise using as input the current values of $T = 2.728$ K and $t = 14 \times 10^9$ $years$, we predict the value of cosmological constant as 2.0×10^{-56} cm^{-2}. Note that Λ is a completely free parameter in General Theory of Relativity. Also it is interesting to note that we obtain not only the value of the cosmological constant but also the sign of the parameter correct though it is a very small number.

4. Mathematical Formulation without Λ

We in this section derive a relation connecting temperature and age of the universe when cosmological constant is zero, by considering a Hamiltonian[1, 2, 3] used by us some time back for the study of a system of self-gravitating particles which is given as:

$$H = -\sum_{i=1}^{N} \left(\frac{\hbar^2}{2m}\right)\nabla_i^2 + \frac{1}{2}\sum_{i=1, i\neq j, j=1}^{N}\sum^{N} v(|\vec{X}_i - \vec{X}_j|) \tag{10}$$

where $v(|\vec{X}_i - \vec{X}_j|) = -g^2/|\vec{X}_i - \vec{X}_j|$, with $g^2 = Gm^2$, G being the universal gravitational constant and m the mass of the effective constituent particles describing the luminous matter and dark matter whose number is $N = \int \rho(\vec{X})d\vec{X}$. Since the measured value for the temperature of the cosmic microwave background radiation is $\approx 2.728K$, it lies in the neighbourhood of almost zero temperature. We,therefore, use the zero temperature formalism for the study of the present problem. Under the situation when N is extremely large, the total kinetic energy of the system is obtained as

$$< KE >= \left(\frac{3\hbar^2}{10m} \right) (3\pi^2)^{2/3} \int d\vec{X} [\rho(\vec{X})]^{5/3} \tag{11}$$

where $\rho(\vec{X})$ denotes the single particle density to account for the distribution of particles (fermions) within the system, which is considered to be a finite one. Eq.(11) has been written in the Thomas-Fermi approximation. The total potential energy of the system, in the Hartree-approximation, is now given as

$$< PE >= -\left(\frac{g^2}{2} \right) \int d\vec{X} d\vec{X}' \frac{1}{|\vec{X} - \vec{X}'|} \rho(\vec{X}) \rho(\vec{X}') \tag{12}$$

Inorder to evaluate the integral in Eq.(11) and Eq.(12), we had chosen a trial single-particle density[3, 2] $\rho(\vec{X})$ which is of the form :

$$\rho(\vec{X}) = \frac{Ae^{-x}}{x^3} \tag{13}$$

where $x = (r/\lambda)^{1/2}$, $r = |\vec{X}|$, λ being the variational parameter and A is the normalization constant given as $A = \frac{N}{16\pi\lambda^3}$. Though $\rho(\vec{X})$ is a trial density, still the physics of the behaviour of the density should be incorporated into it while using one. As one can see from Eq.(13), $\rho(\vec{X})$ is singular at the origin. This looks to be consistent with the concept behind the Big Bang theory of the universe. The early universe was not only known to be super hot, but also it was superdense. To account for the scenario at the time of the Big Bang, we have, therefore, imagined of a single-particle density $\rho(x)$ for the system which is singular at the origin ($r = 0$). This is only true at the microscopic level, which is not so meaningful looking at things macroscopically. Although $\rho(x)$ is singular, the number of particles N, in the system is finite. Since the present universe has a finite size, its present density which is nothing but an average value is finite. At the time of Big Bang ($t = 0$), since the scale factor (identified as the radius of the universe) is supposed to be zero, the average density of the system can assume an infinitely large value, implying its superdense state. Having thought of a singular form of single-particle density at the time of the Big Bang, we have tried with a number of singular form of single particle densities of the kind $\rho(\vec{r}) = B \frac{exp[-(\frac{r}{\lambda})^\nu]}{(\frac{r}{\lambda})^{3\nu}}$ where $\nu = 1, 2, 3, 4...or \frac{1}{2}, \frac{1}{3}, \frac{1}{4}, ...$. Though the normalization constant B here is a function of ν, for $\nu = \frac{1}{2}$, B=A. Integer values of ν are not permissible because they make the normalization constant infinite. Out of the fractional values, $\nu = \frac{1}{2}$ is found to be most appropriate, because, it has been shown in our earlier paper that it gives the expected upper limit for the critical mass of a neutron star[3] beyond which black hole formation takes place and other parameters of the universe[2] satisfactorily correct. Also because, if ν goes to zero (like $1/n$, $n \to \infty$), $\rho(r)$ would tend to the case of a constant density as found in an infinite many-fermion system. In view of the arguments put forth above, one will have to think that the very choice of our $\rho(r)$ is a kind of ansatz in our theory, which is equivalent to the choice of a trial wave function used in the quantum mechanical calculation for the binding energy of a physical system following variational techniques. As mentioned earlier, singularity at $r = 0$ in the single particle distribution has nothing to do with the average particle density in the system, which happens to be finite (because of the fact that N is finite and volume V of the visible universe is finite), and hence it is not going to affect the large scale spatial homogeneity of the observed universe. Having accepted the value $\nu = \frac{1}{2}$, the parameter λ associated with $\rho(r)$ is determined after minimizing $E(\lambda) =< H >$ with respect to λ. This is how, we are able to find the total energy of the system corresponding to its lowest energy state.

After evaluating the integrals in Eq.(11) and Eq.(12), we find the total energy $E(\lambda)$ of the system which is given as

$$E(\lambda) = \frac{\hbar^2}{m} \frac{12}{25\pi} \left(\frac{3\pi N}{16} \right)^{5/3} \frac{1}{\lambda^2} - \frac{g^2 N^2}{16} \frac{1}{\lambda} \tag{14}$$

We minimize the total energy with respect to λ. Differentiating this with respect to λ and then equating it with zero, we obtain the value of λ at which the minimum occurs. This is found as:

$$\lambda_0 = \frac{72}{25} \frac{\hbar^2}{mg^2} \left(\frac{3\pi}{16} \right)^{2/3} \frac{1}{N^{1/3}} \tag{15}$$

Evaluating Eq.(14) at $\lambda = \lambda_0$, the total binding energy of the system is found as

$$E_0 \simeq -(0.015442) N^{7/3} \left(\frac{mg^4}{\hbar^2} \right) \tag{16}$$

Considering the case of the two-particle system (N=2), from Eq.(16), we find

$$E_0 = -(0.077823) \left(\frac{mg^4}{\hbar^2} \right) \tag{17}$$

This is seen to be quite high compared to the actual binding energy of the two-body system whose value is (-0.25) $\left(\frac{mg^4}{\hbar^2} \right)$. Comparing the two results, one should not consider Eq.(16) to be a drawback of the present theory,because it is supposed to be very accurate for very large N. Looking at Eq.(16), we find that E_0 varies as $N^{7/3}$ where N is the particle number. Such a dependence of the binding energy for the system on N was also found by Levy-Leblond[17] by assuming the particles to be fermions and looking at the distribution of N-points on a cubic lattice he was able to obtain both an upper and an lower bound for the binding energy of the system which for large N were given as

$$-(0.5) N^{7/3} \left(\frac{mg^4}{\hbar^2} \right) \le E_0 \le -(0.001055) N^{7/3} \left(\frac{mg^4}{\hbar^2} \right) \tag{18}$$

Anyway, comparing our result, as shown in Eq.(16), with Eq.(18), we find that it does not violate the inequalities established by Levy-Leblond[17].

Following the expression for $< KE >$ evaluated at $\lambda = \lambda_0$, we write down the value of the equivalent temperature T of the system, using the relation

$$T = \frac{2}{3k_B} \left[\frac{< KE >}{N} \right] = \frac{2}{3k_B} (0.015442) N^{4/3} \left(\frac{mg^4}{\hbar^2} \right) \tag{19}$$

The expression for the radius R_0 of the universe, as found by us earlier[3, 2], is given as

$$R_0 = 2\lambda_0 = 4.047528\left(\frac{\hbar^2}{mg^2}\right)/N^{1/3} \tag{20}$$

Our identification of the radius R_0 with $2\lambda_0$ is based on the use of socalled quantum mechanical tunneling[18] effect. Classically, it is well known that a particle has a turning point where the potential energy becomes equal to the total energy. Since the kinetic energy and therefore the velocity are equal to zero at such a point, the classical particle is expected to be turned around or reflected by the potential barrier. From the present theory it is seen that the turning point occurs at a distance $R = 2\lambda_0$.

We now invoke Mach's principle[11] which states that inertial properties of matter are determined by the distribution of matter in the rest of the universe. Mach had the view[7] that the velocity and acceleration of a particle would be meaningless had the particle been alone in the universe. We have to talk of acceleration only with respect to other bodies, just like we talk of velocities with respect to other bodies. This means that the inertial mass of a particle is the result of the particle feeling the presence of other particles in the universe. If we denote the inertial mass of the particle by m_{inert}, it is to be determined by its response to accelerated motion. As far as the universe is concerned, the distance particles beyond the Hubble length which we take as the radius of the visible universe R_0 are unobservable and therefore do not contribute to the determination of local inertial mass. If M denotes the gravitational mass of the observable universe, the gravitational energy of the particle is given by $E_{gr} = \frac{GMm_{grav}}{R_0}$, where m_{grav} is the gravitational mass of the particle, that is, the mass determined by its response to gravity. In accordance with the spirit of Mach's principle, one must have $E_{gr} = \frac{GMm_{grav}}{R_0} = m_{inert}c^2$, where $m_{inert}c^2$ is the intrinsic energy of the particle. Since m_{inert} and m_{grav} are taken to be equal both in Newtonian theory and in the General Theory of Relativity, we have Mach's principle[11] expressed through the relation as $\left(\frac{GM}{R_0c^2}\right) = 1$, and using the fact that the total mass of the universe $M = Nm$, we are able to obtain the total number of particles N constituting the universe, as

$$N = 2.8535954\left(\frac{\hbar c}{Gm^2}\right)^{3/2} \tag{21}$$

Now, substituting Eq.(21) in Eq.(20), we arrive at the expression for R_0, as

$$R_0 = 2.8535954\left(\frac{\hbar}{mc}\right)\left(\frac{\hbar c}{Gm^2}\right)^{1/2} \tag{22}$$

As one can see from above, R_0 is of a form which involves only the fundamental constants like $\hbar, c, G$ and the effective mass m which is ofcourse not fundamental. Now, eliminating N from Eq.(19), by virtue of Eq.(21), we have

$$T = \frac{2}{3}(0.0625019)\left(\frac{mc^2}{k_B}\right) \tag{23}$$

Since we are considering the visible universe, which is actually a patch with a horizon size determined by the speed of light and time that has passed since the Big Bang, we now assume that the radius R_0 of the visible universe is approximately given by the relation

$$R_0 \simeq ct \tag{24}$$

where t denotes the age of the universe at any instant of time.

Following Eq.(22) and Eq.(24), we write m as

$$m = \left(\frac{\hbar^3}{Gc^3}\right)^{1/4} (2.8535954)^{1/2}\frac{1}{\sqrt{t}} \tag{25}$$

It is interesting to see (as shown in Table-1) this variation of mass with time gives approximately the energy and hence the temperature scale of formation of elementary particles in different epochs of nucleosynthesis. We calculate temperatures in different epochs using our Eq.(27) to be derived shortly. This is in good agreement with the calculated values of temperature otherwise known from nucleosynthesis calculations[7, 13]. The period between $t = 7 \times 10^{-5}$ sec and 5 sec is called lepton era, while period before $t = 7 \times 10^{-5} sec$ is hadron era and the early era corresponding to the period $t < 10^{-43}$ sec is known as Planck era.

Table 1.

Age of the universe (t) in sec.	Temperature (T) in K as calculated from Eq. (27)	Temperature (T)in K for the formation of elementary particles[7, 13]
5	$\approx 1 \times 10^9$	$\approx 6 \times 10^9 (e^+, e^-)$
1.2×10^{-4}	$\approx 2.1 \times 10^{11}$	$\approx 1.2 \times 10^{12} (\mu^+, \mu^-$ and their antiparticles)
7×10^{-5}	$\approx 2.8 \times 10^{11}$	$\approx 1.6 \times 10^{12} (\pi^0, \pi^+, \pi^-$ and their antiparticles)
1.5×10^{-6}	$\approx 1.9 \times 10^{12}$	$\approx 10^{13}$ (protons, neutron and their antiparticles)
10^{-43}	$\approx 0.73 \times 10^{31}$	$\approx 10^{32}$ (planck mass)

A substitution of m, from Eq.(25), in Eq.(23), enables us to write

$$\begin{aligned} T &= 0.070388\left(\frac{1}{k_B}\right)\left(\frac{c^5\hbar^3}{G}\right)^{1/4} t^{-1/2} \\ &= 0.06339\left[\frac{c^2}{G\,a_B}\right]^{1/4} t^{-1/2} \end{aligned} \tag{26}$$

This is exactly the Gamow's relation[13, 1] apart from the fact that Gamow's relation had the coefficient 0.41563 instead of 0.06339 as in our expression. Substituting the numerical value of a_B, which is equal to 7.56×10^{-15} $erg\ cm^{-3}K^{-4}$, and the present value for the universal gravitational constant G $[G = 6.67 \times 10^{-8} dyn.cm^2.gm^{-2}]$, in Eq.(26),we obtain

$$T = (0.23172 \times 10^{10})t_{sec}^{-1/2} K \tag{27}$$

If we accept the age of the universe to be close to $14 \times 10^9 year$, which we have used here, with the help of Eq.(27), we arrive at a value for the Cosmic Microwave Background Temperature (CMBT) equal to $\approx 3.5K$. This is very close to the measured value of 2.728 K as reported from the most recent Cosmic Background Explorer (COBE) satellite measurements[19, 20]. However, if we use Gamow's relation, $t = 956$ billion years is required to obtain the exact value of 2.728 K for the cosmic background temperature from. Using our expression, Eq.(27), we would require an age of $22.832 \times 10^9 \ year$ for the universe to get the exact value of 2.728 K. Long back a correction was made to Gamow's relation by multiplying it with a factor of $(\frac{2}{g_d})^{1/4}$ by taking into account the degeneracies of the particles, where $g_d = 9$. This correction effectively multiplies Gamow'relation with a factor of 0.68 and brings back the age of the universe to 425 billion years for the present CMBT. If we multiply our expression by the same factor to correct for the degeneracy of particles, we obtain a value of 2.4 K, which is less than the value of present CMBT. In the next section we see that by including the cosmic repulsion by the part given by cosmological constant we get back 2.728 K, This is physically correct since the cosmological term[11] has the meaning of negative pressure, it adds energy to the system by its tension when the universe expands, though the over all temperature decreases as the universe expands.

5. Entropy, Number of Photons and the Ratio $(\bar{N}_\gamma/N_n)$

In this section we estimate total entropy due to the CMBR, total number of photons and the ratio of number of photons to number of baryons. By virtue of the expression given in Eq. (26), we can rewrite T as

$$T = 0.070388 \left[\left(\frac{c^3}{G} \right) \left(\frac{\pi^2}{60 \, \sigma} \right) \right]^{1/4} t^{-1/2} \tag{28}$$

where $\sigma = \frac{\pi^2 k_B^4}{60 \hbar^3 c^2}$ is the Stefan-Boltzman constant and its numerical value is equal to $5.669 \times 10^{-5} erg/cm^2.K^4.sec$, and we have

$$\sigma T^4 \simeq 2.4547 \times 10^{-5} \left(\frac{\pi^2 c^3}{60G} \right) \frac{1}{t^2} \tag{29}$$

The very form of the above equation suggests that the factor in its right hand side (rhs) can be identified as the energy density of the electromagnetic radiation at a time t. The radiation of this form is belived to follow the black- body law. The very agreement of our calculated result with the most accurate value for the temperature of the background radiation shows that age of the universe is very close to $\approx 14 \times 10^9 yr$. This also creates a kind of confidence in us regarding the correctness of our theory compared to others, inspite of its basic difference from the conventional approaches, relating to the evolution of the universe.

Having evaluated the expression in the rhs of Eq. (29), the energy of the electromagnetic radiation radiated per unit area per unit time is given as

$$u = 1.6345 \times 10^{33} \left(\frac{1}{t^2} \right) \tag{30}$$

where t is the age of the universe in sec at any instant of time. The entropy S associated with the microwave back-ground radiation is obtained as[21]

$$S = \frac{16Vu}{3cT} = 2.9058 \left(\frac{V}{T} \right) \times 10^{23} \left(\frac{1}{t^2} \right) \tag{31}$$

Assuming the present universe to be spherical, its volume V is given as $V = (\frac{4\pi}{3})R_0^3$, where R_0 denotes its radius. Taking $R_0 \simeq 1.325 \times 10^{28}$ cm, which corresponds to the age $t = 14 \times 10^9 yr$, since $(R_0 \approx ct)$, the photonic entropy of the present universe is calculated to give

$$S = 1.45 \times 10^{73} \left(\frac{1}{T}\right) erg/deg \tag{32}$$

For $T = 2.728K$, it becomes,

$$S = 0.5 \times 10^{73} \approx 10^{73} erg/deg \tag{33}$$

The equilibrium number of photons[21] associated with the microwave background radiation is given as

$$\overline{N}_\gamma = \frac{V2\zeta(3)}{\pi^2\hbar^3c^3}k_\beta^3T_0^3 \simeq (410.0)V \tag{34}$$

Following this, the photon density is found to be $(\frac{\overline{N}_\gamma}{V}) \simeq 410$, which is in very good agreement with the estimated value of 400 found[22] by doing a calculation of the total energy density carried by the cosmic microwave background radiation. Using Eq. (34), we have calculated the total number of photons in the present universe, which becomes

$$\overline{N}_\gamma = 0.4 \times 10^{88} \tag{35}$$

Considering the fact that the number of nucleons[23], N_n, in the present universe is $\approx 6.30 \times 10^{78}$, we obtain

$$\frac{\overline{N}_\gamma}{N_n} \simeq 0.063 \times 10^{10} \tag{36}$$

This agrees with the value $(0.14 \sim 0.33) \times 10^{10}$ as speculated by several earlier workers[24] following calculations on baryogenesis. So we find that the theory reproduces the temperature of the cosmic background radiation correctly. Besides, it also succeeds to reproduce the photon density associated with the background radiation, and the value of the ratio $(\overline{N}_\gamma/N_n)$, which nicely match with the results predicted by others.

6.   A Relation Connecting t , T and Λ

The cosmological constant term[10, 11] Λ associated with vacuum energy density was originally introduced by Einstein as a repulsive component in his field equation and when translated from the relativistic to Newtonian picture gives rise to a repulsive harmonic oscillator force per unit mass as $\sim (\Lambda c^2)\vec{r}$ between points in space when Λ is positive. The one-body operator corresponding to the potential can be written as $H_\Lambda = -\Lambda c^2|\vec{X}|^2\rho(\vec{X})$ where $\rho(X)$ here is measured in the unit of mass density and this term also contains a unit volume. Hence the energy corresponding to this repulsive potential can be written as:

$$< H_\Lambda >= -\int \Lambda c^2|\vec{X}|^2\rho^2(\vec{X})\, d\vec{X} \tag{37}$$

By including this contribution of H_Λ in the energy, we have the total energy

$$E(\lambda) = \frac{\hbar^2}{m} \frac{12}{25\pi} \left(\frac{3\pi N}{16} \right)^{5/3} \frac{1}{\lambda^2} - \frac{g_\Lambda^2 N^2}{16} \cdot \frac{1}{\lambda} \tag{38}$$

where $g_\Lambda^2 = g^2 + \frac{3\Lambda c^2}{16\pi}$ and dimension of first term is same as the second term since the second term contains implicitely a product of unit mass and unit volume as can be easily seen by checking the single particle Hamiltonian H_Λ. Calculating as before, we have

$$N = 2.8535954 \left(\frac{1}{Gm^2} \right)^{3/4} \left(\frac{\hbar c}{g_\Lambda} \right)^{3/2} \tag{39}$$

and

$$R_0 = 2.8535954 \left(\frac{\hbar}{mc} \right)^{1/2} \left(\frac{\hbar G^{1/4}}{g_\Lambda^{3/2}} \right) \tag{40}$$

Now equating this R_0 with ct we have

$$Gm^{8/3} + \frac{3\Lambda c^2}{16\pi} m^{2/3} - Q = 0 \tag{41}$$

where $Q = \frac{4.0475279 \hbar^2 G^{1/3}}{c^2} \times \frac{1}{t^{4/3}}$. Using $m' = m^{2/3}$, the above equation can be cast as a quartic equation in m'. We find[25] four analytic solutions for m' and hence for m. Three of the solutions are unphysical and the only solution which is physically correct is given as

$$m = \left(\frac{u^{1/2} + \sqrt{u - 4(u/2 - [(u/2)^2 + Q/G]^{1/2})}}{2} \right)^{3/2} \tag{42}$$

where

$$u = [r + (q^3 + r^2)^{1/2}]^{1/3} + [r - (q^3 + r^2)^{1/2}]^{1/3} \tag{43}$$

and $r = \frac{9\Lambda^2 c^4}{2(16\pi G)^2}$, $q = \frac{4Q}{3G}$. Now the Kinetic energy with the degeneracy factor as discussed in the previous section, is given as

$$T = \left(\frac{2}{g_d} \right)^{\frac{1}{4}} \frac{2}{3k_B} \left[\frac{<KE>}{N} \right] = \left(\frac{2}{g_d} \right)^{\frac{1}{4}} \frac{2}{3k_B} (0.015442) N^{4/3} \left(\frac{mg_\Lambda^4}{\hbar^2} \right) \tag{44}$$

Using Eq.(39) and Eq.(42) in Eq.(44), we finally have the relation,

$$T = 0.0417 \left(\frac{2}{g_d} \right)^{1/4} \frac{c^2}{k_B} \frac{[(\{u^{1/2} + \sqrt{4[(u/2)^2 + Q/G]^{1/2} - u}\}/2)^3 + \frac{3\Lambda c^2}{16\pi G}]}{(\{u^{1/2} + \sqrt{4[(u/2)^2 + Q/G]^{1/2} - u}\}/2)^{3/2}} \tag{45}$$

This is the central result of our paper. This relation connects temperature T with time t and cosmological constant Λ since Q is a function of t and u is also a function of t and Λ. When $\Lambda=0$, we get back the relation Eq.(26) connecting T and t. Since we know the current values of $T = 2.728K$ and $t = 14 \times 10^9 year$, using Eq.(45), we solve for Λ. We do that in Fig.2 by plotting the left hand side and right hand side of Eq. (45) and finding the crossing point. This gives $\Lambda = 2.0 \times 10^{-56}$ cm^{-2} which is the value that has been derived dynamically here.

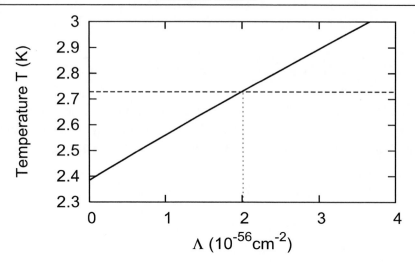

Figure 2. Determination of Λ by plotting the right hand side of Eq.(45) as a function of Λ (solid line) and left hand side as 2.728 K (thin broken line). The vertical dotted line indicates the value of $\Lambda = 2.0 \ 10^{-56} cm^{-2}$

7.   Conclusion

We study the universe as an expanding self-gravitating system using a many-particle Hamiltonian which is recently derived as representing the exactly soluble sector of quantum gravity and studied by us from a condensed matter point of view by using a quantum mechanical variational approach. This can also be viewed as a novel way of looking at the self-gravitating systems and it not only reproduces the results known from Einstein's General Theory of Relativity but also goes beyond by predicting certain relations and specifically the value of the cosmological constant. Instead of looking at the systems through the space-time dynamics, this theory treats the energy of the system directly. Infact from the quantum gravity point of view after quantising the GTR in the $c \to \infty$ (Newton-Cartan theory with spacial vanishing curvature), we have explicitly the above known Hamiltonian. We then special relativise the problem by using Mach's principle in case of the universe .

The Hamiltonian describes the expanding universe as made up of gravitationally interacting particles such as particles of luminous matter and dark matter. The dark energy which is responsible for the accelerated expansion of the universe is taken as the a repulsive harmonic potential among the points in the flat 3-space as the nonrelativistic limit of the potential attributed to the cosmolgical constant. We derive a quantum mechanical relation connecting, temperature of the cosmic microwave background radiation, age, and cosmological constant of the universe. When the cosmological constant is zero, we get back Gamow's relation with a much better coefficient. Otherwise, our theory predicts a value of the cosmological constant 2.0×10^{-56} cm^{-2} when the present values of cosmic microwave background temperature of 2.728 K and age of the universe 14 billion years are taken as input. It is interesting to note that in this flat universe, our method dynamically determines the value of the cosmological constant reasonably well compared to General Theory of Relativity where the cosmological is a free parameter.

References

[1] S. Mishra, *Int. J. Theor. Phys.* **47**, 2655 (2008).

[2] D. N. Tripathy and S. Mishra, *Int. J. Mod. Phys. D* **7**, 6, 917 (1998).

[3] D. N. Tripathy and S. Mishra, *Int. J. Mod. Phys. D* **7**, 3 , 431 (1998).

[4] J. Christian, *Phys. Rev. D* **56**, 4844 (1997)

[5] E. Cartan, *Ann. Ecole Norm. Sup.* **40**, 325 (1923)

[6] Alan H. Guth, in *Bubbles,voids and bumps in time: the new cosmology* ed. James Cornell (Cambridge University Press,Cambridge,1989).

[7] G. Contopoulos and D. Kotsakis, *Cosmology*, (Springer-Verlag,Heidelberg, 1987).

[8] A. G. Riess *et al*[Supernova Search Team Collaboration], *Astron. J.* **116** 1009 (1998).

[9] S. Perlmutter *et al.* [Supernova Cosmology Project Collaboration], *Astrophys. J.* **517** 565 (1999).

[10] P. J. E. Peebles and B. Ratra, *Rev. Mod. Phys.* **75**, 559 (2003)

[11] E. Harrison,*Cosmology*,2nd Ed. (Cambridge: Cambridge University Press,2000).

[12] S. M. Carroll, *Living Reviews in Relativity,* **4**, 1 (2001).

[13] J. V. Narlikar, *An introduction to cosmology, 3rd Ed.* (Cambridge University Press,Cambridge,2002).

[14] G. E. Volovik, *The Universe in a Helium Droplet* (Clarendon, Oxford,2003).

[15] R. B. Laughlin, *Int. J. Mod. Phys. A,* **18**, 831 (2003).

[16] J. P. Hu and S. C. Zhang, *Phys. Rev. B.* **66**, 125301 (2002).

[17] J. M. Levy-Leblond, *J. Math. Phys.* **10**, 806, (1969)

[18] A. Karplus and R. N. Porter, *Atoms and Molecules* (Reading Massachusetts, W. A. Benjamin Inc. 1970).

[19] A. A. Penzias and R. W. Wilson,*Astrophys.Jour.* **142**, 419 (1965).

[20] D. J. Fixen, E. S. Cheng, J. M. Gales, J. C. Mather, R. A. Shafer and E. L. Wright, *Astrophysics. J.* **473**, 576 (1996); A. R. Liddle, *Contemporary Physics,* **39**, no 2, 95,(1998).

[21] R. K. Pathria, *Statistical Mechanics*, (Pergamon Press, Oxford, 1972).

[22] A. M. Boesgaard and G. Steigman, *Ann. Rev. Astron.* **23**, 319 (1985)

[23] P. S. Wesson, *Cosmology and Geophysics*,(Adam Hilger Ltd,Bristol,1978).

[24] I. Affleck and M. Dine, *Nucl. Phys.* **B249**, 361 (1985).

[25] M. Abramowitz and I. A. Stegun, *Handbook of Mathematical Functions*, (Dover, New York, 1965).

In: Dark Energy: Theories, Developments and Implications   ISBN 978-1-61668-271-2
Editors: K. Lefebvre and R. Garcia, pp. 143-156   © 2010 Nova Science Publishers, Inc.

Chapter 8

DARK PRESSURE IN A NON-COMPACT AND NON-RICCI FLAT 5D KALUZA-KLEIN COSMOLOGY

F. Darabi[*]
Department of Physics, Azarbaijan
University of Tarbiat Moallem, 53714-161, Tabriz, Iran

Abstract

In the framework of noncompact Kaluza-Klein theory, we investigate a $(4 + 1)$-dimensional universe consisting of a $(4 + 1)$ dimensional Robertson-Walker type metric coupled to a $(4 + 1)$ dimensional energy-momentum tensor. The matter part consists of an energy density together with a pressure subject to $4D$ part of the $(4 + 1)$ dimensional energy-momentum tensor. The dark part consists of just a dark pressure $\bar{p}$, corresponding to the extra-dimension endowed by a scalar field, with no element of dark energy. It is shown that for a flat universe, coupled with the non-vacuum states of the scalar field, the reduced Einstein field are free of $4D$ pressure and are just affected by an effective pressure produced by the $4D$ energy density and dark pressure. It is then proposed that the expansion of the universe may be controlled by the equation of state in higher dimension rather than four dimensions. This may reveal inflationary behavior at early universe and subsequent deceleration, and account for the current acceleration at the beginning or in the middle of matter dominant era.

1.   Introduction

The recent distance measurements from the light-curves of several hundred type Ia supernovae [1, 2] and independently from observations of the cosmic microwave background (CMB) by the WMAP satellite [3] and other CMB experiments [4, 5] suggest strongly that our universe is currently undergoing a period of acceleration. This accelerating expansion is generally believed to be driven by an energy source called dark energy. The question of dark energy and the accelerating universe has been therefore the focus of a large amount of

---
[*]E-mail address: f.darabi@azaruniv.edu

activities in recent years. Dark energy and the accelerating universe have been discussed extensively from various point of views over the past few years [6, 7, 8]. In principle, a natural candidate for dark energy could be a small positive cosmological constant. This is usually studied in ΛCDM model. This model attempts to explain: cosmic microwave background observations, large scale structure observations, and supernovae observations of accelerating universe. It uses Friedmann-Robertson-Walker (FRW) metric, the Friedmann equations and the cosmological equations of state to describe the universe from right after the inflationary epoch to the present and future times, according to Einstein equation with a cosmological constant

$$R_{\mu\nu} - \frac{1}{2}Rg_{\mu\nu} = kT_{\mu\nu} + \Lambda g_{\mu\nu}.$$

Alternative approaches have also been pursued, a few example of which can be found in [9, 10, 11]. These schemes aim to improve the quintessence approach overcoming the problem of scalar field potential, generating a dynamical source for dark energy as an intrinsic feature. Quintessence is a scalar field with an equation of state which unlike cosmological constant, varies through the space and time. Many models of quintessence have a tracker behavior. In these models, the quintessence field has a density which closely tracks (but is less than) the radiation density until matter-radiation equality , which triggers the quintessence to start, having characteristics similar to dark energy and eventually dominating the universe and starting the acceleration of the universe. The goal would be to obtain a comprehensive model capable of linking the picture of the early universe to the one observed today, that is, a model derived from some effective theory of quantum gravity which, through an inflationary period would result in today accelerated Friedmann expansion driven by some Ω_Λ-term.

Another approach in this direction is to employ what is known as modified gravity where an arbitrary function of the Ricci scalar, namely $f(R)$, is inserted into the Einstein-Hilbert action instead of Ricci scalar R as

$$S = -\frac{1}{2k} \int d^4x \sqrt{-g} f(R) + S_m(g_{\mu\nu}, \psi),$$

where ψ is the matter field. This is a fully covariant theory based on the principle of least action. One of the main considerations with this modified gravity is to know how can we fit this theory with the local solar system tests as well as cosmological constraints [12, 13]. It has been shown that such a modification may account for the late time acceleration and the initial inflationary period in the evolution of the universe so that positive powers of R lead to inflation while negative powers of R result in current acceleration [14, 15, 16]. A scenario where the issue of cosmic acceleration in the framework of higher order theories of gravity in $4D$ is addressed can be found in [17]. One of the first proposals in this regard was suggested in second reference of [15] where a term of the form R^{-1} was added to the usual Einstein-Hilbert action.

The idea that our world may have more than four dimensions is due to Kaluza [18], who unified Einstein's theory of General Relativity with Maxwell's theory of Electromagnetism in a $5D$ manifold. In 1926, Klein reconsidered Kaluza's idea and treated the extra dimension as a topologically compact small circle [19]. Since then the Kaluza-Klein idea has been studied extensively from different angles. Notable amongst them is the so-called Space-Time-Matter (STM) theory, proposed by Wesson and his collaborators, which is designed to explain the origin of matter in terms of the geometry of the bulk space in which our $4D$ world is embedded, for reviews see [20]. More precisely, in STM theory, our world is a hypersurface embedded in a five-dimensional Ricci-flat ($R_{AB} = 0$) manifold where all the matter in our world can be thought of as being the manifestation of the geometrical properties of the higher dimensional space according to

$$G_{\alpha\beta} = 8\pi T_{\alpha\beta}.$$

Applications of the idea of induced matter or induced geometry can also be found in other situations [22]. The STM theory allows for the metric components to be dependent on the extra dimension and does not require the extra dimension to be compact. The sort of cosmologies stemming from STM theory is studied in [23, 24, 25, 26, 28]. The evolution of the universe has also been studied extensively based on this noncompact *vacuum* Kaluza-Klein theory [29] where a 5D mechanism is developed to explain, by a single scalar field, the evolution of the universe including inflationary expansion and the present day observed accelerated expansion.

Since in a variety of inflationary models the scalar fields have been used in describing the transition from the quasi-exponential expansion of the early universe to a power law expansion, it is natural to try to understand the present acceleration of the universe by constructing models where the matter responsible for such behavior is also represented by a scalar field. Such models are worked out, for example, in Ref [30]. In this chapter, based on the above mentioned ideas on higher dimension and scalar field, a $5D$ cosmological model is introduced which is not Ricci flat, but is extended to be coupled to a higher dimensional energy momentum tensor. It is shown that the higher dimensional sector of this model may induce a dark pressure, through a scalar field, in four dimensional universe so that for a flat universe under specific conditions one may have early inflation and current acceleration even for the non-vacuum states of the scalar field.

2. Space-Time-Matter versus Kaluza-Klein Theory

The Kaluza-Klein theory is essentially general relativity in $5D$ subject to two conditions:
1) the so called "cylinder" condition, introduced by Kaluza, to set all partial derivations with respect to the 5th coordinate to zero
2) the "compactification" condition, introduced by Klein, to set a small size and a closed topology for the 5th coordinate.
Physically, both conditions have the motivation of explaining why we perceive the 4 dimensions of space-time and do not observe the fifth dimension. In perfect analogy with general relativity, one may define a 5×5 metric tensor $g_{AB}(A, B = 0, 1, 2, 3, 4)$ where 4 denotes the extra coordinate which is referred to "internal" coordinate. Moreover, one can form a $5D$ Ricci tensor R_{AB}, a $5D$ Ricci scalar R and a $5D$ Einstein tensor $G_{AB} = R_{AB} - \frac{1}{2}Rg_{AB}$. In Principle, the field equations are expected to be

$$G_{AB} = kT_{AB}, \tag{1}$$

where k is an appropriate coupling constant and T_{AB} is a $5D$ momentum tensor. In Kaluza-Klein theory, however, it is usually assumed that the universe in $5D$ is empty, so we have the *vacuum* field equations which may equivalently be defined as

$$R_{AB} = 0, \tag{2}$$

where R_{AB} is the $5D$ Ricci tensor. These 15 relations serve to determine the 15 components of the metric g_{AB}. To this end, some assumptions are to be made on $g_{AB} = g_{AB}(x^C)$ as the choice of coordinates or gauges. Interested in electromagnetism and its unification with gravity, Kaluza realized that g_{AB} may be expressed in the following form that involves the electromagnetic 4-potentials A_α as well as an scalar field Φ

$$g_{AB} = \begin{pmatrix} (g_{\alpha\beta} - \kappa^2\Phi^2 A_\alpha A_\beta) & -\kappa\Phi^2 A_\alpha \\ \\ -\kappa\Phi^2 A_\beta & -\Phi^2 \end{pmatrix}. \tag{3}$$

The five dimensional Ricci tensor and Christoffel symbols exactly as in four dimension are defined in terms of the metric as

$$R_{AB} = \partial_C \Gamma^C_{AB} - \partial_B \Gamma^C_{AC} + \Gamma^C_{AB}\Gamma^D_{CD} - \Gamma^C_{AD}\Gamma^D_{BC}, \tag{4}$$

$$\Gamma^C_{AB} = \frac{1}{2}g^{CD}(\partial_A g_{DB} + \partial_B g_{DA} - \partial_D g_{AB}). \tag{5}$$

Then the field equations $R_{AB} = 0$ reduces to

$$G_{\alpha\beta} = \frac{\kappa^2\Phi^2}{2}T_{\alpha\beta} - \frac{1}{\Phi}(\nabla_\alpha\nabla_\beta\Phi - g_{\alpha\beta}\Box\Phi), \tag{6}$$

$$\nabla^\alpha F_{\alpha\beta} = -3\frac{\nabla^\alpha\Phi}{\Phi}F_{\alpha\beta}, \tag{7}$$

$$\Box\Phi = -\frac{\kappa^2\Phi^3}{4}F_{\alpha\beta}F^{\alpha\beta}, \tag{8}$$

where $G_{\alpha\beta}$, $F_{\alpha\beta}$ and $T_{\alpha\beta}$ are the usual $4D$ Einstein tensor, electromagnetic field strength tensor and energy-momentum tensor for the electromagnetic field, respectively. The Kaluza's case $\Phi^2 = 1$ together with the identification $\kappa = (16\pi G)^{\frac{1}{2}}$ leads to the Einstein-Maxwell equations

$$G_{\alpha\beta} = 8\pi G T_{\alpha\beta}, \tag{9}$$

$$\nabla^\alpha F_{\alpha\beta} = 0. \tag{10}$$

The STM or "induced matter theory" is also based on the vacuum field equation in $5D$ as $R_{AB} = 0$ or $G_{AB} = 0$. But, it differs mainly from the Kaluza-Klein theory about the cylinder condition so that one may now keep all terms containing partial derivatives with respect to the fifth coordinate. This results in the $4D$ Einstein equations

$$G_{\alpha\beta} = 8\pi G T_{\alpha\beta}, \tag{11}$$

provided an appropriate definition is made for the energy-momentum tensor of matter in terms of the extra part of the geometry. Physically, the picture behind this interpretation is that curvature in $(4 + 1)$ space induces effective properties of matter in $(3 + 1)$ space-time. The fact that such an embedding can be done is supported by Campbell's theorem [21] which states that any analytical solution of the Einstein field equations in N dimensions can be locally embedded in a Ricci-flat manifold in $(N + 1)$ dimensions. Since the matter is induced from the extra dimension, this theory is also called the *induced matter theory*. The main point in the induced matter theory is that these equations are a subset of $G_{AB} = 0$ with an effective or induced 4D energy-momentum tensor $T_{\alpha\beta}$, constructed by the geometry of higher dimension, which contains the classical properties of matter.

Taking the metric (3) and choosing coordinates such that the four components of the gauge fields A_α vanish, then the 5-dimensional metric becomes

$$g_{AB} = \begin{pmatrix} g_{\alpha\beta} & 0 \\ 0 & \epsilon\Phi^2 \end{pmatrix}, \tag{12}$$

where the factor ϵ with the requirement $\epsilon^2 = 1$ is introduced in order to allow a timelike, as well as spacelike signature for the fifth dimension. Using the definitions (4), (5) and keeping derivatives with respect to the fifth dimension, the resultant expressions for the $\alpha\beta$, $\alpha4$ and 44 components of the five dimensional Ricci tensor R_{AB} are obtained

$$\hat{R}_{\alpha\beta} = R_{\alpha\beta} - \frac{\nabla_\beta(\partial_\alpha\Phi)}{\Phi}$$
$$+ \frac{\epsilon}{2\Phi^2}\left(\frac{\partial_4\Phi\partial_4 g_{\alpha\beta}}{\Phi} - \partial_4 g_{\alpha\beta} + g^{\gamma\delta}\partial_4 g_{\alpha\gamma}\partial_4 g_{\beta\delta} - \frac{g^{\gamma\delta}\partial_4 g_{\gamma\delta}\partial_4 g_{\alpha\beta}}{2}\right), \tag{13}$$

$$\hat{R}_{\alpha4} = \frac{g^{44}g^{\beta\gamma}}{4} + (\partial_4 g_{\beta\gamma}\partial_\alpha g_{44} - \partial_\gamma g_{44}\partial_4 g_{\alpha\beta})$$
$$+ \frac{\partial_\beta g^{\beta\gamma}\partial_4 g_{\gamma\alpha}}{2} + \frac{g^{\beta\gamma}\partial_4(\partial_\beta g_{\gamma\alpha})}{2} - \frac{\partial_\alpha g^{\beta\gamma}\partial_4 g_{\beta\gamma}}{2}$$
$$- \frac{g^{\beta\gamma}\partial_4(\partial_\alpha g_{\beta\gamma})}{2} + \frac{g^{\beta\gamma}g^{\delta\epsilon}\partial_4 g_{\gamma\alpha}\partial_\beta g_{\delta\epsilon}}{4} + \frac{\partial_4 g^{\beta\gamma}\partial_\alpha g_{\beta\gamma}}{4}, \tag{14}$$

$$\hat{R}_{44} = -\epsilon\Phi\Box\Phi - \frac{\partial_4 g^{\alpha\beta}\partial_4 g_{\alpha\beta}}{2} - \frac{g^{\alpha\beta}\partial_4(\partial_4 g_{\alpha\beta})}{2}$$
$$+ \frac{\partial_4\Phi g^{\alpha\beta}\partial_4 g_{\alpha\beta}}{2\Phi} - \frac{g^{\alpha\beta}g^{\gamma\delta}\partial_4 g_{\gamma\beta}\partial_4 g_{\alpha\delta}}{4}. \tag{15}$$

Assuming that there is no higher dimensional matter, the Einstein equations take the form $R_{AB} = 0$ which produces the following expressions for the 4-dimensional Ricci tensor

$$R_{\alpha\beta} = \frac{\nabla_\beta(\partial_\alpha\Phi)}{\Phi} - \frac{\epsilon}{2\Phi^2}\left(\frac{\partial_4\Phi\partial_4 g_{\alpha\beta}}{\Phi} - \partial_4 g_{\alpha\beta} + g^{\gamma\delta}\partial_4 g_{\alpha\gamma}\partial_4 g_{\beta\delta} - \frac{g^{\gamma\delta}\partial_4 g_{\gamma\delta}\partial_4 g_{\alpha\beta}}{2}\right), \tag{16}$$

$$\nabla_\beta\left[\frac{1}{2\sqrt{\hat{g}_{44}}}(g^{\beta\gamma}\partial_4 g_{\gamma\delta} - \delta_\alpha^\beta g^{\gamma\epsilon}\partial_4 g_{\gamma\epsilon})\right] = 0, \tag{17}$$

$$\epsilon\Phi\Box\Phi = -\frac{\partial_4 g^{\alpha\beta}\partial_4 g_{\alpha\beta}}{4} - \frac{g^{\alpha\beta}\partial_4(\partial_4 g_{\alpha\beta})}{2} + \frac{\partial_4\Phi g^{\alpha\beta}\partial_4 g_{\alpha\beta}}{2\Phi}. \tag{18}$$

The first equation introduces an induced energy-momentum tensor as

$$8\pi G T_{\alpha\beta} = R_{\alpha\beta} - \frac{1}{2} R g_{\alpha\beta}. \tag{19}$$

The four dimensional Ricci scalar is obtained

$$R = g^{\alpha\beta} R_{\alpha\beta} = \frac{\epsilon}{4\Phi^2} [\partial_4 g^{\alpha\beta} \partial_4 g_{\alpha\beta} + (g^{\alpha\beta} \partial_4 g_{\alpha\beta})^2]. \tag{20}$$

Inserting R and $R_{\alpha\beta}$ into eq.(19) one finds that

$$8\pi G T_{\alpha\beta} = \frac{\nabla_\beta(\partial_\alpha \Phi)}{\Phi}$$

$$-\frac{\epsilon}{2\Phi^2} \left[\frac{\partial_4 \Phi \partial_4 g_{\alpha\beta}}{\Phi} - \partial_4 g_{\alpha\beta} + g^{\gamma\delta} \partial_4 g_{\alpha\gamma} \partial_4 g_{\beta\delta} - \frac{g^{\gamma\delta} \partial_4 g_{\gamma\delta} \partial_4 g_{\alpha\beta}}{2} \right.$$
$$\left. + \frac{g_{\alpha\beta}}{4} (\partial_4 g^{\gamma\delta} \partial_4 g_{\gamma\delta} + (g^{\gamma\delta} \partial_4 g_{\gamma\delta})^2) \right]. \tag{21}$$

Therefore, the 4-dimensional Einstein equations $G_{\alpha\beta} = 8\pi G T_{\alpha\beta}$ are automatically contained in the 5-dimensional vacuum equations $G_{AB} = 0$, so that the matter $T_{\alpha\beta}$ is a manifestation of pure geometry in the higher dimensional world and satisfies the appropriate requirements: it is symmetric and reduces to the expected limit when the cylinder condition is re-applied.

3. The Extended Model

As explained above, both Kaluza-Klein and STM theories use the vacuum field equations in 5 dimensions. In Kaluza-Klein case, the energy-momentum tensor is limited to Electromagnetic and scalar fields. In STM case, $T_{\alpha\beta}$ includes more types of matter but is limited to those obtained for an specific form of the $5D$ metric. We are now interested in a $5D$ model with a general $5D$ energy-momentum tensor which is, in principle, independent of the geometry of higher dimension and can be set by physical considerations on 5-dimensional matter. We start with the $5D$ line element

$$dS^2 = g_{AB} dx^A dx^B. \tag{22}$$

The space-time part of the metric $g_{\alpha\beta} = g_{\alpha\beta}(x^\alpha)$ is assumed to define the Robertson-Walker line element

$$ds^2 = dt^2 - a^2(t) \left(\frac{dr^2}{(1 - Kr^2)} + r^2(d\theta^2 + \sin^2\theta d\phi^2) \right), \tag{23}$$

where K takes the values $+1, 0, -1$ according to a close, flat or open universe, respectively. We also take the followings

$$g_{4\alpha} = 0, \quad g_{44} = \epsilon \Phi^2(x^\alpha),$$

where $\epsilon^2 = 1$ and the signature of the higher dimensional part of the metric is left general. This choice has been made because any fully covariant $5D$ theory has five coordinate degrees of freedom which can lead to considerable algebraic simplification, without loss of generality. Unlike the noncompact vacuum Kaluza-Klein theory, we will assume the fully covariant $5D$ non-vacuum Einstein equation

$$G_{AB} = 8\pi G T_{AB}, \tag{24}$$

where G_{AB} and T_{AB} are the $5D$ Einstein tensor and energy-momentum tensor, respectively. Note that the $5D$ gravitational constant has been fixed to be the same value as the $4D$ one[1]. In the following we use the geometric reduction from $5D$ to $4D$ as appeared in [28]. The $5D$ Ricci tensor is given in terms of the $5D$ Christoffel symbols by

$$R_{AB} = \partial_C \Gamma^C_{AB} - \partial_B \Gamma^C_{AC} + \Gamma^C_{AB}\Gamma^D_{CD} - \Gamma^C_{AD}\Gamma^D_{BC}. \tag{25}$$

The $4D$ part of the $5D$ quantity is obtained by putting $A \rightarrow \alpha$, $B \rightarrow \beta$ in (25) and expanding the summed terms on the r.h.s by letting $C \rightarrow \lambda, 4$ etc. Therefore, we have

$$\begin{aligned}
\hat{R}_{\alpha\beta} &= \partial_\lambda \Gamma^\lambda_{\alpha\beta} + \partial_4 \Gamma^4_{\alpha\beta} - \partial_\beta \Gamma^\lambda_{\alpha\lambda} - \partial_\beta \Gamma^4_{\alpha 4} + \Gamma^\lambda_{\alpha\beta}\Gamma^\mu_{\lambda\mu} + \Gamma^\lambda_{\alpha\beta}\Gamma^4_{\lambda 4} + \Gamma^4_{\alpha\beta}\Gamma^D_{4D} \\
&\quad - \Gamma^\mu_{\alpha\lambda}\Gamma^\lambda_{\beta\mu} - \Gamma^4_{\alpha\lambda}\Gamma^\lambda_{\beta 4} - \Gamma^D_{\alpha 4}\Gamma^4_{\beta D},
\end{aligned} \tag{26}$$

where $\hat{}$ denotes the $4D$ part of the $5D$ quantities. One finds the $4D$ Ricci tensor as a part of this equation which may be cast in the following form

$$\hat{R}_{\alpha\beta} = R_{\alpha\beta} + \partial_4 \Gamma^4_{\alpha\beta} - \partial_\beta \Gamma^4_{\alpha 4} + \Gamma^\lambda_{\alpha\beta}\Gamma^4_{\lambda 4} + \Gamma^4_{\alpha\beta}\Gamma^D_{4D} - \Gamma^4_{\alpha\lambda}\Gamma^\lambda_{\beta 4} - \Gamma^D_{\alpha 4}\Gamma^4_{\beta D}. \tag{27}$$

Evaluating the Christoffel symbols for the metric g_{AB} gives

$$\hat{R}_{\alpha\beta} = R_{\alpha\beta} - \frac{\nabla_\alpha \nabla_\beta \Phi}{\Phi}. \tag{28}$$

Putting $A = 4, B = 4$ and expanding with $C \rightarrow \lambda, 4$ in Eq.(25) we obtain

$$R_{44} = \partial_\lambda \Gamma^\lambda_{44} - \partial_4 \Gamma^\lambda_{4\lambda} + \Gamma^\lambda_{44}\Gamma^\mu_{\lambda\mu} + \Gamma^4_{44}\Gamma^\mu_{4\mu} - \Gamma^\lambda_{4\mu}\Gamma^\mu_{4\lambda} - \Gamma^4_{4\mu}\Gamma^\mu_{44}. \tag{29}$$

Evaluating the corresponding Christoffel symbols in Eq.(29) leads to

$$R_{44} = -\epsilon \Phi \Box \Phi. \tag{30}$$

We now construct the space-time components of the Einstein tensor

$$G_{AB} = R_{AB} - \frac{1}{2}g_{AB}R_{(5)}.$$

In so doing, we first obtain the $5D$ Ricci scalar $R_{(5)}$ as

$$R_{(5)} = g^{AB}R_{AB} = \hat{g}^{\alpha\beta}\hat{R}_{\alpha\beta} + g^{44}R_{44} = g^{\alpha\beta}\left(R_{\alpha\beta} - \frac{\nabla_\alpha \nabla_\beta \Phi}{\Phi}\right) + \frac{\epsilon}{\Phi^2}(-\epsilon \Phi \Box \Phi)$$

---

[1]Note that unlike the brane gravity where due to the compactification of extra dimension the gravitational constant in $5D$ is different from $4D$, here one may keep the same value of G for both $4D$ and $5D$ since this model is non-compact.

$$= R - \frac{2}{\Phi}\Box\Phi, \tag{31}$$

where the $\alpha 4$ terms vanish and R is the $4D$ Ricci scalar. The space-time components of the Einstein tensor is written $\hat{G}_{\alpha\beta} = \hat{R}_{\alpha\beta} - \frac{1}{2}\hat{g}_{\alpha\beta}R_{(5)}$. Substituting $\hat{R}_{\alpha\beta}$ and $R_{(5)}$ into the space-time components of the Einstein tensor gives

$$\hat{G}_{\alpha\beta} = G_{\alpha\beta} + \frac{1}{\Phi}(g_{\alpha\beta}\Box\Phi - \nabla_\alpha\nabla_\beta\Phi). \tag{32}$$

In the same way, the 4-4 component is written $G_{44} = R_{44} - \frac{1}{2}g_{44}R_{(5)}$, and substituting R_{44}, $R_{(5)}$ into this component of the Einstein tensor gives

$$G_{44} = -\frac{1}{2}\epsilon R\Phi^2. \tag{33}$$

We now consider the $5D$ energy-momentum tensor. The form of energy-momentum tensor is dictated by Einstein's equations and by the symmetries of the metric (23). Therefore, we may assume a perfect fluid with nonvanishing elements

$$T_{\alpha\beta} = (\rho + p)u_\alpha u_\beta - pg_{\alpha\beta}, \tag{34}$$

$$T_{44} = -\bar{p}g_{44} = -\epsilon\bar{p}\Phi^2, \tag{35}$$

where ρ and p are the conventional density and pressure of perfect fluid in the $4D$ standard cosmology and $\bar{p}$ acts as a pressure living along the higher dimensional sector. Hence, the field equations (24) are to be viewed as *constraints* on the simultaneous geometric and physical choices of G_{AB} and T_{AB} components, respectively.

Substituting the energy-momentum components (34), (35) in front of the $4D$ and extra dimensional part of Einstein tensors (32) and (33), respectively, we obtain the field equations[2]

$$G_{\alpha\beta} = 8\pi G[(\rho + p)u_\alpha u_\beta - pg_{\alpha\beta}] + \frac{1}{\Phi}[\nabla_\alpha\nabla_\beta\Phi - \Box\Phi g_{\alpha\beta}], \tag{36}$$

and

$$R = 16\pi G\bar{p}. \tag{37}$$

By evaluating the $g^{\alpha\beta}$ trace of Eq.(36) and combining with Eq.(37) we obtain

$$\Box\Phi = \frac{1}{3}(8\pi G(\rho - 3p) + 16\pi G\bar{p})\Phi. \tag{38}$$

This equation infers the following scalar field potential

$$V(\Phi) = -\frac{1}{6}(8\pi G(\rho - 3p) + 16\pi G\bar{p})\Phi^2, \tag{39}$$

whose minimum occurs at $\Phi = 0$, for which the equations (36) reduce to describe a usual $4D$ FRW universe filled with ordinary matter ρ and p. In other words, our conventional $4D$

---

[2]The $\alpha 4$ components of Einstein equation (24) result in

$$R_{\alpha 4} = 0,$$

which is an identity with no useful information.

universe corresponds to the vacuum state of the scalar field Φ. From Eq.(38), one may infer the following replacements for a nonvanishing Φ

$$\frac{1}{\Phi}\Box\Phi = \frac{1}{3}(8\pi G(\rho - 3p) + 16\pi G\bar{p}), \tag{40}$$

$$\frac{1}{\Phi}\nabla_\alpha\nabla_\beta\Phi = \frac{1}{3}(8\pi G(\rho - 3p) + 16\pi G\bar{p})u_\alpha u_\beta. \tag{41}$$

Putting the above replacements into Eq.(36) leads to

$$G_{\alpha\beta} = 8\pi G[(\rho + \tilde{p})u_\alpha u_\beta - \tilde{p}g_{\alpha\beta}], \tag{42}$$

where

$$\tilde{p} = \frac{1}{3}(\rho + 2\bar{p}). \tag{43}$$

This energy-momentum tensor effectively describes a perfect fluid with density ρ and pressure $\tilde{p}$. The four dimensional field equations lead to two independent equations

$$3\frac{\dot{a}^2 + k}{a^2} = 8\pi G\rho, \tag{44}$$

$$\frac{2a\ddot{a} + \dot{a}^2 + k}{a^2} = -8\pi G\tilde{p}. \tag{45}$$

Differentiating (44) and combining with (45) we obtain the conservation equation

$$\frac{d}{dt}(\rho a^3) + \tilde{p}\frac{d}{dt}(a^3) = 0. \tag{46}$$

The equations (44) and (45) can be used to derive the acceleration equation

$$\frac{\ddot{a}}{a} = -\frac{4\pi G}{3}(\rho + 3\tilde{p}) = -\frac{8\pi G}{3}(\rho + \bar{p}). \tag{47}$$

The acceleration or deceleration of the universe depends on the negative or positive values of the quantity $(\rho + \bar{p})$.

From extra dimensional equation (37) (or 4-dimensional Eqs.(43), (44) and (45)) we obtain

$$-\frac{6(k + \dot{a}^2 + \ddot{a}a)}{a^2} = 16\pi G\bar{p}. \tag{48}$$

Using power law behaviors for the scale factor and dark pressure as $a(t) = a_0 t^\alpha$ and $\bar{p}(t) = \bar{p}_0 t^\beta$ in the above equation, provided $k = 0$ in agreement with observational constraints, we obtain $\beta = -2$.

Based on homogeneity and isotropy of the 4D universe we may assume the scalar field to be just a function of time, then the scalar field equation (38) reads as the following form

$$\ddot{\Phi} + 3\frac{\dot{a}}{a}\dot{\phi} - \frac{8\pi G}{3}((\rho - 3p) + 2\bar{p})\Phi. \tag{49}$$

Assuming $\Phi(t) = \Phi_0 t^\gamma$ and $\rho(t) = \rho_0 t^\delta$ ($\rho_0 > 0$) together with the equations of state for matter pressure $p = \omega\rho$ and dark pressure $\bar{p} = \Omega\rho$ we continue to calculate the required parameters for inflation, deceleration and then acceleration of the universe. In doing so, we rewrite the acceleration equation (47), scalar field equation (49) and conservation equation (46), respectively, in which the above assumptions are included as

$$\alpha(\alpha - 1) + \frac{8\pi G}{3}\rho_0(1 + \Omega) = 0, \tag{50}$$

$$\gamma(\gamma - 1) + 3\alpha\gamma - \frac{8\pi G}{3}\rho_0((1 - 3\omega) + 2\Omega) = 0, \tag{51}$$

$$2\rho_0[(2 + \Omega)\alpha - 1] = 0, \tag{52}$$

where $\delta = -2$ has been used due to the consistency with the power law behavior $t^{3\alpha-3}$ in the conservation equation. The demand for acceleration $\ddot{a} > 0$ through Eq.(47) with the assumptions $\rho(t) = \rho_0 t^\delta$ and $\bar{p} = \Omega\rho$, requires $\rho_0(1 + \Omega) < 0$ or $\Omega < -1$ which accounts for a negative dark pressure. This negative domain of Ω leads through the conservation equation (46) to $\alpha > 1$ which indicates an accelerating universe as expected. On the other hand, using Friedmann equation we obtain $\alpha = \frac{1}{2+\Omega}$ which together with the condition $\alpha > 1$ requires that $-2 < \Omega < -1$. Now, one may recognize two options as follows.

The first option is to attribute an intrinsic evolution to the parameter Ω along the higher dimension so that it can produce the $4D$ expansion evolution in agreement with standard model including early inflation and subsequent deceleration, and also current acceleration of the universe. Ignoring the phenomenology of the evolution of the parameter Ω, we may require

$$\begin{cases} \Omega \gtrsim -2 & for \quad \text{inflation} \\ \Omega > -1 & for \quad \text{deceleration} \\ \Omega \lesssim -1 & for \quad \text{acceleration.} \end{cases} \tag{53}$$

The first case corresponds to highly accelerated universe due to a large $\alpha >>> 1$. This can be relevant for the inflationary era if one equate the power law with exponential behavior. The second case corresponds to a deceleration $\alpha < 1$, and the third case represents an small acceleration $\alpha \gtrsim 1$. In this option, there is no specific relation between the physical phase along extra dimension, namely Ω, and the ones defined in $4D$ universe by ω. Therefore, an unexpected acceleration in the *"middle"* of matter dominated phase $\omega = 0$ is justified due to the beginning of a new phase of Ω.

The second option is to assume a typical relation between the parameters Ω and ω as $\Omega = f(\omega)$ so that

$$\begin{cases} \Omega \gtrsim -2 & for \quad \omega = -1 \\ \Omega > -1 & for \quad \omega = \frac{1}{3} \\ \Omega \lesssim -1 & for \quad \omega = 0. \end{cases} \tag{54}$$

The physics of ω is well known in the standard cosmology (see bellow) but that of the parameter Ω clearly needs more careful investigation based on effective representation of higher dimensional theories, for instance string or Brane theory. In fact, at early universe when it is of Plank size, it is plausible that all coordinates including 3-space and higher dimension would be symmetric with the same size. However, during the GUT era when some spontaneous symmetry breakings could have happened to trigger the inflation, one may assume that such symmetry breakings could lead to asymmetric compact 3-space and non-compact higher dimension. This could also result in demarcation between three spatial pressures on 3-space and one higher dimensional pressure. Therefore, it is possible to consider a relation $\Omega = f(\omega)$ which is based on the physics of early universe (phase transitions) along with initial conditions to justify this demarcation.

The case $\omega = -1$ corresponds to the early universe and shows a very high acceleration due to $\alpha >>> 1$. The case $\omega = \frac{1}{3}$ corresponds to the radiation dominant era and shows a deceleration $\alpha < 1$. Finally, the case $\omega = 0$ corresponds to the matter dominant era and shows an small acceleration $\alpha \gtrsim 1$ at the *"beginning"* of this era.

4. Conclusion

A $(4 + 1)$-dimensional universe consisting of a $(4 + 1)$ dimensional metric of Robertson-Walker type coupled with a $(4+1)$ dimensional energy-momentum tensor in the framework of noncompact Kaluza-Klein theory is studied. In the matter part, there is energy density ρ together with pressure p subject to $4D$ part of the $(4 + 1)$ dimensional energy-momentum tensor, and a dark pressure $\bar{p}$ corresponding to the extra-dimensional part endowed by a scalar field. A particular (anisotropic) equation of state in $5D$ is used for the purpose of realizing the $4D$ expansion in agreement with observations. This is done by introducing two parameters ω and Ω which may be either independent or related as $\Omega = f(\omega)$. The reduced $4D$ and extra-dimensional components of $5D$ Einstein equations together with different equations of state for pressure p and dark pressure $\bar{p}$ may lead to a $4D$ universe which represents early inflation, subsequent deceleration and current acceleration. In other words, all eras of cosmic expansion may be explained by a single simple mechanism.

The important point of the present model is that the reduced Einstein field equations are free of $4D$ pressure and are just affected by an effective pressure produced by the $4D$ energy density and dark pressure along the extra dimension. This provides an opportunity to consider the expansion of the universe as a higher dimensional effect and so justify either the *acceleration in the middle of matter dominant era* or *acceleration at the beginning of matter dominant era*. Moreover, there is no longer "coincidence problem" within this model. This is because, in the present model there is no element of "dark energy" at all and we have just one energy density ρ associated with ordinary matter. So, there is no notion of coincidental domination of dark energy over matter densities to trigger the acceleration at the present status of the universe. In fact, a dark pressure with different negative values along the 5th dimension by itself may produce expanding universe including inflation, deceleration and acceleration without involving with the coincidence problem. These stages of the $4D$ universe may occur as well because of negative, positive and zero values of the four dimensional pressure, respectively, which leads to a competition between energy density ρ and dark pressure $\bar{p}$ in the acceleration equation (47). For the same reason that there is no element of dark energy in this model, the apparent *phantom like* equation of state for dark pressure $\Omega < -1$ is free of serious problems like *unbounded from below dark energy* or *vacuum instability* [27].

The above results are independent of the signature ϵ by which the higher dimension takes part in the 5D metric. Moreover, one may give a convincing justification for the non-observability of extra dimension in the present model: the scale factor a as the coefficient of the 3-space line element in the metric grows with cosmic time while the scalar field as the coefficient along the 5th coordinate in the $5D$ metric is very impressed by the role of scale factor such that at the matter dominant era where the scale factor is very large the scalar field and its possible fluctuations are very suppressed and the contribution of 5th coordinate becomes negligible compared with 3-space coordinates. Therefore, due to the dynamical compactification the extra dimension effectively becomes non-observable and leaves an effective $4D$ universe in agreement with present observations.

There are general similarities between the results of our $5D$ *non compact* model and that of multi-dimensional *compact* one [31]. Although the equations of state studied in these models are not the same, however, the cosmological consequences like dynamical compactification of extra dimension along with accelerating behavior of four-dimensional universe are rather similar. A detailed comparison of two models shows a physical similarity: The dynamical character of the higher dimensional sectors in both models lets them to compactify dynamically as $\sim a^{-n}$ (n is a positive parameter) and allow the $5D$ cosmology to dynamically evolve towards an effective four-dimensional universe. This feature is the key point in [31] to express the effective pressure in terms of the components of the higher dimensional energy-momentum tensor, and lets it capable of being negative to explain the current acceleration of the universe.

References

[1] Riess, A. G. et. al., *Astron. J.* 1998, 116, 1006.

[2] Perlmutter, S. et. al., *Astron. J.* 1999, 517, 565 ; Spergel, D. N. et. al., *Astrophys. J. Suppl.* 2003, 148, 175.

[3] Bennett, C. L. et. al., Astrophys. J. Suppl. 2003, 148, 1.

[4] Netterfield, C. B. et. al., *Astrophys. J.* 2002, 571, 604.

[5] Halverson, N. W. et. al., *Astrophys. J.* 2002, 568, 38.

[6] Zlatev, I.; Wang, L. and Steinhardt, P. J. *Phys. Rev. Lett.* 1999, 82, 896; Steinhardt, P. J.; Wang, L.; Zlatev, I. *Phys. Rev. D.* 1999, 59, 123504; Turner, M. S. *Int. J. Mod. Phys. A.* 2002, 17S1, 180; Sahni, V. Class. Quant. Grav. 2002, 19, 3435.

[7] Caldwell, R. R.; Kamionkowski, M.; Weinberg, N. N. *Phys. Rev. Lett.* 2003, 91, 071301; Caldwell, R. R. *Phys. Lett. B.* 2002, 545, 23; Singh, P.; Sami, M.; Dadhich, N. *Phys. Rev. D.* 2003, 68, 023522; Hao, J.G.; Li, X.Z. *Phys. Rev. D.* 2003, 67, 107303.

[8] Armendáriz-Picón; Damour, T. ;Mukhanov, *V. Phy. Lett. B.* 1999, 458, 209; Malquarti, M. *et. al., Phys. Rev. D.* 2003, 67, 123503; Chiba, T. *Phys. Rev. D.* 2002, 66, 063514.

[9] Ujjaini, A.; Sahni, V.; Sahni, V.; Shtanov, Y. *JCAP*, 2003, 0311, 014.

[10] Freese, K.; Lewis, M. *Phys. Lett. B.* 2002, 540, 1; Wang, Y.; Freese, K.; Gondolo, P.; Lewis, P.M. *Astrophys. J.* 2003, 594, 25.

[11] Bilic, N.; Tupper, G. B.; Viollier, R. D. *Phys. Lett. B.*, 2002, 535, 17; Fabris, J. C.; Goncalves, S. V. B.; de Souza, P. E. [astro-ph/0207430]; Dev, A.; Alcanitz, J. S.; Jain, D. *Phys. Rev. D.* 2003, 67, 023515.

[12] Chiba, T. *Phys. Lett. B.* 2003, 575, 1; Erickcek, A. L.; Smith, T. L.; Kamionkowski, M.; *Phys. Rev. D* 74, 2006, 121501; Chiba, T.; Smith, T. L.; Erickcek, A. L. *Phys. Rev. D.* 2007, 75, 124014.

[13] Khoury, J.; Weltman, A. *Phys. Rev. Lett.* 2004, 93, 171104; Khoury, J.; Weltman, A. *Phys. Rev. D.* 2004, 69, 044026; Mota, D. F. Barrow, J. D. *Mon. Not. Roy. Astron. Soc.* 2004, 349, 291; Clifton, T.; Mota, D. F.; Barrow, J. D. *Mon. Not. Roy. Astron. Soc.* 2005, 358, 601; Mota, D. F.; Shaw, D. J. *Phys. Rev. Lett.* 2006, 97, 151102; Mota, D. F.; Shaw, D. J. *Phys. Rev. D.* 2007, 75, 063501.

[14] Nojiri, S.; Odintsov, S. D. *Phys. Rev. D.* 2003, 68, 123512; *J. Phys. Conf. Ser.* 2007, 66, 012005; *Int. J. Geom. Meth. Mod. Phys.* 2007, 4, 115; *J. Phys. A.* 2007, 40, 6725; Nojiri, S.; Odintsov, S. D.; Stefancic, H. *Phys. Rev. D.* 2006, 74, 086009; Capozziello, S. *Int. J. Mod. Phys. D.* 2002, 11, 483; Capozziello, S. *et. al.*, *Int. Mod. Phys. D.* 2003, 12, 1969; *Phys. Lett. B.* 2006, 639, 135; Capozziello, S.; Troisi, A.; Phys. Rev. D. 2005, 72, 044022; Capozziello, S.; Stabile, A.; Troisi, A.; *Mod. Phys. Lett. A.* 2006, 21, 2291; Briscese, F. *et. al.*, Phys. Lett. B. 2007, 646, 105; Soussa, M. E.; Woodard, R. P. *Gen. Rel. Grav.* 2004, 36, 855; Dick, R. *Gen. Rel. Grav.* 2004, 36, 217; Dominguez, A. E.; Barraco, D. E. *Phys. Rev. D.* 2004, 70, 043505; Faraoni, V. *Phys. Rev. D.* 2007, 75, 067302; de Souza, J. C. C.; Faraoni, V. Class. Quant. Grav. 2007, 24, 3637; Easson, D. A. *Int. J. Mod. Phys. A.* 2004, 19, 5343; Olmo, G. J. *Phys. Rev. Lett.* 2005, 95, 261102; Allemandi, G. et. al., *Gen. Rel. Grav.* 2005, 37, 1891; Clifton, T.; Barrow, J. D. *Phys. Rev. D.* 2005, 72, 103005; Sotiriou, T. P. *Gen. Rel. Grav.* 2006, 38, 1407; Dolgov, A.; Pelliccia, D. N. *Nucl. Phys. B.* 2006, 734, 208; Atazadeh, K.; Sepangi, H. R. *Int. J. Mod. Phys. D.* 2007, 16, 687.

[15] Carroll, S. M. *et. al.*, *Phys. Rev. D.* 2004, 70, 043528.

[16] Deffayet, C. *Phys. Lett. B.* 2001, 502, 199; Deffayet, C.; Dvali, G. R.; Gabadadze, G. *Phys. Rev. D.* 2002, 65 044023; Deffayet, C.; Landau, S. J.; Raux, J.; Zaldarriaga, M.; Astier, P. Phys. Rev. D. 2002, 66, 024019; Alcaniz, J. S. *Phys. Rev. D.* 2002, 65, 123514; Jain, D.; Dev, A.; Alcaniz, J. S. *Phys. Rev. D.* 2002, 66 083511; Lue, A.; Scoccimarro, R.; Starrkman, G. *Phys. Rev. D,* 2004, 69 044005.

[17] Capozziello, S. *Int. J. Mod. Phys. D.* 2002, 11 483, Capozziello, S. *Int. J. Mod. Phys. D.* 2003, 12, 1969.

[18] Kaluza, T. *On The Problem Of Unity In Physics*, Sitzungsber. Preuss. Akad. Wiss. Berlin *Math. Phys.* K1 (1921), 966.

[19] Klein, O. *Z. Phys.* 1926, 37, 895.

[20] Wesson, P. S. *Space-Time-Matter*; World Scientific, Singapore 1999; Overduin, J. M.; Wesson, P. S. *Phys. Rept.* 1997, 283, 303.

[21] Campbell, J. E. *A Course of Differential Geometry*; Clarendon Oxford, 1926.

[22] Frolov, V.; Snajdr, M.; Stojkovic, D. *Phys. Rev.D.* 2003, 68, 044002.

[23] Liu, H. Y.; Wesson, P. S. *Astrophys. J.* 2001, 562, 1.

[24] Liko, T.; Wesson, P. S. *Int. J. Mod. Phys. A.* 2005, 20, 2037; Seahra, S. S.; Wesson, P. S. *J. Math. Phys.* 2003, 44, 5664; Ponce de Leon, *J. Gen. Rel. Grav.* 1988, 20, 539; Xu, L. X.; Liu, H. Y.; Wang, B. L. *Chin. Phys. Lett.* 2003, 20, 995; Liu, H. Y. *Phys. Lett. B.* 2003, 560, 149.

[25] Wang, B. L.; Liu, H. Y.; Xu, L.X. *Mod. Phys. Lett. A.* 2004, 19, 449; Xu, L. X.; Liu, H. Y. *Int. J. Mod. Phys. D.* 2005, 14, 883.

[26] Liu, H. Y.; Mashhoon, B. *Ann. Phys.* 1995, 4, 565.

[27] Cline, J. M.; Jeon, S.; Moore, G. D. *Phys. Rev. D.* 2004, 70, 043543.

[28] Wesson, P. S.; Ponce de Leon, J. *J. Math. Phys.* 1992, 33, 3883; Ponce de Leon, J.; Wesson, P. S. *J. Math. Phys.* 1993, 34, 4080.

[29] Madriz Aguilar, J. E.; Bellini, M. *Phys. Lett. B.* 2004, 596, 116; *Eur. Phys. J. C.* 2004, 38, 367; Bellini, M. *Nucl. Phys. B.* 2003, 660, 389; Ponce de Leon, J. Int. J. *Mod. Phys. D.* 2006, 15, 1237.

[30] Starobinsky, A. A. *JETP Lett.* 1998, 68, 757; Saini, T. D.; Raychaudhury, S.; Sahni, V.; Starobinsky, A. A. *Phys. Rev. Lett.* 2000, 85, 1162.

[31] N. Mohammedi, *Phys. Rev. D* 65 104018 (2002).

In: Dark Energy: Theories, Developments and Implications   ISBN 978-1-61668-271-2
Editors: K. Lefebvre and R. Garcia, pp. 157-213   © 2010 Nova Science Publishers, Inc.

Chapter 9

FALSIFYING FIELD-BASED DARK ENERGY MODELS

Genly Leon[*], *Yoelsy Leyva*[†], *Emmanuel N. Saridakis*[‡],
Osmel Martin[§] *and Rolando Cardenas*[¶]
Department of Mathematics,
Universidad Central de Las Villas, Santa Clara   CP 54830, Cuba
Department of Physics,
Universidad Central de Las Villas, Santa Clara   CP 54830, Cuba and
Department of Physics,
University of Athens, GR-15771 Athens, Greece

Abstract

We survey the application of specific tools to distinguish amongst the wide variety of dark energy models that are nowadays under investigation. The first class of tools is more mathematical in character: the application of the theory of dynamical systems to select the better behaved models, with appropriate attractors in the past and future. The second class of tools is rather physical: the use of astrophysical observations to crack the degeneracy of classes of dark energy models. In this last case the observations related with structure formation are emphasized both in the linear and non-linear regimes. We exemplify several studies based on our research, such as quintom and quinstant dark energy ones. Quintom dark energy paradigm is a hybrid construction of quintessence and phantom fields, which does not suffer from fine-tuning problems associated to phantom field and additionally it preserves the scaling behavior of quintessence. Quintom dark energy is motivated on theoretical grounds as an explanation for the crossing of the phantom divide, i.e. the smooth crossing of the dark energy state equation parameter below the value -1. On the other hand, quinstant dark energy is considered to be formed by quintessence and a negative cosmological constant, the inclusion of this later component allows for a viable mechanism to halt

[*]E-mail address: genly@uclv.edu.cu
[†]E-mail address: yoelsy@uclv.edu.cu
[‡]E-mail address: msaridak@phys.uoa.gr
[§]E-mail address: osmel@uclv.edu.cu
[¶]E-mail address: rcardenas@uclv.edu.cu

acceleration. We comment that the quinstant dark energy scenario gives good predictions for structure formation in the linear regime, but fails to do that in the non-linear one, for redshifts larger than one. We comment that there might still be some degree of arbitrariness in the selection of the best dark energy models.

1. Introduction

The current accelerated expansion of our universe has been one of the most active fields in modern cosmology. Many cosmological models have been proposed to interpret this mysterious phenomenon, see e.g. [1, 2] for recent reviews. The simplest candidate is a positive cosmological constant Λ [3, 4]. It is well-known that its interpretation as the vacuum energy is problematic because of its exceeding smallness [5]. Notwithstanding its observational merits, the ΛCDM scenario is seriously plagued by the well known coincidence and fine tuning problems [6] which are the main motivations to look for alternative models.

Dark energy (DE) models with two scalar fields (quintessence and phantom) have settled out explicitly and named quintom models [7, 8, 9, 10, 11, 12, 13, 14, 15, 16, 22, 23, 24, 25, 26, 27, 28, 29, 30]. The quintom paradigm is a hybrid construction of a quintessence component, usually modelled by a real scalar field that is minimally coupled to gravity, and a phantom field: a real scalar field –minimally coupled to gravity– with negative kinetic energy. Let us define the equation of state parameter of any cosmological fluid as $w \equiv$ pressure/density. The simplest model of dark energy (vacuum energy or cosmological constant) is assumed to have $w = -1$. A key feature of quintom-like behavior is the crossing of the so called phantom divide, in which the equation of state parameter crosses through the value $w = -1$. [1] Quintom behavior (i.e., the $w = -1$ crossing) has been investigated in the context of h-essence cosmologies [14, 15]; in the context of holographic dark energy [17, 18, 19, 20, 21]; inspired by string theory [22, 23, 24]; derived from spinor matter [25]; for arbitrary potentials [27, 28, 29, 30]; using isomorphic models consisting of three coupled oscillators, one of which carries negative kinetic energy (particularly for investigating the dynamical behavior of massless quintom)[31]. The crossing of the phantom divide is also possible in the context of scalar tensor theories [32, 33, 34, 35, 36] as well as in modified theories of gravity [37].

The cosmological evolution of quintom model with exponential potential has been examined, from the dynamical systems viewpoint, in [12] and [13, 16]. The difference between [12] and [13, 16] is that in the second case the potential considers the interaction between the conventional scalar field and the phantom field. In [13] it had been proven that in the absence of interactions, the solution dominated by the phantom field should be the attractor of the system and the interaction does not affect its attractor behavior. In [16] the case in which the interaction term dominates against the mixed terms of the potential, was studied. It was proven there, that the hypothesis in [13] is correct only in the cases in which the existence of the phantom phase excludes the existence of scaling attractors (in which the energy density of the quintom field and the energy density of DM are proportional). Some of this results were extended in [27], for arbitrary potentials. There it was settled down under what conditions on the potential it is possible to obtain scaling regimes. It was

---

[1] In section 2. we refer briefly to observational evidence in favor the quintom DE model.

proved there, that for arbitrary potentials having asymptotic exponential behavior, scaling regimes are associated to the limit where the scalar fields diverge. Also it has been proven that the existence of phantom attractors in this framework is not generic and consequently the corresponding cosmological solutions lack the big rip singularity.

In the first part of the chapter we investigate basic cosmological observables of quintom paradigm. We perform the cosmological perturbations analysis of quintom model for independent quadratic potentials. We investigate the evolution of quintom cosmology with exponential potentials in a background of a comoving perfect fluid. First, we review the flat FRW subcase (with dust background). Then, we consider both negative and positive curvature FRW models. We construct two dynamical systems, one adapted to negative curvature and the other adapted to positive curvature. We characterize the critical points of the resulting systems. By devising well-defined monotonic functions we get global results for ever expanding and contracting models. We find the existence of orbits starting from and recollapsing to a singularity (given by a massless scalar field cosmology) for positive curvature models. There is also a closed FRW solution with no scalar field starting from a big-bang and recollapsing to a "big-crunch". We have determined conditions for the existence of different types of global attractors. Furthermore, our monotonic functions rule out periodic orbits, recurrent orbits or homoclinic orbits. We comment about the interplay between dynamical analysis and observational checking as tools for discriminate among different quintom proposals.

A large variety of dark energy models suffers from the eternal acceleration problem, due to the exponential de-Sitter expansion. One of the consequences of the eternal acceleration is that a cosmic horizon appears (see e.g. [39] for a further discussion). This problem is not strictly related to Λ, should we replace Λ with a quintessence scalar field, the universe should still be eternally accelerated finally reaching a de Sitter phase and hence again a finite cosmic horizon. In the second part of the chapter we explore from both observational testing and dynamical systems perspective a theoretical model to address the horizon problem. We consider an effective dark energy fluid as a source of the accelerated expansion. We follow a model presented by some of us [38, 39] whose dark energy component is the sum of a negative cosmological constant and a quintessence scalar field evolving under the action of an exponential potential [2]. As a result, although the model is presently accelerating, eternal acceleration disappears and the universe ends in a Big Crunch like singularity in a finite time. Motivated by these theoretical virtues, we further explore this model from the observational point of view in order to see whether a negative Λ is indeed compatible with the astrophysical data at hand. We conclude that a negative Λ is indeed allowed and could represent a viable mechanism to halt eternal acceleration. We also explore the predictions of this class of model concerning the structure formation in the Universe. We conclude that this model give good predictions for structure formation in the linear regime, but fail to do so in the non-linear.

---

[2]Another point of view of the composite dark energy models can be found in [40, 41, 42, 43] where the cosmon model is introduced.

2. Observational Evidence for Quintom Dark Energy Paradigm

In this section we are going to refer briefly on the observational evidence that favor the quintom DE model.

2.1. Basic Observables

In this subsection we examine the basic observational quantity, which is the dark energy (DE) Equation-of-State (EoS) parameter. In 2004, supernovae Ia data were accumulated, opening the road to constraint imposition on the time variation of DE EoS. In [44] un-correlated and nearly model independent band power estimates (basing on the principal component analysis [45]) of the EoS of DE and its density as a function of redshift were presented, by fitting to the SNIa data. Quite unexpectedly, they found marginal (2σ) evidence for $w(z) < -1$ at $z < 0.2$, which is consistent with other results in the literature [46, 47, 48, 49, 50, 51].

The aforementioned result implied that the EoS of DE could indeed vary with time. Therefore, one could use a suitable parametrization of w_{DE} as a function of the redshift z, in order to satisfactory describe such a behavior. There are two well-studied parametrizations. The first (ansatz A) is:

$$w_{DE} = w_0 + w'z \, , \tag{1}$$

where w_0 the DE EoS at present and w' an additional parameter. However, this parametrization is only valid at low redshift, since it suffers from severe divergences at high ones, for example at the last scattering surface $z \sim 1100$. Therefore, a new, divergent-free ansatz (ansatz B) was proposed [52, 53]:

$$w_{DE} = w_0 + w_1(1 - a) = w_0 + w_1\frac{z}{1+z} \, , \tag{2}$$

where a is the scale factor and $w_1 = -dw/da$. This parametrization exhibits a very good behavior at high redshifts.

In [54] the authors used the "gold" sample of 157 SNIa, the low limit of cosmic ages and the HST prior, as well as the uniform weak prior on $\Omega_m h^2$, to constrain the free parameters of above two DE parameterizations. As can be seen in Fig.1 they found that the data seem to favor an evolving DE with the EoS being below -1 around the present epoch, while it was in the range $w > -1$ in the near cosmological past. This result holds for both parametrizations (1),(2), and in particular the best fit value of the EoS at present is $w_0 < -1$, while its "running" coefficient is larger than 0.

Apart from the SNIa data, CMB and LSS data can be also used to study the variation of EoS of DE. In [55], the authors used the first year WMAP, SDSS and 2dFGRS data to constrain different DE models. They indeed found that evidently the data favor a strongly time-dependent w_{DE}, and this result is consistent with similar project of the literature [56, 57, 58, 59, 60, 61, 62, 63, 64, 65]. Using the latest 5-year WMAP data, combined with SNIa and BAO data, the constraints on the DE parameters of ansatz B are: $w_0 = -1.06 \pm 0.14$ and $w_1 = 0.36 \pm 0.62$ [68, 66, 67], and the corresponding contour plot is presented in Fig.2.

In conclusion, as can be observed, the current observational data mildly favor w_{DE} crossing the phantom divide during the evolution of universe.

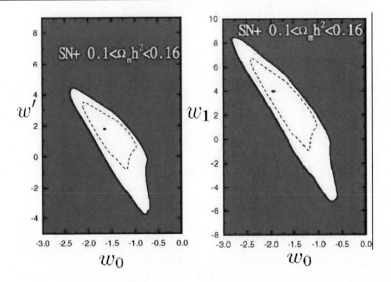

Figure 1. Two-dimensional contour-plots of the DE equation-of-state parameters, in two different parameterizations and using SNIa data. The left graph corresponds to ansatz A (expression (1)) and the right graph are to ansatz B (expression (2)). From Ref. [54].

Let us make some comments here. First of all, we mention that the above results can also fit the basic ΛCDM paradigm, where dark energy is attributed to the simple cosmological constant. Thus, many authors believe that according to data resolution we can still trust the ΛCDM paradigm, and thus there is no need to introduced additional and more complex mechanisms. The second comment is the following: even if we accept that the results seem to favor a DE EoS below -1 at present, this does not necessarily means that a two field explanation (one canonical and one phantom, i.e the basic quintom model) is automatically justified. One can still result to $w_0 < -1$ through many different frameworks including modified gravity, braneworld constructions, stringy or strong-inspired models, spinor models, etc. [69]. Thus, in order to distinguish between these alternatives, one has to find more complicated signatures of the two-field quintom model, apart from the simple observable of DE EoS. One step towards this direction is to investigate the perturbation spectrum of two-field quintom model, and then examine its relation to observations.

2.2. Perturbation Theory and Current Observational Constraints

In this subsection we study the perturbations of two-field quintom DE paradigm and the effects of these perturbations on the current observations. Additionally, since it is important to check the consistency of this model at the classical level, it requires us to analyze the behavior of perturbations when the EoS crosses the cosmological constant boundary [70].

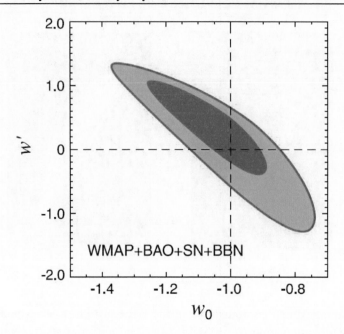

Figure 2. Two-dimensional contour-plot of the DE equation-of-state parameters, in parameterization ansatz B (expression (2)), and using WMAP, BAO, SNIa data. From Ref. [68].

2.2.1.   Analysis of Perturbations in Quintom Cosmology

In the following discussion on the quintom perturbations we will restrict ourselves to the two-field quintom model, with a Lagrangian:

$$\mathcal{L} = \mathcal{L}_Q + \mathcal{L}_P, \tag{3}$$

where

$$\mathcal{L}_Q = \frac{1}{2}\partial_\mu\phi_1\partial^\mu\phi_1 - V_1(\phi_1) \tag{4}$$

describes the canonical (quintessence) component, and

$$\mathcal{L}_P = -\frac{1}{2}\partial_\mu\phi_2\partial^\mu\phi_2 - V_2(\phi_2) \tag{5}$$

the phantom one. The equations of motion for the two scalar fields $\phi_i (i = 1, 2)$ read

$$\ddot{\phi}_i + 2\mathcal{H}\dot{\phi}_i \pm a^2\frac{\partial V_i}{\partial\phi_i} = 0, \tag{6}$$

where the positive sign is for the quintessence and the minus sign for the phantom field. Although in general the two scalar fields could be coupled with each other, here for simplicity we neglect these interactions.

Now, for a complete study on the perturbations, apart from the fluctuations of the fields, one has to consider also the metric perturbations. In the conformal Newtonian gauge the perturbed metric writes

$$ds^2 = a^2(\tau)[(1 + 2\Psi)d\tau^2 - (1 - 2\Phi)dx^i dx_i]. \tag{7}$$

Using the notation of [71], the perturbation equations satisfied by each of the quintom components are:

$$\dot{\delta}_i = -(1+w_i)(\theta_i - 3\dot{\Phi}) - 3\mathcal{H}\left(\frac{\delta P_i}{\delta\rho_i} - w_i\right)\delta_i, \tag{8}$$

$$\dot{\theta}_i = -\mathcal{H}(1-3w_i)\,\theta_i - \frac{\dot{w}_i}{1+w_i}\,\theta_i + k^2\left(\frac{\delta P_i/\delta\rho_i}{1+w_i}\delta_i - \sigma_i + \Psi\right), \tag{9}$$

where

$$\theta_i = (k^2/\dot{\phi}_i)\delta\phi_i, \quad \sigma_i = 0, \tag{10}$$

$$w_i = \frac{P_i}{\rho_i}, \tag{11}$$

and

$$\delta P_i = \delta\rho_i - 2V_i'\delta\phi_i = \delta\rho_i + \frac{\rho_i\theta_i}{k^2}\left[3\mathcal{H}(1-w_i^2) + \dot{w}_i\right]. \tag{12}$$

Thus, combining Eqs. (8), (9) and (12), we obtain

$$\dot{\theta}_i = 2\mathcal{H}\theta_i + \frac{k^2}{1+w_i}\delta_i + k^2\Psi, \tag{13}$$

$$\dot{\delta}_i = -(1+w_i)(\theta_i - 3\dot{\Phi}) - 3\mathcal{H}(1-w_i)\delta_i - 3\mathcal{H}\left[\frac{\dot{w}_i + 3\mathcal{H}(1-w_i^2)}{k^2}\right]\theta_i. \tag{14}$$

Since the simple two-field quintom model is essentially a combination of a quintessence and a phantom field, one obtains the perturbation equations by combining the aforementioned equations. The corresponding variables for the quintom system are

$$w_{quintom} = \frac{\sum_i P_i}{\sum_i \rho_i}, \tag{15}$$

$$\delta_{quintom} = \frac{\sum_i \rho_i\delta_i}{\sum_i \rho_i}, \tag{16}$$

and

$$\theta_{quintom} = \frac{\sum_i(\rho_i + p_i)\theta_i}{\sum_i(\rho_i + P_i)}. \tag{17}$$

Note that for the quintessence component, $-1 \le w_1 \le 1$, while for the phantom component, $w_2 \le -1$.

The two-field quintom model is characterized by the potentials V_i. Lets us consider the simplified case of quadratic potentials $V_i(\phi_i) = \frac{1}{2}m_i^2\phi_i^2$. In general the perturbations of ϕ_i arise from two origins, namely from the adiabatic and the isocurvature modes. Using instead of δ_i the gauge invariant variable $\zeta_i = -\Phi - \mathcal{H}\frac{\delta\rho_i}{\dot{\rho}_i}$, and in addition the relation $\Phi = \Psi$ in a universe without anisotropic stress, the equations (14) and (13) can be rewritten as,

$$\dot{\zeta}_i = -\frac{\theta_i}{3} - C_i\left(\zeta_i + \Phi + \frac{\mathcal{H}}{k^2}\theta_i\right) \tag{18}$$

$$\dot{\theta}_i = 2\mathcal{H}\theta_i + k^2(3\zeta_i + 4\Phi), \tag{19}$$

where

$$C_i = \frac{\dot{w}_i}{1 + w_i} + 3\mathcal{H}(1 - w_i) = \partial_0[\ln(a^6|\rho_i + p_i|)]. \tag{20}$$

In these expressions ζ_i is the curvature perturbation on the uniform-density hypersurfaces for the i-component of the universe [72]. Usually, the isocurvature perturbations of ϕ_i are characterized by the differences between the curvature perturbation of the uniform-ϕ_i-density hypersurfaces and that of the uniform-radiation-density hypersurfaces,

$$S_{ir} \equiv 3(\zeta_i - \zeta_r), \tag{21}$$

where the subscript r stands for radiation. Here we assume there are no matter isocurvature perturbations, and thus $\zeta_m = \zeta_r$. Eliminating ζ_i in equations (18) and (19), we obtain a second order equation for θ_i, namely

$$\ddot{\theta}_i + (C_i - 2\mathcal{H})\dot{\theta}_i + (C_i\mathcal{H} - 2\dot{\mathcal{H}} + k^2)\theta_i = k^2(4\dot{\Phi} + C_i\Phi). \tag{22}$$

The general solutions of this inhomogeneous differential equation, is the sum of the general solution of its homogeneous part with a special integration. In the following, we will show that the special integration corresponds to the adiabatic perturbation.

As it assumed, before the era of DE domination, the universe was dominated by either radiation or dark matter. The perturbation equations for these background fluids read:

$$\begin{aligned}\dot{\zeta}_f &= -\theta_f/3\,, \\ \dot{\theta}_f &= -\mathcal{H}(1 - 3w_f)\theta_f + k^2[3w_f\zeta_f + (1 + 3w_f)\Phi]\,. \end{aligned} \tag{23}$$

From the Poisson equation

$$-\frac{k^2}{\mathcal{H}^2}\Phi = \frac{9}{2}\sum_{\alpha}\Omega_\alpha(1 + w_\alpha)\left(\zeta_\alpha + \Phi + \frac{\mathcal{H}}{k^2}\theta_\alpha\right) \simeq \frac{9}{2}(1 + w_f)\left(\zeta_f + \Phi + \frac{\mathcal{H}}{k^2}\theta_f\right)\,, \tag{24}$$

on large scales we approximately acquire:

$$\Phi \simeq -\zeta_f - \frac{\mathcal{H}}{k^2}\theta_f. \tag{25}$$

Therefore, combining the equations above with $\mathcal{H} = 2/[(1 + 3w_f)\tau]$, we get (note that numerically $\theta_f \sim \mathcal{O}(k^2)\zeta_f$)

$$\begin{aligned}\zeta_f &= -\frac{5 + 3w_f}{3(1 + w_f)}\Phi = \text{const.}\,, \\ \theta_f &= \frac{k^2(1 + 3w_f)}{3(1 + w_f)}\Phi\tau\,. \end{aligned} \tag{26}$$

Therefore, from (22) we observe that there is a special solution which on large scales it is given approximately by

$$\theta_i^{ad} = \theta_f, \tag{27}$$

while (19) leads to

$$\zeta_i^{ad} = \zeta_f.$$ (28)

This indicates that the special integration of (22) corresponds to the adiabatic perturbation. Hence, concerning the isocurvature perturbations of ϕ_i, we can consider only the solution to the homogeneous part of (22),

$$\ddot{\theta}_i + (C_i - 2\mathcal{H})\dot{\theta}_i + (C_i\mathcal{H} - 2\dot{\mathcal{H}} + k^2)\theta_i = 0 .$$ (29)

These solutions are represented by θ_i^{iso} and ζ_i^{iso}. The relation between them is

$$\zeta_i^{iso} = \frac{\dot{\theta}_i^{iso} - 2\mathcal{H}\theta_i^{iso}}{3k^2} .$$ (30)

Since the general solution of ζ_i is

$$\zeta_i = \zeta_i^{ad} + \zeta_i^{iso} = \zeta_r + \zeta_i^{iso} ,$$ (31)

the isocurvature perturbations are simply $S_{ir} = 3\zeta_i^{iso}$.

In order to solve (29), we need to know the forms of C_i and $\mathcal{H}$ as functions of time τ. For this purpose, we solve the background equations (6). During the radiation dominated period, $a = A\tau$, $\mathcal{H} = 1/\tau$ and we thus we have

$$\phi_1 = \tau^{-1/2} \left[A_1 J_{1/4}\left(\frac{A}{2}m_1\tau^2\right) + A_2 J_{-1/4}\left(\frac{A}{2}m_1\tau^2\right) \right] ,$$ (32)

and

$$\phi_2 = \tau^{-1/2} \left[\tilde{A}_1 I_{1/4}\left(\frac{A}{2}m_2\tau^2\right) + \tilde{A}_2 I_{-1/4}\left(\frac{A}{2}m_2\tau^2\right) \right] ,$$ (33)

respectively, where A, A_i and $\tilde{A}_i$ are constants, $J_\nu(x)$ is the νth order Bessel function and $I_\nu(x)$ is the νth order modified Bessel function. Since the masses are usually small in comparison with the expansion rate of the early universe $m_i \ll \mathcal{H}/a$, we can approximate the (modified) Bessel functions as $J_\nu(x) \sim x^\nu(c_1 + c_2 x^2)$ and $I_\nu(x) \sim x^\nu(\tilde{c}_1 + \tilde{c}_2 x^2)$. We mention that $J_{-1/4}$ and $I_{-1/4}$ are divergent when $x \to 0$. Given these arguments we can see that it requires large initial values of ϕ_i and $\dot{\phi}_i$ if A_2 and $\tilde{A}_2$ are not vanished. Imposing small initial values, which is the natural choice if the DE fields are assumed to survive after inflation, only A_1 and $\tilde{A}_1$ modes exist, so $\dot{\phi}_i$ will be proportional to τ^3 in the leading order. Thus, the parameters C_i in (20) will be $C_i = 10/\tau$ (we have used $|\rho_i + p_i| = \dot{\phi}_i^2/a^2$). So, we acquire the solution of (29),

$$\theta_i^{iso} = \tau^{-4}[D_{i1}\cos(k\tau) + D_{i2}\sin(k\tau)].$$ (34)

Therefore, θ_i^{iso} presents an oscillatory behavior, with an amplitude damping with the expansion of the universe. This fact leads the isocurvature perturbations ζ_i^{iso} to decrease rapidly. If we choose large initial values for ϕ_i and $\dot{\phi}_i$, A_2 and $\tilde{A}_2$ modes are present, $\dot{\phi}_i$ will be proportional to τ^{-2} in the leading order and $C_i = 0$. Now the solution of (29) is:

$$\theta_i^{iso} = \tau[D_{i1}\cos(k\tau) + D_{i2}\sin(k\tau)] .$$ (35)

That is, θ_i^{iso} will oscillate with an increasing amplitude, so ζ_i^{iso} remains constant on large scales.

Similarly, during matter dominated era, $a = B\tau^2$, $\mathcal{H} = 2/\tau$, and thus the solutions for the fields ϕ_i respectively read

$$\phi_1 = \tau^{-3}\left[B_1\sin\left(\frac{B}{3}m_1\tau^3\right) + B_2\cos\left(\frac{B}{3}m_1\tau^3\right)\right] \tag{36}$$

and

$$\phi_2 = \tau^{-3}\left[\tilde{B}_1\sinh\left(\frac{B}{3}m_2\tau^3\right) + \tilde{B}_2\cosh\left(\frac{B}{3}m_2\tau^3\right)\right]. \tag{37}$$

Therefore, we do get the same conclusions with the analysis for the radiation dominated era. Firstly, choosing small initial values at the beginning of matter domination, we deduce that the isocurvature perturbations in ϕ_i will decrease with time. On the contrary, for large initial values, the isocurvature perturbations remain constant on large scales. This behavior was expected, since in the case of large initial velocity the energy density of the scalar field is dominated by the kinetic term and it behaves like a fluid with $w = 1$, and thus its isocurvature perturbation remains constant on large scales. However, on the other hand, the energy density of the scalar field will be dominated by the potential energy due to the slow rolling, that is it will behave like a cosmological constant and thus there are only tiny isocurvature perturbations in it.

In summary, we have seen that the isocurvature perturbations in quintessence-like or phantom-like field under quadratical potentials decrease or remain constant at large scales, depending on the initial velocities. In other words, the isocurvature perturbations are stable on large scales, with their amplitude being proportional to the value of Hubble parameter evaluated during the period of inflation H_{inf} (if indeed their quantum nature originates from inflation). In the case of a large H_{inf}, the isocurvature dark energy perturbations can be non-negligible and thus they will contribute to the observed CMB anisotropy [73, 74]. However, in the cases analyzed in this subsection, these isocurvature perturbations are negligible. Firstly, as mentioned above, large initial velocities are not possible if we desire the quintom fields to survive after inflation. Furthermore, even if the initial velocities are large at the beginning of the radiation domination, they will be reduced to a small value due to the small masses and the damping effect of Hubble expansion.

In conclusion, we deduce that the contributions of DE isocurvature perturbations are not very large [75] and thus for simplicity we assume that H_{inf} is small enough in order to make the isocurvature contributions negligible. Therefore, it is safe to focus only in the effects of the adiabatic perturbations of the quintom model.

2.2.2. Signatures of Perturbations in Quintom Scenario

Let us now investigate the observational signatures of perturbations in quintom scenario. For this shake we use the perturbation equations (16) and (17), and we are based on the code of CAMB [76]. For simplicity we impose a flat geometry as a background, although this is not necessary. Moreover, we assume the fiducial background parameters to be $\Omega_b = 0.042, \Omega_{DM} = 0.231, \Omega_{DE} = 0.727$, where b stands for baryons, DM for dark matter and

DE for dark energy, while today's Hubble constant is fixed at $H_0 = 69.255$ km/s Mpc^{-1}. We will calculate the effects of perturbed quintom on CMB and LSS.

In the two-field quintom model there are two parameters, namely the quintessence and phantom masses. When the quintessence mass is larger than the Hubble parameter, the field starts to oscillate and consequently one obtains an oscillating quintom. In the numerical analysis we will fix the phantom mass to be $m_P \sim 2.0 \times 10^{-60} M_{pl}$, and we vary the quintessence mass with the typical values being $m_Q = 10^{-60} M_{pl}$ and $4 \times 10^{-60} M_{pl}$ respectively.

Oscillatory Quintom

In Fig. 3 we depict the equation-of-state parameters as a function of the scale factor, for the aforementioned two parameter-sets, and additionally their corresponding effects on observations. We clearly observe the quintom oscillating behavior as the mass of quintessence component increases. After reaching the $w = -1$ pivot for several times, w crosses -1 consequently with the phantom-component domination in dark energy. As a result, the quintom fields modifies the metric perturbations: $\delta g_{00} = 2a^2 \Psi, \delta g_{ii} = 2a^2 \Phi \delta_{ij}$ and consequently they contribute to the late-time Integrated Sachs-Wolfe (ISW) effect. The ISW effect is an integrant of $\dot{\Phi} + \dot{\Psi}$ over conformal time and wavenumber k. The above two specific quintom models yield quite different evolving $\Phi + \Psi$ as shown in the right panel of Fig. 3, where the scale is $k \sim 10^{-3}$ Mpc^{-1}. As we can see, the late time ISW effects differ significantly when DE perturbations are taken into account (solid lines) or not (dashed lines).

ISW effects constitute an important part on large angular scales of CMB and on the matter power spectrum of LSS. For a constant EoS of phantom it has been shown that the low multipoles of CMB will get significantly enhanced when DE perturbations are neglected [77]. On the other hand for a matter-like scalar field, where the EoS is around zero, perturbations will also play an important role on the large scales of CMB [78]. Our results on CMB and LSS reflect the two combined effects of phantom and oscillating quintessence. We mention that while in the early studies of quintessence effects on CMB, one could consider a constant w_{eff} instead:

$$w_{eff} \equiv \frac{\int da \Omega(a) w(a)}{\int da \Omega(a)} , \qquad (38)$$

this is not enough for the study of effects on SNIa, nor for CMB, when the EoS of DE has a very large variation with redshift, such as the model of oscillating quintom considered above.

To analyze the oscillating quintom-model under the current observations, we perform a preliminary fitting to the first year WMAP TT and the TE temperature–polarization cross-power spectrum as well as the recently released 157 "Gold" SNIa data [79]. Following [80, 81] in all the fittings below we fix $\tau = 0.17$, $\Omega_m h^2 = 0.135$ and $\Omega_b h^2 = 0.022$, setting the spectral index as $n_S = 0.95$, and using the amplitude of the primordial spectrum as a continuous parameter. In the fittings of oscillating quintom we've fixed the phantom-mass to be $m_P \sim 6.2 \times 10^{-61} M_{pl}$. Fig. 4 delineates 3σ WMAP and SNIa constraints on the two-field quintom model, and in addition it shows the corresponding best fit values. The parameters m_Q and m_P stand for the masses of quintessence and phantom respectively.

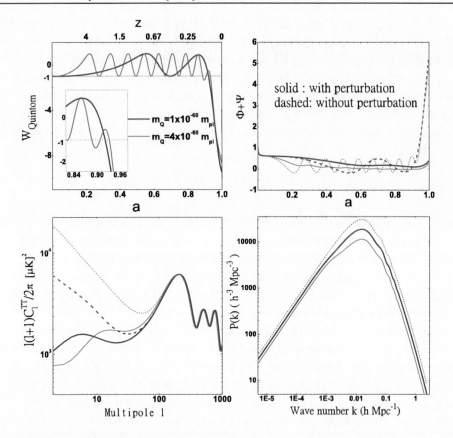

Figure 3. **Effects of the two-field oscillating quintom on the observables. The phantom mass is fixed at** $2.0 \times 10^{-60} M_{pl}$ **and the quintessence mass at** $10^{-60} M_{pl}$ **(thicker line) and** $4.0 \times 10^{-60} M_{pl}$ **(thinner line) respectively. The upper right graph depicts the evolution of the metric perturbations** $\Phi + \Psi$ **of the two models, with (solid lines) and without (dashed lines) DE perturbations. The scale is** $k \sim 10^{-3}$ Mpc^{-1}**. The lower left graph shows the CMB effects and the lower right one delineates the effects on the matter power-spectrum, with (solid lines) and without (dashed lines) DE perturbations. From Ref. [70].**

In the left graph of Fig.4 we present the separate WMAP and SNIa constraints. The green (shaded) area is WMAP constraints on models where DE perturbations have been included, while the blue area (contour with solid lines) is the corresponding area without DE perturbations. The perturbations of DE have no effects on the geometric constraint of SNIa. The right graph shows the combined WMAP and SNIa constraints on the two-field quintom model with perturbations (green/shaded region) and without perturbations (red region/contour with solid lines). We conclude that the confidence regions indeed present a large difference, if the DE perturbations have been taken into account or not.

Non-oscillatory Quintom

As we have mentioned, the basic observables could also described by the simple cosmological constant. Thus, in order to distinguish the quintom model from the cosmological

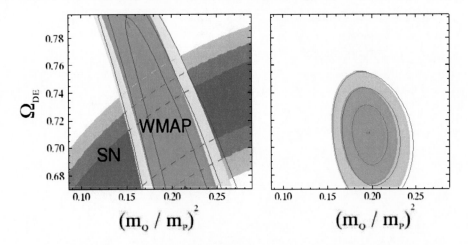

Figure 4. 3σ **WMAP and SNIa constraints on two-field quintom model, shown together with the best fit values.** m_Q and m_P denote the quintessence and phantom mass respectively. **We have fixed** $m_P \sim 6.2 \times 10^{-61} M_{pl}$ and we have varied the value of m_Q. Left graph: separate WMAP and SNIa constraints. The green (shaded) area marks the WMAP constraints on models where DE perturbations have been included, while the blue area (contour with solid lines) corresponds to the case where DE perturbations have not been **taken into account. Right graph: combined WMAP and SNIa constraints on the two-field** quintom model with perturbations (green/shaded region) and without perturbations (red region/contour with solid lines). From Ref. [70].

constant, we now consider a quintom scenario where w crosses -1 smoothly without oscillations. It is interesting to study the effects of this type of quintom model, with its effective EoS defined in (38) exactly equal to -1, on CMB and matter power spectrum. Indeed, we have realized such a quintom model in the lower right panel of Fig. 5, which can be easily given in the two-field model with a lighter quintessence mass. In this example we have set $m_Q \sim 2.6 \times 10^{-61} M_{pl}, m_P \sim 6.2 \times 10^{-61} M_{pl}$. Additionally, we assume that there is no initial kinetic energy. The initial value of the quintessence component is set to $\phi_{1i} = 0.226 M_{pl}$, while for the phantom part we impose $\phi_{2i} = 6.64 \times 10^{-3} M_{pl}$. We find that the EOS of quintom crosses -1 at $z \sim 0.15$, which is consistent with the latest SNIa results.

The model of quintom, which is mainly favored by current SNIa only, needs to be confronted with other observations in the framework of concordance cosmology. Since SNIa offer the only direct detection of DE, this model is the most promising to be distinguished from the cosmological constant and other dynamical DE models which do not get across -1, by future SNIa projects on the low redshift (for illustrations see e.g. [44]). This is also the case for the quintom model in the full parameter space: it can be most directly tested in low redshift Type Ia supernova surveys.

In the upper left panel of Fig. 5 we delineate the different ISW effects among the cosmological constant (red/light solid), the quintom model which gives $w_{eff} = -1$ with (blue/dark solid) and without (blue dashed) perturbations. Similarly to the previous oscil-

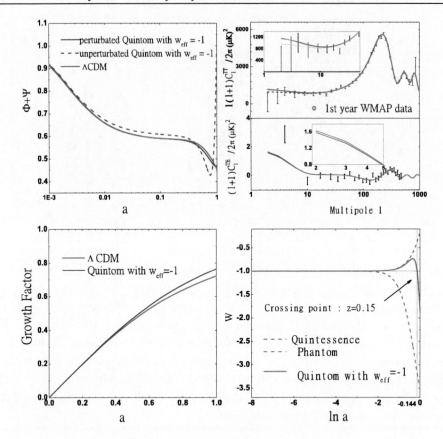

Figure 5. Comparison of the effects of the two-field quintom model with $w_{eff} = -1$ and of the simple cosmological constant, in CMB (WMAP), the metric perturbations $\Phi + \Psi$ (the scale is $k \sim 10^{-3}$ Mpc^{-1}) and the linear growth factor. The binned error bars in the upper right graph are WMAP TT and TE data [82, 83]. From Ref. [70].

lating case, the difference is very large when switching off quintom perturbations and much smaller when including the perturbations. In the upper right panel we find that the quintom model cannot be distinguished from a cosmological constant in light of WMAP. The two models give almost exactly the same results in CMB TT and TE power spectra when including the perturbations. We deduce that the difference in CMB is hardly distinguishable even by cosmic variance.

2.2.3.   Breaking the Degeneracy between Quintom and Cosmological Constant Scenarios

So far we have see that CMB observations cannot distinguish between a quintom model with $w_{eff} = -1$ and a cosmological constant. Thus, in order to acquire distinctive signatures, we have to rely in other observations. To achieve that we need to consider the physical observables which can be affected by the evolving w sensitively. In comparison with the cosmological constant, such a quintom model exhibits a different evolution of the universe's expansion history, and in particular it gives rise to a different epoch of matter-radiation equality. The Hubble expansion parameter becomes:

$$H \equiv \frac{\dot{a}}{a^2} = H_0[\Omega_m a^{-3} + \Omega_r a^{-4} + X]^{1/2} \tag{39}$$

where X, the energy density ratio of DE between the early times and today, is quite different between the *quintom*-CDM and ΛCDM. In the ΛCDM scenario X is simply a constant, while in general for DE models with varying energy density or EoS we obtain

$$X = \Omega_{DE} a^{-3} e^{-3 \int w(a) d \ln a}. \tag{40}$$

Therefore, the two models will give different Hubble expansion rates. This is also the case between the quintom model with $w_{eff} = -1$ in the left panel of Fig. 5 and a cosmological constant.

Finally, we mention that different H leads directly to different behaviors of the growth factor. In particular, according to the linear perturbation theory all Fourier modes of the matter density perturbations grow at the same rate, that is the matter density perturbations are independent of k:

$$\ddot{\delta}_k + \mathcal{H}\dot{\delta}_k - 4\pi G a^2 \rho_{\mathrm{M}} \delta_k = 0. \tag{41}$$

The growth factor $D_1(a)$ characterizes the growth of the matter density perturbations, namely $D_1(a) = \delta_k(a)/\delta_k(a = 1)$, and it is normalized to unity today. In the matter-dominated epoch we have $D_1(a) = a$. Analytically $D_1(a)$ is often approximated by the Meszaros equation [84]:

$$D_1(a) = \frac{5\Omega_m H(a)}{2H_0} \int_0^a \frac{da'}{(a'H(a')/H_0)^3}. \tag{42}$$

Therefore, we can easily observe the difference between the quintom and cosmological constant scenarios, due to the different Hubble expansion rates. In particular, one needs to solve (41) numerically. In the lower left graph of Fig. 5 we show the difference of $D_1(a)$ between the quintom model with $w_{eff} = -1$ and the cosmological constant one. The difference in the linear growth function is considerably large in the late time evolution and possibly distinguishable in future LSS surveys and in weak gravitational lensing (WGL) observations. WGL has emerged with a direct mapping of cosmic structures and it has been recently shown that the method of cosmic magnification tomography can be extremely efficient [85, 86, 87], which leaves a promising future for breaking the degeneracy between quintom and a cosmological constant.

3. Exponential Quintom: Phase Space Analysis

In the following discussion on the quintom phase space analysis we restrict ourselves to the two-field quintom model, with a Lagrangian:

$$\mathcal{L} = \frac{1}{2}\partial_\mu \phi \partial^\mu \phi - \frac{1}{2}\partial_\mu \varphi \partial^\mu \varphi - V(\phi, \varphi), \tag{43}$$

and we include, also, ordinary matter (a comoving perfect fluid) in the gravitational action. As in [16] we consider here the efective two-field potential

$$V = V_0 e^{-\sqrt{6}(m\phi + n\varphi)}, \tag{44}$$

where the scalar field ϕ represents quintessence and φ represents a phantom field. For simplicity, we assume $m > 0$ and $n > 0$.

Quintom (non-conventional) cosmologies with exponential potentials has been investigated, from the dynamical systems approach, for instance, in references [12, 13, 16] (see section 3.1. for a brief review).

We shall consider the Friedmann-Lemaitre-Robertson-Walker (FLRW) line element:

$$ds^2 = -dt^2 + a^2(t) \left(\frac{dx^2}{1 - kx^2} + x^2 \left(d\vartheta^2 + \sin^2 \vartheta \, d\varphi^2 \right) \right), \tag{45}$$

where $k = 1, 0, -1$, identifies the three types of FRW universes: closed, flat, and open, respectively.

The field equations derived from (45), are

$$H^2 - \tfrac{1}{6} \left(\dot{\phi}^2 - \dot{\varphi}^2 \right) - \tfrac{1}{3} V_{\text{eff}} - \tfrac{1}{3} \rho_{\text{M}} = -\tfrac{k}{a^2}, \tag{46}$$

$$\dot{H} = -H^2 - \tfrac{1}{3} \left(\dot{\phi}^2 - \dot{\varphi}^2 \right) + \tfrac{1}{3} V_{\text{eff}} - \tfrac{1}{6} \rho_{\text{M}}, \tag{47}$$

$$\dot{\rho}_{\text{M}} = -3H \rho_{\text{M}}, \tag{48}$$

$$\ddot{\phi} + 3H \dot{\phi} - \sqrt{6} m V = 0, \tag{49}$$

$$\ddot{\varphi} + 3H \dot{\varphi} + \sqrt{6} n V = 0, \tag{50}$$

where $H = \frac{\dot{a}(t)}{a(t)}$ denotes de Hubble expansion scalar.

The dot denotes derivative with respect the time t. We consider a pressureless perfect fluid (dust) as the background matter.

3.1. Flat FRW Subcase

To investigate the flat models we introduced the same normalized variables as in [16]: (x_ϕ, x_φ, y), defined by

$$x_\phi = \frac{\dot{\phi}}{\sqrt{6} H}, \quad x_\varphi = \frac{\dot{\varphi}}{\sqrt{6} H}, y = \frac{\sqrt{V}}{\sqrt{3} H}. \tag{51}$$

They are related through the Friedman equation 46 by $x_\phi^2 - x_\varphi^2 + y^2 = 1 - \frac{\rho_{\text{M}}}{3H^2} \le 1$.

The dynamics in the space is given by the ordinary differential equations [16]:

$$x_\phi' = \frac{1}{3} \left(3my^2 + (q - 2)x_\phi \right) \tag{52}$$

$$x_\varphi' = -\frac{1}{3} \left(3ny^2 - (q - 2)x_\varphi \right) \tag{53}$$

$$y' = \frac{1}{3}(1 + q - 3(mx_\phi + nx_\varphi))y \tag{54}$$

Table 1. Location, existence and deceleration factor of the critical points for $m > 0$, $n > 0$ and $y \geq 0$. We use the notation $\delta = m^2 - n^2$ (from reference [16]).

Name	x_ϕ	x_φ	y	Existence	q
O	0	0	0	All m and n	$\dfrac{1}{2}$
$C_\pm$	$\pm\sqrt{1 + x_\varphi^{*\,2}}$	x_φ^*	0	All m and n	2
P	m	$-n$	$\sqrt{1 - \delta}$	$\delta < 1$	$-1 + 3\delta$
T	$\dfrac{m}{2\delta}$	$-\dfrac{n}{2\delta}$	$\dfrac{1}{2\sqrt{\delta}}$	$\delta \geq 1/2$	$\dfrac{1}{2}$

defined in the phase space given by

$$\Psi = \{\mathbf{x} = (x_\phi, x_\varphi, y) : 0 \leq x_\phi^2 - x_\varphi^2 + y^2 \leq 1\}. \tag{55}$$

Here the prime denotes differentiation with respect to a new time variable $\tau = \log a^3$, where a is the scale factor of the space-time. The deceleration factor $q \equiv -\ddot{a}a/\dot{a}^2$ can then be written

$$q = \frac{1}{2}\left(3\left(x_\phi^2 - x_\varphi^2 - y^2\right) + 1\right). \tag{56}$$

The system (52-54) admits the critical points points $O, C_\pm, T, P$ reported in [16]). In the table 1 the location, existence and deceleration factor of the critical points for $m > 0$, $n > 0$ and $y \geq 0$. We use the notation $\delta = m^2 - n^2$.

By analyzing the sign of the real part of the normally-hyperbolic curves $C_\pm$ we get the following results (we are assuming $m > 0$ and $n > 0$):

- If $m < n$, C_+ contains an infinite arc parameterized by x_φ^* such that $x_\varphi^* < \frac{-n - m\sqrt{1-\delta}}{\delta}$ that is a local source. C_- contains an infinite arc parameterized by x_φ^* such that $x_\varphi^* < \frac{-n + m\sqrt{1-\delta}}{\delta}$ that is a local source.

- If $m = n$, C_+ contains an infinite arc parameterized by x_φ^* such that $x_\varphi^* < \frac{1-m^2}{2n}$ that is a local source. All of C_- is a local source.

- If $m > n$ there are two possibilities

  - If $\delta < 1$, all of C_- is a local source. A finite arc of C_+ parameterized by x_φ^* such that $\frac{-n - m\sqrt{1-\delta}}{\delta} < x_\varphi^* < \frac{-n + m\sqrt{1-\delta}}{\delta}$ is a local source.

  - If $\delta \geq 1$, no part of C_+ is a local source and all of C_- is a local source.

Perhaps, the most appealing result in [16] is that, by introducing properly defined monotonic functions and by making some numerical integrations, it was possible to identify heteroclinic sequences

Table 2. Location and existence conditions for the critical points at infinity.

Name	θ_1	θ_2	Existence
$P_1^\pm$	0	$\pm\frac{\pi}{2}$	always
$P_2^\pm$	π	$\pm\frac{\pi}{2}$	always
$P_3^\pm$	$\frac{\pi}{4}$	$\pm\cos^{-1}\left(-\frac{m}{n}\right)$	$-\pi < \pm\cos^{-1}\left(-\frac{m}{n}\right) \leq \pi, n \neq 0$
$P_4^\pm$	$\frac{3\pi}{4}$	$\pm\cos^{-1}\left(\frac{m}{n}\right)$	$-\pi < \pm\cos^{-1}\left(\frac{m}{n}\right) \leq \pi, n \neq 0$
P_5	$\theta_1^\star$	0	$0 \leq \theta_1^\star \leq \pi$
P_6	$\theta_1^\star$	π	$0 \leq \theta_1^\star \leq \pi$

- Case i) For $m < \sqrt{n^2 + 1/2}$, the point P is a stable node, whereas the point T does not exist. The heteroclinic sequence in this case is $C_\pm \longrightarrow O \longrightarrow P$.

- Case ii) For $\sqrt{n^2 + 1/2} < m \leq \sqrt{n^2 + 4/7}$, the point T is a stable node and the point P is a saddle. For these conditions the heteroclinic sequence is $C_\pm \longrightarrow O \longrightarrow T \longrightarrow P$.

- Case iii) For $\sqrt{n^2 + 4/7} < m < \sqrt{1 + n^2}$, the point T is a spiral node and the point P is a saddle. For these conditions the heteroclinic sequence is the same as in the former case.

- Case iv) For $m > \sqrt{1 + n^2}$ the point T is a spiral node whereas the point P does not exist. The heteroclinic sequence in this case is $C_- \longrightarrow O \longrightarrow T$.

From the possibilities listed above, there still the possibility of other attractors different from phantom ones in the exponential quintom scenario, particularly scaling attractors (T). This fact is a counterexample of one of the result in [13]. However, we are aware on the small probability that T represents the actual stage on the universe evolution due this solution is matter-dominated. Another compelling result in quintom cosmology is that the existence of monotonic functions in the state space can rule out periodic orbits, homoclinic orbits, and other complex behaviour in invariant sets. If so, the dynamics is dominated by critical points (and possibly, heteroclinic orbits joining it). Additionally, some global results can be obtained. A similar approach (i.e., that of devising monotonic functions) for multiple scalar field cosmologies with matter was used in [88, 89, 90]. However, in that work they do not consider phantom-like scalar fields, as we do here.

3.1.1. Analysis at Infinity

The numerical experiments in [16] suggest that there is an open set of orbits that tends to infinity. Let us investigate the dynamics at infinity. In order to do that we will use

Table 3. Stability of the critical points at infinity. We use the notation $\delta = m^2 - n^2$ and $\lambda^{\pm} = n \cos \theta_1^{\star} \pm m \sin \theta_1^{\star}$.

Name	(λ_1, λ_2)	ρ'	Stability
$P_1^{\pm}$	$(-n, n)$	> 0	saddle
$P_2^{\pm}$	$(-n, n)$	> 0	saddle
$P_3^{\pm}$	$\left(\frac{\sqrt{2}\delta}{n}, \frac{\delta}{\sqrt{2}n}\right)$	$\begin{cases} > 0, & \delta < 0 \\ < 0, & \delta > 0 \end{cases}$	source if $n < 0, n < m < -n$ saddle otherwise
$P_4^{\pm}$	$\left(-\frac{\sqrt{2}\delta}{n}, -\frac{\delta}{\sqrt{2}n}\right)$	$\begin{cases} > 0, & \delta < 0 \\ < 0, & \delta > 0 \end{cases}$	source if $n > 0, -n < m < n$ saddle otherwise
P_5	$(0, \lambda^+)$	$\begin{cases} < 0, & \frac{\pi}{4} < \theta_1^{\star} < \frac{3\pi}{4} \\ > 0, & \text{otherwise} \end{cases}$	nonhyperbolic
P_6	$(0, \lambda^-)$	$\begin{cases} < 0, & \frac{\pi}{4} < \theta_1^{\star} < \frac{3\pi}{4} \\ > 0, & \text{otherwise} \end{cases}$	nonhyperbolic

the central Poincaré projection method. Thus, to obtain the critical points at infinity we introduce spherical coordinates (ρ is the inverse of $r = \sqrt{x_\phi^2 + x_\varphi^2 + y^2}$, then, $\rho \to 0$ as $r \to \infty$):

$$x_\phi = \frac{1}{\rho} \sin \theta_1 \cos \theta_2, \tag{57}$$

$$y = \frac{1}{\rho} \sin \theta_1 \sin \theta_2, \tag{58}$$

$$x_\varphi = \frac{1}{\rho} \cos \theta_1 \tag{59}$$

where $0 \leq \theta_1 \leq \pi$ and $-\pi < \theta_2 \leq \pi$, and $0 < \rho < \infty$.

Defining the time derivative $f' \equiv \rho \dot{f}$, the system (52-54), can be written as

$$\rho' = \frac{1}{2} \left(\cos^2 \theta_1 - \cos(2\theta_2) \sin^2 \theta_1\right) + 2n \cos \theta_1 \sin^2 \theta_1 \sin^2 \theta_2 \rho + O\left(\rho^2\right). \tag{60}$$

and

$$\theta_1' = n \cos(2\theta_1) \sin \theta_1 \sin^2 \theta_2 - \cos \theta_1 \sin \theta_1 \sin^2 \theta_2 \rho + O\left(\rho^2\right),$$
$$\theta_2' = (n \cos \theta_1 \cos \theta_2 + m \sin \theta_1) \sin \theta_2 - \cos \theta_2 \sin \theta_2 \rho + O\left(\rho^2\right). \tag{61}$$

Since equation (60) does not depends of the radial component at the limit $\rho \to 0$, we can obtain the critical points at infinity by solving equations (61) in the limit $\rho \to 0$. Thus, the critical points at infinite must satisfy the compatibility conditions

$$\cos(2\theta_1)\sin\theta_1\sin^2\theta_2 = 0,$$
$$(n\cos\theta_1\cos\theta_2 + m\sin\theta_1)\sin\theta_2 = 0. \tag{62}$$

First, we examine the stability of the pairs $(\theta_1^\star, \theta_2^\star)$ satisfying the compatibility conditions (62) in the plane θ_1-θ_2, and then, we examine the global stability by substituting in (60) and analyzing the sign of $\rho'(\theta_2^\star, \theta_2^\star)$. In table 2 it is offered information about the location and existence conditions of these critical points. In table 3 we summarize the stability properties of these critical points.

Let us describe the cosmological solutions associated with the critical points at infinity. The cosmological solutions associated to the critical points $P_1^\pm$ and $P_2^\pm$ have the evolution rates $\dot{\phi}^2/V = 0$, $\dot{\phi}/\dot{\varphi} = 0$ and $H/\dot{\varphi} \equiv \rho/\sqrt{6} \to 0$.[3] These solutions are always saddle points at infinity. The critical points $P_3^\pm$ and $P_4^\pm$ are sources provided $n < 0, n < m < -n$ or $n > 0, -n < m < n$, respectively. They are saddle points otherwise. The associated cosmological solutions to $P_3^\pm$ have the evolution rates $\dot{\phi}^2/V = \frac{2m^2}{n^2-m^2}$, $\dot{\phi}/\dot{\varphi} = -m/n$, and $H/\dot{\phi} \equiv -n\rho/(\sqrt{3}m) \to 0$, and $H/\dot{\varphi} \equiv \rho/\sqrt{3} \to 0$, whereas the associated cosmological solutions to $P_4^\pm$ have the evolution rates $\dot{\phi}^2/V = \frac{2m^2}{n^2-m^2}$, $\dot{\phi}/\dot{\varphi} = -m/n$, and $H/\dot{\phi} \equiv n\rho/(\sqrt{3}m) \to 0$, and $H/\dot{\varphi} \equiv -\rho/\sqrt{3} \to 0$. The curves of critical points P_5 and P_6 are nonhyperbolic. The associated cosmological solutions have expansion rates (valid for $\theta_1^\star \neq \pi/4$) $V/\dot{\phi}^2 = 0, \dot{\phi}/\dot{\varphi} = \tan\theta_1^\star, H/\dot{\varphi} = \rho\sec\theta_1^\star/\sqrt{6} \to 0$, and $V/\dot{\phi}^2 = 0, \dot{\phi}/\dot{\varphi} = -\tan\theta_1^\star, H/\dot{\varphi} = \rho\sec\theta_1^\star/\sqrt{6} \to 0$, respectively.

3.2. Models with Negative Curvature

In this section we investigate negative curvature models.

3.2.1. Normalization, State Space and Dynamical System

For the investigation of negative curvature models we shall use the normalized variables: $(x_\phi, x_\varphi, y, \Omega)$, defined by

$$x_\phi = \frac{\dot{\phi}}{\sqrt{6}H}, \quad x_\varphi = \frac{\dot{\varphi}}{\sqrt{6}H}, \quad y = \frac{\sqrt{V}}{\sqrt{3}H}, \quad \Omega = \frac{\rho_M}{3H^2}. \tag{63}$$

This choice allows to recast the Friedmann equation (46) as

$$1 - \left(x_\phi^2 - x_\varphi^2 + y^2 + \Omega\right) = \Omega_k \geq 0, \tag{64}$$

where

$$\Omega_k = -\frac{k}{a^2H^2}, \quad k = -1, 0. \tag{65}$$

Thus,

$$0 \leq x_\phi^2 - x_\varphi^2 + y^2 + \Omega \leq 1. \tag{66}$$

---

[3]Do not confuse ρ with the matter energy density, the latter denoted by ρ_M.

Let us introduce the new time variable, τ, such that $\tau \to -\infty$ as $t \to 0$ and $\tau \to +\infty$ as $t \to +\infty$. Since the time direction must be preserved we can choose $d\tau = 3\epsilon H dt$ where $\epsilon = \pm 1 = \text{sign}(H)$.

The evolution equations for the variables (63) are

$$
\begin{aligned}
x'_\phi &= \epsilon \left(\tfrac{1}{3}\,(q-2)\,x_\phi + my^2\right), \\
x'_\varphi &= \epsilon \left(\tfrac{1}{3}\,(q-2)\,x_\varphi - ny^2\right), \\
y' &= \epsilon \left(\tfrac{1}{3}(1+q) - m\,x_\phi - n\,x_\varphi\right) y, \\
\Omega' &= \tfrac{1}{3}\epsilon\,(2\,q-1)\,\Omega,
\end{aligned}
\tag{67}
$$

where the prime denotes derivative with respect to τ and $q = 2\left(x_\phi^2 - x_\varphi^2\right) - y^2 + \tfrac{1}{2}\Omega$, is the expression for the deceleration parameter. The DE EoS parameter, w, can be rewritten, in terms of the phase variables, as

$$
w = \frac{x_\phi^2 - x_\varphi^2 - y^2}{x_\phi^2 - x_\varphi^2 + y^2}.
\tag{68}
$$

Notice that the evolution equation (67 c) is form invariant under the coordinate transformation $y \to \epsilon y$. Then, the sign of ϵy is invariant by proposition 4.1 in [91], in such way that we can assume, without lost generality, for fixed ϵ, $\epsilon y \geq 0$. Hence, for each choice of sign of ϵ, the equations (67) define a flow in the phase space

$$
\begin{aligned}
\Psi^\pm = \{(x_\phi, x_\varphi, y, \Omega) : {}& 0 \leq x_\phi^2 - x_\varphi^2 + y^2 + \Omega \leq 1, \\
& x_\phi^2 - x_\varphi^2 + y^2 \geq 0, \Omega \geq 0, \epsilon y \geq 0\}.
\end{aligned}
\tag{69}
$$

3.2.2. Form Invariance under Coordinate Trasformations

First recall that the positive "branch" ($\epsilon = +1$) describe the dynamics of models ever expanding and the negative "branch" ($\epsilon = -1$) describes the dynamics for contracting models. The system is form invariant under the change $\epsilon \to -\epsilon$, i.e., the system is symmetric under time-reversing. In this way it is enough to characterize de dynamics in Ψ^+.

3.2.3. Monotonic Functions

Let be defined in the phase space Ψ^+ (or Ψ^-, depending of the choice of ϵ) the function

$$
M = \frac{(n\,x_\phi + m\,x_\varphi)^2\,\Omega^2}{(1 - x_\phi{}^2 + x_\varphi{}^2 - y^2 - \Omega)^3}, \quad M' = -2\epsilon M.
\tag{70}
$$

This is a monotonic function for $\Omega > 0$ and $n\,x_\phi + m\,x_\varphi \neq 0$. Then, the existences of such monotonic function rule out periodic orbits, recurrent orbits, or homoclinic orbits in the phase space and also, there is possible global results from the local stability analysis of critical points. Additionally, from the expresion of M one can see inmediatly that $\Omega \to 0$, or $n\,x_\phi + m\,x_\varphi \to 0$ o $|n\,x_\phi + m\,x_\varphi| \to +\infty$ (implying x_ϕ or x_φ or both diverge) or $\Omega_k \to 0$ asymptotically.

Table 4. Coordinates and existence conditions for the critical points of the system 67. We have used the notation $\delta = m^2 - n^2$. The subindexes in the labels have the following meaning: the left subindex (denoted by $\epsilon = \pm 1$) indicates when the model is expanding (+) or contracting (−); the right subindex denotes the sign of x_ϕ (i.e., the sign of $\dot\phi$) and it is displayed by the sign $\pm$.

Label	Coordinates: $(x_\phi, x_\varphi, y, \Omega)$	Existence
$\pm K_\pm$	$\left(\pm\sqrt{1 + x_\varphi^{\star 2}}, x_\varphi^\star, 0, 0\right)$	All m and n
$\pm M$	$(0, 0, 0, 0)$	All m and n
$\pm F$	$(0, 0, 0, 1)$	All m and n
$\pm SF$	$(m, -n, \epsilon\sqrt{1-\delta}, 0)$	$\delta < 1$
$\pm CS$	$\left(\frac{m}{3\delta}, -\frac{n}{3\delta}, \frac{\epsilon\sqrt{2}}{3\sqrt{\delta}}, 0\right)$	$\delta > \frac{1}{3}$
$\pm MS$	$\left(\frac{m}{2\delta}, -\frac{n}{2\delta}, \frac{\epsilon\sqrt{1}}{2\sqrt{\delta}}, \sqrt{1-\frac{1}{2\delta}}\right)$	$\delta > \frac{1}{2}$

3.2.4.   Local Analysis of Critical Points

By the discussion about the invariance of the system, it is sufficient characterize dynamically the critical points $_+K_\pm$, $_+M$, $_+F$ $_+SF$ $_+CS$ y $_+MS$, in the phase space Ψ^+. In tables 4 and 5, it is offered information about the location, existence and eigenvalues of the critical points of the system (67) in the phase space (69) (for each choice of ϵ) and also, it is displayed the values of some cosmological parameters associated to the corresponding cosmological solutions.

Now we shall investigate the local stability of the critical points (and curves of critical points). We shall characterize de associated cosmological solutions.

The set of critical points $_+K_\pm$ and the isolated critical points $_+M$ are located in the invariant set of massless scalar field (MSF) cosmologies without matter. The isolated critical points $_+F$ are located in the invariant set of MSF cosmologies with matter.

The arcs of hyperbolae $_+K_\pm$ parameterized by the real value $x_\varphi^\star$ denote cosmological models dominated by the energy density of DE ($\Omega_{de} \rightarrow 1$), particularly by its kinetic energy. DE mimics a stiff fluid solution. Since this are a set of critical points, then necessarily, they have a zero eigenvalue. They are local sources (and in general they constitute the past attractor in the phase space Ψ^+) provided $nx_\varphi^\star \pm m\sqrt{1 + x_\varphi^{\star 2}} < 1$.

The isolated critical points $_+M$ denote the Milne's universe. They are non-hyperbolic. The critical points $_+F$ represent flat FRW solutions (dominated by matter). They are hyperbolic. For this points the quintom field vanishes, then, the DE's cosmological parameters are not applicable to this points.

The stable manifold of $_+M$ is 3-dimensional and it is tangent at the point to the 3-

Table 5. DE EoS parameter (w), deceleration parameter (q), fractional energy densities, and eigenvalues of the perturbation matrix associated to the critical points of the system 67. We use the notation $\lambda^{\pm} = nx_{\varphi}^{\star} \pm m\sqrt{1 + x_{\varphi}^{\star 2}}$. $\pm K_{\pm}, \pm F, \pm SF \pm MS$ corresponds to $k = 0$, the eigenvalues of these points in the invariant set of zero-curvature models are the same as displayed in the table but the first from the left.

Label	w	q	$\Omega_m, \Omega_{de}, \Omega_k$	Eigenvalues
$\pm K_{\pm}$	1	2	$0, 1, 0$	$\frac{4}{3}\epsilon, 0, \epsilon\left(1 - \lambda^{\pm}\right), \epsilon$
$\pm M$	-	0	$0, 0, 1$	$-\frac{2}{3}\epsilon, -\frac{2}{3}\epsilon, \frac{1}{3}\epsilon, -\frac{1}{3}\epsilon$
$\pm F$	-	$\frac{1}{2}$	$1, 0, 0$	$\frac{1}{2}\epsilon, -\frac{1}{2}\epsilon, -\frac{1}{2}\epsilon, \frac{1}{3}\epsilon$
$\pm SF$	$-1 + 2\delta$	$-1 + 3\delta$	$0, 1, 0$	$2\left(\delta - \frac{1}{3}\right)\epsilon, (\delta - 1)\epsilon, (\delta - 1)\epsilon, (2\delta - 1)\epsilon$
$\pm CS$	$-\frac{1}{3}$	0	$0, \frac{1}{3\delta}, 1 - \frac{1}{3\delta}$	$-\frac{2}{3}\epsilon, -\frac{1}{3}\epsilon, -\frac{1}{3}\left(\epsilon \pm \sqrt{\frac{4}{3\delta} - 3}\right)$
$\pm MS$	0	$\frac{1}{2}$	$1 - \frac{1}{2\delta}, \frac{1}{2\delta}, 0$	$\frac{1}{3}\epsilon, -\frac{1}{2}\epsilon, -\frac{1}{4}\left(\epsilon \pm \sqrt{\left(-7 + \frac{4}{\delta}\right)}\right)$

dimensional space $(x_{\phi}, x_{\varphi}, \Omega)$ whereas the unstable one is 1-dimensional and tangent to the axis y. This means the the critical point $_{+}M$ is unstable to perturbations in y. The critical point $_{+}F$ have a 2-dimensional stable manifold tangent at the point to the plane (x_{ϕ}, x_{φ}) and a 2-dimensional unstable manifold tangent at the critical point to the plane (y, Ω).

The isolated critical points $_{\pm}SF$ and $_{\pm}CS$ denotes cosmological solutions dominated by quintom dark energy and curvature scaling solutions, respectively. These are located in the invariant set of MSF cosmologies without matter ($\Omega = 0$). The critical points $_{\pm}MS$ (belonging to the invariant set of MSF cosmologies with matter ($\Omega > 0$)) represent flat matter scaling solutions.

The stable manifold of $_{+}SF$ in Ψ^{+} is 4-dimensional provided $\delta < 1/3$. In this case $_{+}SF$ is the global attractor on Ψ^{+}. $_{+}SF$ is a saddle with a 3-dimensional stable manifold, if $\frac{1}{3} < \delta < \frac{1}{2}$ or 2-dimensional if $\frac{1}{2} < \delta < 1$.

The isolated critical points $_{\pm}CS$ are non-hyperbolic if $\delta = \frac{1}{3}$. On the other hand, the critical points $_{\pm}MS$ are non-hyperbolic if $\delta = \frac{1}{2}$.

$_{+}CS$ is stable (with a 4-dimensional stable manifold) and then, it is a global attractor provided $\frac{1}{3} < \delta \leq \frac{4}{9}$ (in this case all the eigenvalues are real) or if $\delta > \frac{4}{9}$ (in which case there exists two complex conjugated eigenvalues in such way that the orbits initially at the subspace spanned by the corresponding eigenvectors spiraling toward the critical point).

Let us notice that $_{+}MS$ is the global attractor of the system (it have a 4-dimensional stable manifold) only if $0 < \gamma < \frac{2}{3}$, $\delta > \frac{7}{2}$ (where γ denotes the barotropic index of the perfect fluid). Since we are assuming $\gamma = 1$ (i.e., dust background) then, the critical point $_{+}MS$ is a saddle. It have a 3-dimensional stable manifold if $\frac{1}{2} < \delta \leq \frac{4}{7}$ (in which case all the eigenvalues are real) or if $\delta > \frac{4}{7}$ (in which case there are two complex conjugated eigenvalues and then the orbits initially at the subspace spanned by the corresponding eigenvalues spiral in towards the critical point).

3.2.5. Bifurcations

Observe that the critical points $_{\pm}MS$ and $_{\pm}SF$ are the same as $\delta \to \frac{1}{2}^{+}$. $_{+}SF$ ($_{-}SF$) coincide with a point in the arc $_{+}K_{+}$ ($_{-}K_{-}$) as $\delta \to 1^{-}$. This values of δ where the critical points coincide correspond to bifurcations since the stability changes.

3.2.6. Typical Behavior

Once the attractors have been identified one can give a quantitative description of the physical behaviour of a typical open ($k = -1$) quintom cosmology. For example, for ever expanding cosmologies, near the initial singularity the model behave as de flat FRW with stiff fluid (DE mimics a stiff fluid) represented by a critical point in $_{+}K_{+}$ or in $_{+}K_{-}$, depending on the selection of the free parameters m, n and $x_{\varphi}^{\star}$ (see table 6). Whenever $_{+}CS$ exists (i.e., provided $\delta > \frac{1}{3}$) it is the global attractor of the system. In absence of this type of points, i.e., if $\delta < \frac{1}{3}$, the late time dynamics is determined by the critical point $_{+}SF$, i.e., the universe will be accelerated, almost flat ($\Omega_k \to 0$) and dominated by DE ($\Omega_{de} \to 1$). DE behaves like quintessence ($-1 < q < 0$, i.e., $-1 < w < -\frac{1}{3}$) or a phantom field ($q < -1$, i.e., $w < -1$) if $\delta > 0$ or $\delta < 0$, respectively. This means that, typically, the ever expanding open quintom model crosses the phantom divide (DE EoS parameter have values less than -1). [4] The intermediate dynamics will be governed by the critical points $_{+}CS$, $_{+}MS$, y $_{+}M$, which have the highest lower-dimensional stable manifold.

For contracting models, the typical behavior, is in some way, the reverse of the above. If $\delta < \frac{1}{3}$ the early time dynamics is dominated by $_{-}CS$. Otherwise, if $\delta > \frac{1}{3}$, the past attractor is $_{-}SF$, i.e., the model is accelerating, close to flatness ($\Omega_k \to 0$) and dominated by DE. The intermediate dynamics is dominated at large extent by the critical points $_{-}CS$, $_{-}MS$, y $_{-}F$, which have the highest lower-dimensional stable manifold. A typical model behaves at late times as a flat FRW universe with stiff fluid (i.e., ME mimics a stiff fluid) represented by the invariant sets $_{-}K_{+}$ or $_{-}K_{-}$, depending on the choice of the values of the free parameters m, n y $x_{\varphi}^{\star}$.

3.3. Models with Positive Curvature

In this section we investigate positive curvature models we shall make use of the variables similar but not equal to those defined in [88] section VI.A.

---

[4]For flat models, is well known the, whenever it exists, (i.e., provided $\delta > \frac{1}{2}$) the attractor is $_{+}MS$ (denoted by T in [16]). When we include curvature, the stability of the matter scaling solution is transferred to the curvature scaling solution, as we prove here.

Table 6. Summary of attractors of the system 67. Observe that, whenever exists, the solution dominated by curvature $_-CS$ ($_+CS$) is the past (the future) attractor for $\epsilon = -1$, i. e., for contracting models ($\epsilon = 1$, i.e., for expanding models). We use the notation $\lambda^\pm = nx_\varphi^\star \pm m\sqrt{1 + x_\varphi^{\star\,2}}$.

Restrictions	Past attractor	Future attractor
$\epsilon = -1$	$_-SF$ if $\delta < \frac{1}{3}$ $_-CS$ if $\delta > \frac{1}{3}$	$_-K_\pm$ if $\lambda^\pm > -1$
$\epsilon = 1$	$_+K_\pm$ if $\lambda^\pm < 1$	$_+SF$ if $\delta < \frac{1}{3}$ $_+CS$ if $\delta > \frac{1}{3}$

3.3.1. Normalization, State Space and Dynamical System

Let us introduce the normalization factor

$$\hat{D} = 3\sqrt{H^2 + a^{-2}}. \tag{71}$$

Observe that

$$\hat{D} \to 0 \Leftrightarrow H \to 0,\, a \to +\infty$$

(i.e., at a singularity). This means that it is not possible that $\hat{D}$ vanishes at a finite time.
Let us introduce the following normalized variables $(Q_0, \hat{x}_\phi, \hat{x}_\varphi, \hat{y}, \hat{\Omega})$, given by

$$Q_0 = \frac{3H}{\hat{D}}, \ \hat{x}_\phi = \sqrt{\frac{3}{2}}\frac{\dot{\phi}}{\hat{D}}, \hat{x}_\varphi = \sqrt{\frac{3}{2}}\frac{\dot{\varphi}}{\hat{D}}, \hat{y} = \frac{\sqrt{3V}}{\hat{D}}, \hat{\Omega} = \frac{3\rho_M}{\hat{D}^2}. \tag{72}$$

From the Friedmann equation we find

$$0 \le \hat{x}_\phi^2 - \hat{x}_\phi^2 + \hat{y}^2 = 1 - \hat{\Omega} \le 1 \tag{73}$$

and by definition

$$-1 \le Q_0 \le 1. \tag{74}$$

By the restrictions (73, 74), the state variables are in the state space

$$\hat{\Psi} = \{(Q_0, \hat{x}_\phi, \hat{x}_\varphi, \hat{y}) : 0 \le \hat{x}_\phi^2 - \hat{x}_\phi^2 + \hat{y}^2 \le 1, -1 \le Q_0 \le 1\}. \tag{75}$$

As before, this state space is not compact.
Let us introduce the time coordinate

$$' \equiv \frac{d}{d\hat{\tau}} = \frac{3}{\hat{D}}\frac{d}{dt}.$$

$\hat{D}$ has the evolution equation

$$\hat{D}' = -3Q_0\hat{D}\left(\hat{x}_\phi^2 - \hat{x}_\varphi^2 + \frac{1}{2}\hat{\Omega}\right)$$

where

$$\hat{\Omega} = 1 - \left(\hat{x}_\phi^2 - \hat{x}_\varphi^2 + \hat{y}\right).$$

This equation decouples from the other evolution equations. Thus, a reduced set of evolution equations is obtained.

$$\begin{aligned}
Q_0' &= \left(1 - Q_0^2\right)\left(1 - 3\Xi\right),\\
\hat{x}_\phi' &= 3\,m\,\hat{y}^2 + 3\,Q_0\,\hat{x}_\phi\left(-1 + \Xi\right),\\
\hat{x}_\varphi' &= -3\,n\,\hat{y}^2 + 3\,Q_0\,\hat{x}_\varphi\left(-1 + \Xi\right),\\
\hat{y}' &= -3\,\hat{y}\left(m\,\hat{x}_\phi + n\,\hat{x}_\varphi - Q_0\,\Xi\right).
\end{aligned} \tag{76}$$

Where $\Xi = \hat{x}_\phi^2 - \hat{x}_\varphi^2 + \frac{1}{2}\hat{\Omega}$.

There is also an auxiliary evolution equation

$$\hat{\Omega}' = -Q_0\left(-2\left(\hat{x}_\phi^2 - \hat{x}_\varphi^2\right) + \left(1 - \hat{\Omega}\right)\right)\hat{\Omega}. \tag{77}$$

It is useful to express some cosmological parameters in terms of our state variables. [5]

$$(\Omega_m, \Omega_{de}, \Omega_k, q) = \left(\hat{\Omega}, 1 - \hat{\Omega}, Q_0^2 - 1, -1 + 3\Xi\right)/Q_0^2,$$

and

$$w = \frac{\hat{x}_\phi^2 - \hat{x}_\varphi^2 - \hat{y}^2}{\hat{x}_\phi^2 - \hat{x}_\varphi^2 + \hat{y}^2}.$$

3.3.2.   Invariance under Coordinate Transformations

Observe that the system (76, 77) is invariant under the transformation of coordinates

$$\left(\hat{\tau}, Q_0, \hat{x}_\phi, \hat{x}_\varphi, \hat{y}, \hat{\Omega}\right) \rightarrow \left(-\hat{\tau}, -Q_0, -\hat{x}_\phi, -\hat{x}_\varphi, \hat{y}, \hat{\Omega}\right). \tag{78}$$

Thus, it is sufficient to discuss the behaviour in one part of the phase space, the dynamics in the other part being obtained via the transformation (78). In relation with the possible attractors of the system we will characterize those corresponding to the "positive" branch. The dynamical behavior of the critical points in the "negative" branch is determined by the transformation (78).

---

[5] We have defined $\Omega_k \equiv \frac{k}{a^2H^2} = \frac{1}{a^2H^2}$.

3.3.3. Monotonic Functions

The function

$$N = \frac{(n\,\hat{x}_\phi + m\,\hat{x}_\varphi)^2 \,\hat{\Omega}^2}{\left(1 - Q_0^2\right)^3}, \quad N' = -6\,Q_0\,N \tag{79}$$

is monotonic in the regions $Q_0 < 0$ and $Q_0 > 0$ for $Q_0^2 \neq 1$, $n\,\hat{x}_\phi + n\,\hat{x}_\varphi \neq 0$, $\hat{\Omega} > 0$. Hence, there can be no periodic orbits or recurrent orbits in the interior of the phase space. Furthermore, it is possible to obtain global results. From the expression N we can immediately see that asymptotically $Q_0^2 \to 1$ or $n\hat{x}_\phi + m\hat{x}_\varphi \to 0$ or $\hat{\Omega} \to 0$.

3.3.4. Local Analysis of Critical Points

In the tables 7 and 8 it is summarized the location, existence conditions, some properties of the critical points and the eigenvalues of the linearized system around each critical point.

Table 7. Critical points of the system (76). We use the same notation as in table 4.

Label	Coordinates: $(Q_0, \hat{x}_\phi, \hat{x}_\varphi, \hat{y})$	Existence
$\pm\hat{K}_\pm$	$(\epsilon, \pm\sqrt{1 + x_\varphi^{\star\,2}}, x_\varphi^\star, 0)$	All m and n
$\pm\hat{F}$	$(\epsilon, 0, 0, 0)$	All m and n
$\pm\hat{SF}$	$(\epsilon, m\epsilon, -n\epsilon, \sqrt{1-\delta})$	$\delta < 1$
$\pm\hat{CS}$	$(\sqrt{3\delta}\epsilon, \frac{m\epsilon}{\sqrt{3\delta}}, -\frac{n\epsilon}{\sqrt{3\delta}}, \sqrt{\frac{2}{3}})$	$0 < \delta < \frac{1}{3}$
$\pm\hat{MS}$	$(\epsilon, \frac{m}{2\delta}, -\frac{n}{2\delta}, \frac{\sqrt{1}}{2\sqrt{\delta}}, \sqrt{1 - \frac{1}{2\delta}})$	$\delta > \frac{1}{2}$

In the following we will characterize the dynamical behavior of the cosmological solutions associated with them.

The critical points $\pm\hat{K}, \pm\hat{F}, \pm\hat{SF}$ and $\pm\hat{MS}$ represents flat FRW solutions.

The set of critical points $_+\hat{K}_\pm$ parameterized by the real value $x_\varphi^\star$ represents stiff fluid cosmological solutions (DE mimics a stiff fluid). It is the past attractor for ever expanding models provided $nx_\varphi^\star \pm m\sqrt{1 + x_\varphi^{\star\,2}} < 1$. As we proceed before, a simple application of the symmetry (78), allows to the identification of the future attractor for collapsing models: the typical orbits tends asymptotically to $_-\hat{K}_\pm$ as $\hat{\tau} \to \infty$ provided $nx_\varphi^\star \pm m\sqrt{1 + x_\varphi^{\star\,2}} > -1$, and $-1 \leq Q_0 < 0$. This fact has interesting consequences. If $x_\varphi^\star$ is a fixed value and n and m are such that $-1 < nx_\varphi^\star + m\sqrt{1 + x_\varphi^{\star\,2}} < 1$, then, there exists one orbit of the type $_+\hat{K}_+ \to {}_-\hat{K}_+$. If n and m are such that $-1 < nx_\varphi^\star - m\sqrt{1 + x_\varphi^{\star\,2}} < 1$, then, there is one orbit of the type $_+\hat{K}_- \to {}_-\hat{K}_-$. These are solutions starting from and recollapsing to a singularity given by a MSF cosmology (see figure 6 b).

Table 8. DE EoS parameter (w), deceleration parameter (q), fractional energy densities, and eigenvalues of the perturbation matrix associated to the critical points of the system (76). We use the notation $\lambda^{\pm} = nx_{\varphi}^{\star} \pm m\sqrt{1 + x_{\varphi}^{\star\,2}}$. When the flow is restricted to the invariant sets $Q_0 = \pm 1$, the eigenvalues associated to the critical points $_{\pm}\hat{F}$, $_{\pm}\hat{SF}$ and $_{\pm}\hat{MS}$ and to the critical sets $_{\pm}\hat{K}_{\pm}$, are, in each case, the same as those displayed, but the first from the left.

Label	w	q	$\Omega_m, \Omega_{de}, \Omega_k$	Eigenvalues
$_{\pm}\hat{K}_{\pm}$	1	2	$0, 1, 0$	$4\epsilon, 0, 3\left(\epsilon - \lambda^{\pm}\right), 3\epsilon$
$_{\pm}\hat{F}$	-	$\frac{1}{2}$	$1, 0, 0$	$\epsilon, \frac{3}{2}\epsilon, -\frac{3}{2}\epsilon, -\frac{3}{2}\epsilon$
$_{\pm}\hat{SF}$	$-1 + 2\delta$	$-1 + 3\delta$	$0, 1, 0$	$2\left(3\delta - 1\right)\epsilon, 3\left(\delta - 1\right)\epsilon, 3\left(\delta - 1\right)\epsilon, 3\epsilon$
$_{\pm}\hat{CS}$	$-\frac{1}{3}$	0	$0, \frac{1}{3\delta}, 1 - \frac{1}{3\delta}$	$-2\sqrt{3\delta}\epsilon, -\sqrt{3\delta}\epsilon \pm \sqrt{4 - 9\delta}, -\sqrt{3\delta}\epsilon$
$_{\pm}\hat{MS}$	0	$\frac{1}{2}$	$1 - \frac{1}{2\delta}, \frac{1}{2\delta}, 0$	$\epsilon, -\frac{3}{2}\epsilon, -\frac{3}{4}\left(\epsilon \pm \sqrt{\left(-7 + \frac{4}{\delta}\right)}\right)\epsilon$

The critical points $_{\pm}\hat{F}$ represent flat FRW solutions. They are hyperbolic. For these points the scalar fields vanish, so the cosmological parameters associated to DE are not applicable to these points. If $\delta > \frac{2}{3}$, the unstable (stable) manifold of $_{+}\hat{F}$ ($_{-}\hat{F}$) is tangent to the critical point and parallel to the plane $\hat{y} - Q_0$. This means that there is an orbit connecting $_{+}\hat{F}$ and $_{-}\hat{F}$ pointing towards $_{-}\hat{F}$ in the direction of the Q_0-axis. It represents the closed FRW solution with no scalar field starting from a big-bang at $_{+}\hat{F}$ and recollapsing to a "big-crunch" at $_{-}\hat{F}$ (see figure 6 a).

The critical point $_{+}\hat{SF}$ represents a solution dominated by the scalar field (with non-vanishing potential). It can be the global attractor in the sets $0 < Q_0 < 1$ or $Q_0 = 1$ (i.e., for ever expanding models, or flat models) for the values of the parameters displayed in table 9. It can be a phantom dominated solution provided $\delta < 0$. It also can represent quintessence dominated or de Sitter solutions.

The critical point $_{+}\hat{MS}$ exist if $\delta > \frac{1}{2}$. They represent flat matter scaling solutions, for which both the fluid and quintom are dynamically important. It is a saddle point.

For $0 < \delta < \frac{1}{3}$ there exists the critical points $_{\pm}\hat{CS}$ for which the matter is unimportant, but curvature is non-vanishing ($Q_0^2 \neq 1$) and tracks the scalar field. These are called curvature scaling solutions. The values of its cosmological parameters are the same as for $_{\pm}CS$ (displayed in table 5), but it represents a different cosmological solution with positive curvature. These critical points are typically saddle points.

In table 9, where we present a summary of attractors for the quintom model with $k = 1$.

Table 9. Summary of attractors for the quintom model with $k = 1$ (system (76)). We use the notation $\lambda^{\pm} = nx^{\star}_{\varphi} \pm m\sqrt{1 + x^{\star\,2}_{\varphi}}$.

Restrictions	Past attractor	Future attractor
$Q_0 = -1$	$_{-}\hat{SF}$ if $\delta < \frac{1}{2}$ $_{-}\hat{MS}$ if $\delta > \frac{1}{2}$	$_{-}\hat{K}_{\pm}$ if $\lambda^{\pm} > -1$
$-1 < Q_0 < 0$	$_{-}\hat{SF}$ if $\delta < \frac{1}{3}$	as above
$0 < Q_0 < 1$	$_{+}\hat{K}_{\pm}$ if $\lambda^{\pm} < 1$	$_{+}\hat{SF}$ if $\delta < \frac{1}{3}$
$Q_0 = 1$	as above	$_{+}\hat{SF}$ if $\delta < \frac{1}{2}$ $_{+}\hat{MS}$ if $\delta > \frac{1}{2}$

3.3.5. Bifurcations

Observe that the critical points $_{\pm}\hat{MS}$ y $_{\pm}\hat{SF}$ coincides as $\delta \to \frac{1}{2}^{+}$. $_{\pm}\hat{CS}$ y $_{\pm}\hat{SF}$ coincides as $\delta \to \frac{1}{3}^{-}$. Additionally, $_{+}\hat{SF}$ ($_{-}\hat{SF}$) coincides with a point at the arc $_{+}\hat{K}_{+}$ ($_{-}\hat{K}_{-}$) as $\delta \to 1^{-}$. For this values of δ a bifurcation occurs.

3.3.6. Typical Behaviour

Once the attractors have been identified one can give a quantitative description of the physical behaviour of a typical closed quintom cosmology. For example, for ever expanding cosmologies, near the big-bang a typical model behaves like a flat FRW model with stiff fluid represented by the critical set $_{+}\hat{K}_{+}$ or by $_{+}\hat{K}_{-}$, depending on the choice of the values of the free parameters m, n and $x^{\star}_{\varphi}$. If $\delta < \frac{1}{3}$ and $0 < Q_0 < 1$ the late time dynamics is determined by $_{+}\hat{SF}$, (with the same physical properties as $_{+}SF$). The intermediate dynamics will be governed to a large extent by the fixed points $_{+}\hat{CS}$, $_{+}\hat{MS}$, and $_{+}\hat{F}$, which have the highest lower-dimensional stable manifold. For flat models (i.e., in the invariant set $Q_0 = 1$), the late time dynamics is determined by the critical point $_{+}\hat{SF}$ provided $\delta < \frac{1}{2}$ or $_{+}\hat{MS}$ provided $\delta > \frac{1}{2}$.

For contracting models, the typical behavior is, in one sense, the reverse of the above. If $\delta < \frac{1}{3}$ and $-1 < Q_0 < 0$ the early time dynamics is determined by $_{-}\hat{SF}$. The intermediate dynamics will be governed to a large extent by the fixed points $_{-}\hat{CS}$, $_{-}\hat{MS}$, and $_{-}\hat{F}$, which have the highest lower-dimensional stable manifold. For flat models (i.e., in the invariant set $Q_0 = -1$), the early time dynamics is determined by the critical point $_{-}\hat{SF}$ ($_{-}\hat{MF}$)

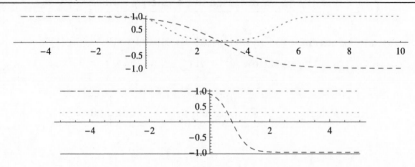

Figure 6. The collapse of quintom cosmologies with positive curvature for the values for the parameters $m = 0.7$, $n = 0.3$ and $\gamma = 1$. In (a) we have selected the initial conditions $Q_0(0) = \hat{\Omega}(0) = 0.9$, and $\hat{x}_\phi(0) = \hat{x}_\varphi(0) = \hat{y}(0) = 0$. The dashed line represents the evolution of Q_0 vs τ (observe that Q_0 evolves from 1 to -1, and eventually takes zero value). The dotted line represents $\hat{\Omega}$ vs τ. This illustrates the existence of a closed FRW solution with no scalar field starting from a big-bang at $_+\hat{F}$ and recollapsing to a "big-crunch" at $_-\hat{F}$. In (b) we have selected the initial conditions $Q_0(0) = 0.9$, $\hat{\Omega}(0) = 0$ and $\hat{x}_\phi(0) = -\sqrt{1 + \hat{x}_\varphi(0)^2}$ with $\hat{x}_\varphi(0) = 0.3$. The dashed line denotes Q_0 vs τ. Observe that Q_0 goes from the value 1 to -1 (i.e., the model collapses). The dot-dashed line denotes the evolution of $\hat{x}_\phi^2 - \hat{x}_\varphi^2$ vs τ (which is identically equal to 1). The dotted line denotes the value of $\hat{x}_\varphi$ vs τ and the straight line denotes the value of $\hat{x}_\phi$ vs τ. This illustrates the existence of orbits of the type $_+\hat{K}_- \rightarrow \ _-\hat{K}_-$. By choosing $\hat{x}_\phi(0) = \sqrt{1 + \hat{x}_\varphi(0)^2}$, with the same initial conditions for the other variables, we obtain orbits of the type $_+\hat{K}_+ \rightarrow \ _-\hat{K}_+$. These are solutions starting from and recollapsing to a singularity (given by a massless scalar field cosmology).

provided $\delta < \frac{1}{2}$ ($\delta > \frac{1}{2}$). A typical model behaves at late times like a flat FRW model with stiff fluid (i.e. the DE mimics a stiff fluid) represented by the critical set $_-\hat{K}_+$ or by $_-\hat{K}_-$ depending on the choice of the values of the free parameters m, n and $x_\varphi^\star$.

4.   Observational Evidence for Quinstant Dark Energy Paradigm

4.1.   The Model

Looking at the impressive amount of papers addressing the problem of cosmic acceleration clearly shows that two leading candidates to the dark energy throne are the old cosmological constant Λ and a scalar field ϕ evolving under the influence of its self-interaction potential $V(\phi)$.

In the usual approach, one adds either a scalar field or a cosmological constant term to the field equations. However, since what we see is only the final effect of the dark energy components, in principle nothing prevents us to add more than one single component provided that the effective dark energy fluid coming out is able to explain the data at hand. Moreover, as we have hinted upon above, a single scalar field, while explaining cosmic speed up, leads to a problematic eternal acceleration. A possible way out of this problem has been proposed by some of us [38, 39, 92] through the introduction of a negative cosmological term.

Motivated by those encouraging results, we therefore consider a spatially flat universe filled by dust matter, radiation, scalar field and a (negative) cosmological constant term. The Einstein equations thus read:

$$H^2 = \frac{1}{3}\left[\rho_M + \rho_r + \rho_\Lambda + \frac{1}{2}\dot{\phi}^2 + V(\phi)\right] , \qquad (80)$$

$$2\dot{H} + 3H^2 = -\left[\frac{1}{3}\rho_r - \rho_\Lambda + \frac{1}{2}\dot{\phi}^2 - V(\phi)\right] , \qquad (81)$$

where we have used natural units with $8\pi G = c = 1$.

4.2. Matching with the Data

Notwithstanding how well motivated it is, a whatever model must be able to reproduce what is observed. This is particularly true for the model we are considering because the presence of a negative cosmological constant introduces a positive pressure term potentially inhibiting the cosmic speed up. Moreover, contrasting the model against the data offers also the possibility to constrain its characteristic parameters and estimate other derived interesting quantities, such as q_0, the transition redshift z_T and the age of the universe t_0. Motivated by these considerations, we will therefore fit our model to the dataset described below parametrizing the model itself with the matter density Ω_M, the scalar field quantities (Ω_ϕ, w_0) and the dimensionless Hubble constant h (i.e., H_0 in units of 100 km/s/Mpc), while we will set the radiation density parameter as $\Omega_r = 10^{-4.3}$ as in [93] from a median of different values reported in literature.

4.2.1. The Method and the Data

In order to constrain the model parameters we will consider several observational test: (a) the distance modulus $\mu = m - M$, i.e. the difference between the apparent and absolute magnitude of an object at redshift z, (b) the gas mass fraction in galaxy clusters, (c) the measurement of the baryonic acoustic oscillation (BAO) peak in the large scale correlation function at $100\ h^{-1}$ Mpc separation detected by Eisenstein et al. [94] using a sample of 46748 luminous red galaxies (LRG) selected from the SDSS Main Sample [95], (d) the shift parameter [96] [6] and we maximize the following likelihood taking into account the above test:

$$\mathcal{L} \propto \exp\left[-\frac{\chi^2(\mathbf{p})}{2}\right] \qquad (82)$$

where $\mathbf{p} = (\Omega_M, \Omega_\phi, w_0, h)$ denotes the set of model parameters and the pseudo-χ^2 merit function reads:

---

[6] A complete discussion about this observational test and the quinstant model can be found in [39].

$$\chi^2(\mathbf{p}) = \sum_{i=1}^{N} \left[\frac{\mu^{th}(z_i, \mathbf{p}) - \mu_i^{obs}}{\sigma_i} \right]^2 + \sum_{i=1}^{N} \left[\frac{f_{gas}^{th}(z_i, \mathbf{p}) - f_{gas,i}^{obs}}{\sigma_i} \right]^2$$
$$+ \left[\frac{\mathcal{A}(\mathbf{p}) - 0.474}{0.017} \right]^2 + \left[\frac{\mathcal{R}(\mathbf{p}) - 1.70}{0.03} \right]^2 + \left(\frac{h - 0.72}{0.08} \right)^2. \tag{83}$$

4.2.2. Results

Table 1 shows the best fit model parameters, median values and 1 and 2σ ranges for the parameters $(\Omega_M, \Omega_\Lambda, w_0, h, \Omega_\phi)$.

Table 10. Best fit (bf) and median (med) values and 1σ and 2σ ranges of the parameters $(\Omega_M, \Omega_\Lambda, w_0, h, \Omega_\phi)$) as obtained from the likelihood analysis (from [39]).

Par	bf	med	1σ	2σ
Ω_M	0.283	0.307	$(0.272, 0.352)$	$(0.246, 0.410)$
Ω_Λ	-0.072	-0.298	$(-0.54, -0.11)$	$(-0.92, -0.02)$
w_0	-0.72	-0.67	$(-0.74, -0.60)$	$(-0.79, -0.53)$
h	0.632	0.620	$(0.588, 0.654)$	$(0.554, 0.692)$
Ω_ϕ	0.789	0.989	$(0.799, 1.226)$	$(0.700, 1.574)$

Figs.7 shows how well our best fit model reproduce the data on the SNeIa Hubble diagram and gas mass fraction. The best fit model is in quite good agreement with both the SNeIa and gas data. Actually, the χ^2 values are respectively 206 and 48 to be contrasted with the number of datapoints, being 192 and 42 respectively. Besides the predicted values for the acoustic peak and shift parameters are in satisfactory agreement with the observed ones:

$$\mathcal{A} = 0.45 \ , \ \mathcal{R} = 1.67 \ . \tag{84}$$

Because of these results, we can therefore conclude that including a negative Λ leads to a model still in agreement with the data so that this approach to halting eternal acceleration is a viable one from an observational point of view. [7]

Another interesting tools to study the viability of a dark energy model is the point of view of structure formation. This kind of analysis can break beetwen models with similar prediction from the cosmic expansion history, in this sense the growth of the large scale structure in the universe provide and important companion test. Following this line in [97] the authors showed that the quinstant model makes reasonable predictions for the formation of linear large scale structure of the Universe but it fails in the non linear regime because of the density contrast at virialisation increase with the value of virialisation redshif.

Concerning the predictions of the cluster abundances, the quinstant model is capable of reproducing the results of the other models in a satisfactory way backwards in time up to redshifts a bit larger than $z = 1$ for the three range of mass values [8]. Then, it shows

---

[7] See [39] for a further discussion about the observational results and implications.
[8] The same behavior is obtained for other mass range [97].

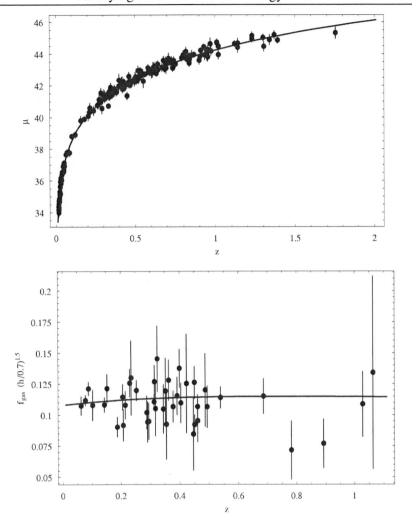

Figure 7. (a) Best fit curve superimposed to the data on the SNeIa Hubble diagram. (b) Best fit curve superimposed to the data on the gas mass fraction. Note that the theoretical curve plots indeed $f_{gas}(z) \times (h/0.7)^{1.5}$ with h set to its best fit value (from [39]).

abrupt peaks of structure formation, in a serious departure of the hierarchical model for large scale structure. This seems to be caused by the unusual equation of state of quinstant dark energy, which behaves as stiff matter for redshifts a bit larger than one. This would result in enhanced accretion of the forming structures, both because of gravitational and viscous forces.

5. Exponential Quinstant: Phase Space Analysis

In this section we will investigate, from the dynamical systems viewpoint the quinstant dark energy model with exponential potential $V(\phi) = V_0 e^{\lambda \phi}$. We do not consider radiation fluid here but a background of a perfect fluid with equation of state $w = \gamma - 1$. The cosmological equations reads:

$$H^2 - \tfrac{1}{6}\dot{\phi}^2 - \tfrac{1}{3}V(\phi) - \tfrac{1}{3}\rho_{\mathrm{M}} - \tfrac{\Lambda}{3} = -\tfrac{k}{a^2}, \quad k = -1, 0, 1, \tag{85}$$

$$\dot{H} = -H^2 - \tfrac{1}{3}\dot{\phi}^2 + \tfrac{1}{3}V(\phi) + \tfrac{\Lambda}{3} - \tfrac{1}{6}\left(3\gamma - 2\right)\rho_{\mathrm{M}}, \tag{86}$$

$$\dot{\rho}_{\mathrm{M}} = -3\gamma H \rho_{\mathrm{M}}, \tag{87}$$

$$\ddot{\phi} + 3H\dot{\phi} + \tfrac{dV(\phi)}{d\phi} = 0. \tag{88}$$

Our purpose is to re-express the former equations as an autonomous dynamical system.

5.1. Flat FRW Case

5.1.1. Normalization, State Space, and Dynamical System

In order to get a first order autonomous system of ordinary differential equations (ODEs) is is convenient introduce normalized variables an a new convenient (monotonic) time variable. If the phase space is compact the flow of the system admits both past and future attractors. Let us introduce the normalization factor $D = \sqrt{H^2 - \Lambda/3} > 0$, the new time variable $d\tau = D dt$, and the phase space variables:

$$x = \frac{\dot{\phi}}{\sqrt{6}D}, \quad y = \frac{\sqrt{V(\phi)}}{\sqrt{3}D}, \quad \mathcal{H} = \frac{H}{D}. \tag{89}$$

The former variables lies in the compact phase space

$$\left\{(x, y, \mathcal{H}) : x^2 + y^2 \le 1, \ y \ge 0, \ -1 \le \mathcal{H} \le 1\right\}.$$

The variables (89) satisfy the ODEs (the prime denoting derivative with respect to τ):

$$x' = -\frac{3}{2}x\mathcal{H}\left((\gamma - 2)x^2 + \left(y^2 - 1\right)\gamma + 2\right) - \sqrt{\frac{3}{2}}\lambda y^2, \tag{90}$$

$$y' = \frac{3}{2}y\left(\frac{\sqrt{6}\lambda x}{3} - \mathcal{H}\left((\gamma - 2)x^2 + \left(y^2 - 1\right)\gamma\right)\right), \tag{91}$$

$$\mathcal{H}' = -\frac{3}{2}\left(\mathcal{H}^2 - 1\right)\left((\gamma - 2)x^2 + \left(y^2 - 1\right)\gamma\right) \tag{92}$$

For convenience, let us express some cosmological quantities in terms of the variables (89). The deceleration parameter is explicitly

$$q \equiv -\ddot{a}a/\dot{a}^2 = -1 + \frac{3}{2}\left[\frac{x^2\left(2 - \gamma\right) + \left(1 - y^2\right)\gamma}{\mathcal{H}^2}\right]; \tag{93}$$

the fractional energy density of the scalar field is

$$\Omega_\phi = \frac{x^2 + y^2}{\mathcal{H}^2}; \tag{94}$$

and the 'effective' EoS parameter is given by

$$\omega_{eff} \equiv \frac{P_{tot}}{\rho_{tot}} \equiv \frac{\tfrac{1}{2}\dot{\phi}^2 - V(\phi) + (\gamma - 1)\rho_{\mathrm{M}} - \Lambda}{\tfrac{1}{2}\dot{\phi}^2 + V(\phi) + \rho_{\mathrm{M}} + \Lambda} = -1 + \frac{(2 - \gamma)x^2 + \left(1 - y^2\right)\gamma}{\mathcal{H}^2}. \tag{95}$$

5.1.2. Form Invariance under Coordinate Transformations

The system (90-92) is form invariant under the coordinate transformation and time reversal

$$(\tau, x, y, \mathcal{H}) \to (-\tau, -x, y, -\mathcal{H}). \tag{96}$$

Thus, it is sufficient to discuss the behaviour in one part of the phase space, the dynamics in the other part being obtained via the transformation (96). Observe also, that equations (90-92) are form invariant under the coordinate transformation $y \to -y$. Then, (90-92) is form invariant under its composition with (96).

From equation (92) follows that $\mathcal{H} = \pm 1$ are invariant sets of the flow. From equation (91) follows that the sign of y is invariant.

5.1.3. Monotonic Functions

Let be defined

$$Z(x, y, \mathcal{H}) = \left(\frac{\mathcal{H}+1}{\mathcal{H}-1}\right)^2 \tag{97}$$

in the invariant set

$$S = \left\{(x, y, \mathcal{H}) : x^2 + y^2 < 1, \ y > 0, \ -1 < \mathcal{H} < 1\right\}.$$

Then, Z is monotonic decreasing in S since

$$Z' \equiv \nabla Z \cdot f = -6 Z \left(x^2 \left(2 - \gamma\right) + \left(1 - y^2\right) \gamma\right) < 0$$

in S. The existence of this monotonic allows to state that there can be no periodic orbits or recurrent orbits in the interior of the phase space. Furthermore, it is possible to obtain global results. The range of Z is the semi-interval $(0, +\infty)$, and $Z \to 0$ as $\mathcal{H} \to -1$ (since $\mathcal{H}$ is bounded) and $Z \to +\infty$ as $\mathcal{H} \to 1$. By applying the Monotonicity Principle (theorem 4.12 [91]) we find that, for all $p \in S$, the past asymptotic attractor of p (the α-limit) belongs to $\mathcal{H} = 1$ and the future asymptotic attractor of p (the ω-limit) belongs to $\mathcal{H} = -1$.

5.1.4. Local Analysis of Critical Points

The system (90-91) admits ten critical points with the labels $P_i^{\pm}$ with $i = 1 \ldots 5$. In table 11 we offer some partial information about the location, conditions for existence and some additional properties of them. All the critical points satisfy $\mathcal{H} = \pm 1$. In other words, they are solutions with $H = \pm D$ (i.e. with $H \to \pm\infty$). If sign$(\mathcal{H}) = -1$ the associated solutions ends in a collapse (since $H < 0$), whereas, if sign$(\mathcal{H}) = 1$ we have ever expanding cosmological solutions. The expected cosmological behavior of our model is that the attractor solutions represent collapsing solutions due the negative value of the cosmological constant.

Now, let us make some comments about the cosmological solutions associated to these critical points.

The critical points $P_1^{\pm}$ and $P_3^{\pm}$ represent stiff-matter solutions which are associated with massless scalar field cosmologies (the kinetic energy density of the scalar field dominated

Table 11. Location and existence conditions of the critical points of the system (90-92)

Label	Coordinates: $(x, y, \mathcal{H})$	Existence
$P_1^\pm$	$(-1, 0, \pm 1)$	All λ
$P_2^\pm$	$(0, 0, \pm 1)$	All λ
$P_3^\pm$	$(1, 0, \pm 1)$	All λ
$P_4^\pm$	$\left(\mp \frac{\lambda}{\sqrt{6}}, \sqrt{1 - \frac{\lambda^2}{6}}, \pm 1 \right)$	$-\sqrt{6} < \lambda < \sqrt{6}$
$P_5^\pm$	$\left(\mp \sqrt{\frac{3}{2}} \frac{\gamma}{\lambda}, \sqrt{\frac{3}{2}} \sqrt{\frac{(2-\gamma)\gamma}{\lambda^2}}, \pm 1 \right)$	$\gamma = 0, \ \lambda \neq 0$ $0 < \gamma \leq 2, \ \|\lambda\| \geq \sqrt{3\gamma}$

against the potential energy density). In the former case the scalar field is a monotonic decreasing function of t (since its time-derivative is negative). In the last case the scalar field is an increasing function of t since its time-derivative is positive. These solutions are always decelerated. The critical points $P_2^\pm$ represent a flat FRW solution fuelled by perfect fluid. They represent accelerating solutions for $\gamma < \frac{2}{3}$. The critical points $P_4^\pm$ represent solutions dominated by the scalar field ($\Omega_\phi = 1$, and $H \to \pm \infty$) which are accelerating if $\lambda^2 < 2$. Our models does not devoid of scaling phases: the critical points $P_5^\pm$ are such that neither the scalar field nor the perfect fluid dominates the evolution. There $\Omega_m / \Omega_\phi = $ const., and $\gamma_\phi = \gamma$.

Before proceed to make some numerical experiments let us discuss some aspects concerning the symmetry (96). Observe that the critical points $P_3^\mp$, $P_2^\mp$, $P_4^\mp$ and $P_5^\mp$ are related by the transformation (96) with $P_1^\pm$, $P_2^\pm$, $P_4^\pm$ and $P_5^\pm$ respectively. In order to analyze the local stability of $P_1^+, P_2^+, P_3^+, P_4^+, P_5^+$ it is sufficient analyse the local stability of $P_3^-, P_2^-, P_1^-, P_4^-, P_5^-$ respectively, and then, infer the stability of the points in the "positive" branch by using (96). In table 12 we offer partial information about the dynamical character of the critical points corresponding to the "negative" branch.

The critical point P_1^- is nonhyperbolic provided $\gamma = 2$ or $\lambda = -\sqrt{6}$. It is a stable node (future attractor) provided $\lambda > -\sqrt{6}$ and $\gamma \neq 2$. It is a saddle otherwise with a 2D stable manifold and a 1D unstable manifold tangent to the y-axis. P_2^- is nonhyperbolic provided $\gamma = 0$ or $\gamma = 2$. It is a saddle point otherwise with a 2D stable manifold and a 1D unstable manifold tangent to the x-axis. The critical point P_3^- is nonhyperbolic provided $\gamma = 2$ or $\lambda = \sqrt{6}$. It is a stable node (future attractor) provided $\lambda < \sqrt{6}$ and $\gamma \neq 2$. It is a saddle otherwise with a 2D stable manifold and a 1D unstable manifold tangent to the y-axis. P_4^- is nonhyperbolic if $\lambda^2 \in \{0, 3\gamma, 6\}$. Saddle otherwise, with a 2D stable manifold provided $\lambda^2 > 3\gamma$ or 1D if $\lambda^2 < 3\gamma$. The critical point P_5^- is nonhyperbolic if $\gamma = 0$ or $\lambda^2 = 3\gamma$. It is a saddle point, otherwise, with a 2D unstable manifold provided $0 < \gamma < 2$ and $\lambda^2 > 3\gamma$.

Table 12. Eigenvalues, and dynamical character of the fixed points of (90-92). We use the notation $\Delta = (2 - \gamma)(24\gamma^2 + \lambda^2(2 - 9\gamma))$.

Label	Eigenvalues	Dynamical character
P_1^-	$-6, -3 - \sqrt{\frac{3}{2}}\lambda, -3(2 - \gamma)$	nonhyperbolic if $\gamma = 2$ or $\lambda = -\sqrt{6}$; stable (node) if $\lambda > -\sqrt{6}$ and $\gamma \neq 2$; saddle, otherwise.
P_2^-	$\frac{3}{2}(2 - \gamma), -3\gamma, -\frac{3\gamma}{2}$	nonhyperbolic if $\gamma = 0$ or $\gamma = 2$; saddle, otherwise.
P_3^-	$-6, -3 + \sqrt{\frac{3}{2}}\lambda, -3(2 - \gamma)$	nonhyperbolic if $\gamma = 2$ or $\lambda = \sqrt{6}$; stable (node) if $\lambda < \sqrt{6}$ and $\gamma \neq 2$; saddle, otherwise.
P_4^-	$-\lambda^2, \frac{1}{2}(6 - \lambda^2), -\lambda^2 + 3\gamma$	nonhyperbolic if $\lambda = 0$ or $\lambda^2 = 3\gamma$; saddle, otherwise.
P_5^-	$-3\gamma, \frac{3}{4}\left(2 - \gamma \pm \frac{1}{\lambda}\sqrt{\Delta}\right)$	nonhyperbolic if $\gamma = 0$ or $\lambda^2 = 3\gamma$; saddle, otherwise.

P_4^- (P_5^-) is the past attractor in the invariant set $\mathcal{H} = -1$ provided $0 < \gamma < 2$, $\lambda^2 < 3\gamma$ ($0 < \gamma < 2$, $\lambda^2 > 3\gamma$).

In table 14, where we present a summary of attractors (both past and future) for the quinstant model with $k = 0$.

In figure 8 we show some orbits in the phase space for the values $\lambda = -\sqrt{\frac{3}{2}}$ and $\gamma = 1$. For this choice $\lambda^2 < 3\gamma$ and $-\sqrt{6} < \lambda < \sqrt{6}$. Thus the critical points $P_5^{\pm}$ do not exist. By the linear analysis (see table 12) we find that the critical points P_1^- and P_3^- have a 3-dimensional stable manifold, P_3^- having a stronger attracting manifold tangent to the y-axis (see figure 8), i.e., two global future attractors might coexist (bistability). The critical points P_1^+ and P_3^+ are local sources in the invariant set $\mathcal{H} = 1$. Numerical inspection suggest and analytical results confirm that they are also global sources, P_1^+ having a stronger unstable direction tangent to y-axis. The critical points $P_2^{\pm}$ acts locally as saddles. For P_2^- (resp. P_2^+) the stable (resp. unstable) manifold is 2-dimensional and tangent to the y-$\mathcal{H}$ plane. There are orbits (corresponding to exact cosmological solutions) connecting $P_{1,2,3}^+$ with $P_{1,2,3}^-$ (recollapse occurs). The critical point P_4^+, with coordinates $(1/2, \sqrt{3}/2, 1)$, have eigenvalues $-2.25, -1.5, 1.5$ acting as a local attractor in the invariant set $\mathcal{H} = 1$ and P_4^-, with coordinates $(-1/2, \sqrt{3}/2, 1)$, and eigenvalues $2.25, 1.5, -1.5$ is the local source for the invariant set $\mathcal{H} = -1$. They are saddle points for the 3D dynamics (see figure 8).

Table 13. Some properties of the critical points of the system (90-92)

Label	Deceleration q	Ω_ϕ	ω_{eff}
$P_1^\pm$	2	1	1
$P_2^\pm$	$-1+\frac{3\gamma}{2}$	0	$-1+\gamma$
$P_3^\pm$	2	1	1
$P_4^\pm$	$-1+\frac{\lambda^2}{2}$	1	$-1+\frac{\lambda^2}{3}$
$P_5^\pm$	$-1+\frac{3\gamma}{2}$	$\frac{3\gamma}{\lambda^2}$	$-1+\gamma$

Table 14. Summary of attractors for for the quinstant model with $k = 0$ (system (90-92)).

Restrictions	Past attractor		Future attractor	
$\mathcal{H} = -1$	P_4^-	if $0 < \gamma < 2$, $\lambda^2 < 3\gamma$	P_3^-	if $\lambda < \sqrt{6}$, $\gamma \neq 2$
	P_5^-	$0 < \gamma < 2$, $\lambda^2 > 3\gamma$	P_1^-	if $\lambda > -\sqrt{6}$, $\gamma \neq 2$
$-1 < \mathcal{H} < 1$	P_3^+	if $\lambda > -\sqrt{6}$, $\gamma \neq 2$	as above	
	P_1^+	if $\lambda < \sqrt{6}$, $\gamma \neq 2$		
$\mathcal{H} = 1$	as above		P_4^+	if $0 < \gamma < 2$, $\lambda^2 < 3\gamma$
			P_5^+	$0 < \gamma < 2$, $\lambda^2 > 3\gamma$

5.1.5. Bifurcations

The critical points $\left(P_4^+, P_4^-\right)$ reduce to $\left(P_1^+, P_3^-\right)$ as $\lambda \to (\sqrt{6})^-$. The critical points $\left(P_4^+, P_4^-\right)$ reduce to $\left(P_3^+, P_1^-\right)$ as $\lambda \to (-\sqrt{6})^+$. The critical points $P_5^\pm$ reduce to $P_2^\pm$ as $\gamma \to 0^+$. On the other hand, $P_5^\pm$ reduce to $P_4^\pm$ as $\lambda \to (\sqrt{3\gamma})^+$ or $\lambda \to (-\sqrt{3\gamma})^-$. For these values of the parameters a bifurcation arises.

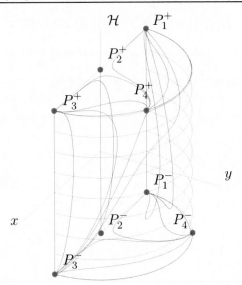

Figure 8. Some orbits of (90-91) in the phase space for the values $\lambda = -\sqrt{\frac{3}{2}}$ and $\gamma = 1$. The critical points P_1^- (resp. P_1^+) and P_3^- (resp. P_3^+) are the future (resp. past) asymptotic attractors, P_3^- (resp. P_1^+) having a stronger attracting (resp. unstable) manifold tangent to the y-axis. P_4^+, acts as a local attractor in the invariant set $\mathcal{H} = 1$ and P_4^-, acts as the local source for the invariant set $\mathcal{H} = -1$ (they are, however, saddle points for the 3D dynamics).

5.1.6.   Typical Behavior

Once the attractors have been identified one can give a quantitative description of the physical behavior of a typical flat quinstant cosmology.

For example, for ever expanding cosmologies with $H > 0, H \to +\infty, \mathcal{H} = 1$, i.e., the standard expanding cosmology near the big-bang, a typical model behaves like a massless scalar field (kinetic dominated energy density) represented by P_3^+ or P_1^+ provided $\lambda > -\sqrt{6}, \gamma \neq 2$ or $\lambda < \sqrt{6}, \gamma \neq 2$, respectively. This types of solutions might coexist in the same phase space. The late time dynamics in $\mathcal{H} = 1$ is given by either a scalar field dominated solution ($\Omega_\phi \to 1$) represented by P_4^+ or by a scaling solution ($\Omega_m/\Omega_\phi = O(1)$) represented by P_5^+ provided $\lambda^2 < 3\gamma$ or $\lambda^2 > 3\gamma$, respectively. For finite values of H, i.e., $-1 < \mathcal{H} < 1$, the early time dynamics is the same as in the previous case but there are subtle differences with respect the late time dynamics. In fact, in the invariant set $-1 < \mathcal{H} < 1$ the future attractors are P_3^- or P_1^- depending if $\lambda < \sqrt{6}, \gamma \neq 2$ or $\lambda > -\sqrt{6}, \gamma \neq 2$. If $|\lambda| < \sqrt{6}$ the system is bistable. Such solutions represent contracting stiff-fluid cosmologies. This means that a typical quinstant cosmologies allows the collapse of matter when the time evolves. For contracting cosmologies with ($H < 0, H \to \infty, \mathcal{H} = -1$), i.e., the standard contracting model near the initial singularity, the late time dynamics is the same as int the previously described case, i.e., the collapse. However, there are subtle differences concerning the early time dynamics. The late time dynamics in $\mathcal{H} = -1$ is given by either a scalar field dominated solution ($\Omega_\phi \to 1$) represented by P_4^- or by a scaling solution ($\Omega_m/\Omega_\phi = O(1)$) represented by P_5^- provided $\lambda^2 < 3\gamma$ or $\lambda^2 > 3\gamma$.

5.2.   Quinstant Cosmology with Negative Curvature

5.2.1.   Normalization, State Space, and Dynamical System

Let us consider the same normalization as in section 5.1.1., i.e, the normalization factor $D = \sqrt{H^2 - \Lambda/3} > 0$ and the time variable $d\tau = D dt$. We will consider the variables (89) augmented by the new variable $z = \frac{1}{aD}$. These variables lies in the compact phase space

$$\left\{ (x, y, z, \mathcal{H}) : x^2 + y^2 + z^2 \leq 1, \ y \geq 0, \ z \geq 0, \ -1 \leq \mathcal{H} \leq 1 \right\}.$$

The evolution equations for the variables x, y, z, and $\mathcal{H}$ with respect to the time variable τ are (the prime denoting derivative with respect to τ):

$$x' = -\frac{3}{2} x \mathcal{H} \left((\gamma - 2)x^2 + (y^2 - 1)\gamma + z^2 \left(\gamma - \frac{2}{3} \right) + 2 \right) - \sqrt{\frac{3}{2}} \lambda y^2, \quad (98)$$

$$y' = \frac{3}{2} y \left(\frac{\sqrt{6}\lambda x}{3} - \mathcal{H} \left((\gamma - 2)x^2 + (y^2 - 1)\gamma + z^2 \left(\gamma - \frac{2}{3} \right) \right) \right), \quad (99)$$

$$z' = \frac{3}{2} z \mathcal{H} \left((\gamma - 2)x^2 + (y^2 - 1)\gamma + z^2 \left(\gamma - \frac{2}{3} \right) - \frac{2}{3} \right) \quad (100)$$

$$\mathcal{H}' = -\frac{3}{2} \left(\mathcal{H}^2 - 1 \right) \left((\gamma - 2)x^2 + (y^2 - 1)\gamma + z^2 \left(\gamma - \frac{2}{3} \right) \right) \quad (101)$$

As before, we will re-express the cosmological magnitudes of interest in terms of the normalized variables.

The deceleration parameter is explicitly

$$q \equiv -\ddot{a}a/\dot{a}^2 = -1 + \frac{3}{2} \left[\frac{x^2 (2 - \gamma) + (1 - y^2) \gamma + z^2 (\frac{2}{3} - \gamma)}{\mathcal{H}^2} \right]; \quad (102)$$

the fractional energy density of the scalar field and curvature are given respectively by

$$\Omega_\phi = \frac{x^2 + y^2}{\mathcal{H}^2}; \Omega_k = \frac{z^2}{\mathcal{H}^2} \quad (103)$$

and the 'effective' EoS parameter is given by

$$\omega_{eff} \equiv \frac{P_{tot}}{\rho_{tot}} \equiv \frac{\frac{1}{2}\dot{\phi}^2 - V(\phi) + (\gamma - 1)\rho_{\mathrm{M}} - \Lambda}{\frac{1}{2}\dot{\phi}^2 + V(\phi) + \rho_{\mathrm{M}} + \Lambda} = -1 + \frac{(2 - \gamma) x^2 + (1 - y^2) \gamma - \gamma z^2}{\mathcal{H}^2 - z^2}. \quad (104)$$

5.2.2.   Form Invariance under Coordinate Transformations

The system (98-101) is form invariant under the coordinate transformation and time reversal

$$(\tau, x, y, z, \mathcal{H}) \to (-\tau, -x, y, z, -\mathcal{H}). \quad (105)$$

Thus, it is sufficient to discuss the behaviour in one part of the phase space, the dynamics in the other part being obtained via the transformation (105). Observe that equations (98-101) are form invariant under the coordinate transformation $y \to -y$ and $z \to -z$. Then, (98-101) is form invariant under they composition with (105).

There are four obvious invariant sets under the flow of (98-101), they are $y = 0$, $z = 0$, and $\mathcal{H} = \pm 1$. They combination defines other invariant sets. The dynamics restricted to the invariant set $z = 0$ is the same described in section 5.1.. It is of course of interest the analysis of the behavior of the 4D orbits near the invariant set $z = 0$ (or the other enumerated above). We will discuss this in next sections.

5.2.3. Monotonic Functions

Let be defined

$$Z_1 = \left(\frac{\mathcal{H} + 1}{\mathcal{H} - 1} \right)^2 \tag{106}$$

in the invariant set

$$\{ (x, y, z, \mathcal{H}) : x^2 + y^2 + z^2 < 1,\ y > 0,\ z > 0,\ -1 < \mathcal{H} < 1 \}.$$

Then, Z is monotonic decreasing in S since

$$Z' \equiv \nabla Z \cdot f = -Z \left(4z^2 + 6x^2 \left(2 - \gamma \right) + 6 \left(1 - y^2 - z^2 \right) \gamma \right) < 0$$

in S. The range of Z is the semi-interval $(0, +\infty)$, and $Z \to 0$ as $\mathcal{H} \to -1$ (since $\mathcal{H}$ is bounded) and $Z \to +\infty$ as $\mathcal{H} \to 1$. By applying the Monotonicity Principle (theorem 4.12 [91]) we find that, for all $p \in S$, the past asymptotic attractor of p (the α-limit) belongs to $\mathcal{H} = 1$ and the future asymptotic attractor of p (the ω-limit) belongs to $\mathcal{H} = -1$.

Let be defined in the same invariant set the function

$$Z_2 = \frac{z^4}{\left(1 - x^2 - y^2 - z^2 \right)^2},\ Z_2' = -2 \left(2 - 3\gamma \right) \mathcal{H} Z_2 \tag{107}$$

This function is monotonic in the regions $\mathcal{H} < 0$ and $\mathcal{H} > 0$ for $\gamma \neq \frac{2}{3}$.

The existence of monotonic functions allows to state that there can be no periodic orbits or recurrent orbits in the interior of the phase space. Futhermore, it is possible to obtain global results. From the expression Z_2 we can immediately see that asymptotically $z \to 0$ or $x^2 + y^2 + z^2 \to 1$.

5.2.4. Local Analysis of Critical Points

The system (98-101) admits fourteen critical points. We will denote the critical points of the system (98-101) located at the invariant set $z = 0$ in the same way as in the table 11 of section 5.1.4.. We submit the reader to this table for the conditions for their existence. In table 17 are summarized the stability properties of the critical points.

Observe that the critical points $P_6^{\mp}$ and $P_7^{\pm}$ are related through (105) with the critical points $P_6^{\pm}$ and $P_7^{\pm}$ respectively.

In order to analyze the local stability of $P_1^+, P_2^+, P_3^+, P_4^+, P_5^+, P_6^+, P_7^+$ it is sufficient analyze the local stability of $P_3^-, P_2^-, P_1^-, P_4^-, P_5^-, P_6^-, P_7^-$ respectively, and then, infer the stability of the points in the "positive" branch by using (105). In table 17 we offer a detailed analysis of the dynamical character of the critical points corresponding to the "negative" branch.

The critical point P_1^- is nonhyperbolic if $\gamma = 2$ or $\lambda = -\sqrt{6}$. It is a local sink provided $\lambda > -\sqrt{6}, \gamma \neq 0$. If $\lambda < -\sqrt{6}$ then there exists a 1D unstable manifold tangent to the y-axis, and a 3D stable manifold. The critical point P_2^- is nonhyperbolic if $\gamma = 0, \gamma = \frac{2}{3}$ or $\gamma = 2$. There exist always at least a 1D unstable manifold tangent to the x axis. The unstable manifold is 2D provided $\gamma < \frac{2}{3}$. In this case there exists a 2D stable manifold tangent to the y-$\mathcal{H}$ plane. The critical point P_3^- is nonhyperbolic if $\gamma = 2$ or $\lambda = \sqrt{6}$. It is a local sink provided $\lambda < \sqrt{6}, \gamma \neq 0$. If $\lambda > \sqrt{6}$ then there exists a 1D unstable manifold tangent to the y-axis, and a 3D stable manifold. Observe that P_1^- and P_3^- coexist provided $\lambda^2 < 6$. In this case both are the future attractors of the system, they attract solutions in its basin of attraction. The critical point P_4^-, is hyperbolic if $\lambda \in \{0, \pm 3\gamma, \pm 2\}$. If not, it is always a saddle point with an unstable manifold at least 1D. The stable manifold is 3D provided $0 < \gamma \leq \frac{2}{3}, 2 < \lambda^2 < 6$ or $\frac{2}{3} < \gamma < 2, 3\gamma < \lambda^2 < 6$. It is 2D provided $0 < \gamma < \frac{2}{3}, 3\gamma < \lambda^2 < 2$ or $\frac{2}{3} < \gamma < 2, 2 < \lambda^2 < 3\gamma$ or 1D provided $0 < \gamma \leq \frac{2}{3}, 0 < \lambda^2 < 3\gamma$ or $\frac{2}{3} < \gamma < 2, 0 < \lambda^2 < 2$. The critical point P_5^- is nonhyperbolic if $\gamma = 0$ or $\lambda^2 = 3\gamma$ or $\gamma = \frac{2}{3}$. If not, it have always two conjugate complex eigenvalues with positive real parts (there exists at least a 2D unstable manifold). The stable manifold is 1D provided $0 < \gamma \leq \frac{2}{9}, \lambda^2 > 3\gamma$ or $\frac{2}{9} < \gamma < \frac{2}{3}, \lambda^2 < \frac{24\gamma^2}{-2+9\gamma}$ or 2D if $\frac{2}{3} < \gamma < 2, \lambda^2 < \frac{24\gamma^2}{-2+9\gamma}$. Thus, P_5^- is always a saddle point.

In the negative curvature case there are two new classes of critical points: Mine's solutions and curvature-scaling solutions, denoted by $P_6^\pm$ and $P_7^\pm$ respectively. In table 15 it is displayed the location, existence and some properties of them.

Table 15. Location and existence conditions of the critical points $P_6^\pm$ and $P_7^\pm$ of the system (98-101).

Label	Coordinates: $(x, y, z, \mathcal{H})$	Existence
$P_6^\pm$	$(0, 0, 1, \pm 1)$	All λ
$P_7^\pm$	$(\mp \frac{1}{\lambda}\sqrt{\frac{2}{3}}, \frac{2}{\sqrt{3}\lambda}, \frac{1}{\lambda}\sqrt{-2+\lambda^2}, \pm 1)$	$\frac{2}{3} < \gamma \leq 2$ and $\|\lambda\| \geq \sqrt{2}$

In table 18, where we present a summary of attractors (both past and future) for the quinstant model with $k = -1$.

In figure 9 are displayed typical orbits of the system (98-101) in the invariant set $\mathcal{H} = 1$ for the values $\lambda = -\sqrt{\frac{3}{2}}$ and $\gamma = 1$. The critical points P_1^+ and P_3^+ are the past asymptotic attractors, P_1^+ having a stronger unstable manifold tangent to the y-axis. P_4^+, acts as a local attractor in the invariant set $\mathcal{H} = 1$. The Milne's universe (P_6^+) is the local future attractor

Table 16. Some properties of of the critical points $P_6^{\pm}$ and $P_7^{\pm}$ of the system (98-101).

Label	Deceleration q	Ω_ϕ	ω_{eff}
$P_6^{\pm}$	0	1	0
$P_7^{\pm}$	$\frac{2}{\lambda^2}$	$1 - \frac{2}{\lambda^2}$	$-\frac{1}{3}$

in the invariant set $y = 0$. **In figure** 10 are drawn some orbits in the invariant set $\mathcal{H} = -1$ for the same choice. The critical points P_1^- and P_3^- are the future asymptotic attractors, P_3^- having a stronger attracting manifold tangent to the y-axis. P_4^- acts as the local source for the invariant set $z = 0$ (they are, however, saddle points for the 4D dynamics). The Milne's universe (P_6^-) is the local past attractor in the invariant set $y = 0$. This is the better we can do numerically since the phase space is actually 4D.

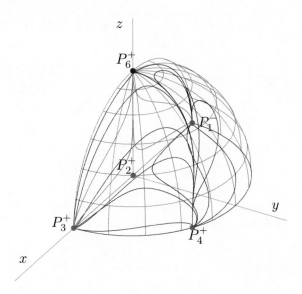

Figure 9. Some orbits of (98-101) in the invariant set $\mathcal{H} = 1$ for the values $\lambda = -\sqrt{\frac{3}{2}}$ and $\gamma = 1$.

5.2.5. Bifurcations

The critical points $\left(P_4^+, P_4^-\right)$ reduce to $\left(P_1^+, P_3^-\right)$ as $\lambda \to (\sqrt{6})^-$. The critical points $\left(P_4^+, P_4^-\right)$ reduce to $\left(P_3^+, P_1^-\right)$ as $\lambda \to (-\sqrt{6})^+$. The critical points $P_5^{\pm}$ reduce to $P_2^{\pm}$ as $\gamma \to 0^+$. On the other hand, $P_5^{\pm}$ reduce to $P_4^{\pm}$ as $\lambda \to (\sqrt{3\gamma})^+$ or $\lambda \to (-\sqrt{3\gamma})^-$. The

Table 17. Eigenvalues, and dynamical character of the fixed points of the system (98-101). We use the notation $\Delta = (2 - \gamma)(24\gamma^2 + \lambda^2(2 - 9\gamma))$.

Label	Eigenvalues	Dynamical character
P_1^-	$-6, -3 - \sqrt{\frac{3}{2}}\lambda, -3(2 - \gamma), -2$	nonhyperbolic if $\gamma = 2$ or $\lambda = -\sqrt{6}$; stable (node) if $\lambda > -\sqrt{6}$ and $\gamma \neq 2$; saddle, otherwise.
P_2^-	$\frac{3}{2}(2 - \gamma), -3\gamma, -\frac{3\gamma}{2}, 1 - \frac{3\gamma}{2}$	nonhyperbolic if $\gamma = 0$ or $\gamma = 2$ or $\gamma = \frac{2}{3}$ saddle, otherwise.
P_3^-	$-6, -3 + \sqrt{\frac{3}{2}}\lambda, -3(2 - \gamma), -2$	nonhyperbolic if $\gamma = 2$ or $\lambda = \sqrt{6}$; stable (node) if $\lambda < \sqrt{6}$ and $\gamma \neq 2$; saddle, otherwise.
P_4^-	$-\lambda^2, \frac{1}{2}(6 - \lambda^2), -\lambda^2 + 3\gamma, 1 - \frac{\lambda^2}{2}$	nonhyperbolic if $\lambda = 0$ or $\lambda^2 = 3\gamma$; saddle, otherwise.
P_5^-	$-3\gamma, \frac{3}{4}\left(2 - \gamma \pm \frac{1}{\lambda}\sqrt{\Delta}\right), 1 - \frac{3\gamma}{2}$	nonhyperbolic if $\gamma = 0$ or $\lambda^2 = 3\gamma$ or $\gamma = \frac{2}{3}$; saddle, otherwise.
P_6^-	$-2, 2, -1, -2 + 3\gamma$	nonhyperbolic if $\gamma = \frac{2}{3}$; saddle, otherwise.
P_7^-	$-2, 1 \pm \sqrt{\frac{8}{\lambda^2} - 3}, -2 + 3\gamma$	nonhyperbolic if $\lambda^2 = 2$ or $\gamma = \frac{2}{3}$; unstable, otherwise.

critical points $P_7^\pm$, $P_5^\pm$ and $P_4^\pm$ coincide as $\gamma \to (\frac{2}{3})^+$ and $\lambda \to (\sqrt{2})^+$ simultaneously. For these values of the parameters a bifurcation arises.

5.2.6. Typical Behavior

Once the attractors have been identified one can give a quantitative description of the physical behavior of a typical negatively curved quinstant cosmology.

For example, for ever expanding cosmologies with $H > 0, H \to +\infty, \mathcal{H} = 1$, i.e., the standard expanding cosmology near the big-bang, a typical model behaves like a massless scalar field (kinetic dominated energy density) represented by P_3^+ or P_1^+ provided $\lambda > -\sqrt{6}, \gamma \neq 2$ or $\lambda < \sqrt{6}, \gamma \neq 2$, respectively. This types of solutions might coexist in the same phase space. The late time dynamics in $\mathcal{H} = 1$ is given by either a scalar field dominated solution ($\Omega_\phi \to 1$) represented by P_4^+ provided $0 < \gamma < 2, \lambda^2 < 3\gamma$ or $\frac{2}{3} < \gamma \leq 2, \lambda^2 < 2$; or by a scaling solution ($\Omega_m/\Omega_\phi = O(1)$) represented by P_5^+ for $0 < \gamma < \frac{2}{3}, \lambda^2 > 2$; or by a curvature scaling solution represented by P_7^+ provided $\frac{2}{3} < \gamma \leq 2, \lambda^2 > 2$. For finite values of H, i.e., $-1 < \mathcal{H} < 1$, the early time dynamics

Table 18. Summary of attractors for for the quinstant model with $k = -1$ (system (98-101)).

Restrictions	Past attractor	Future attractor
$\mathcal{H} = -1$	P_4^- if $0 < \gamma < 2$, $\lambda^2 < 3\gamma$ or if $\frac{2}{3} < \gamma \leq 2$, $\lambda^2 < 2$ P_5^- $\quad 0 < \gamma < \frac{2}{3}$, $\lambda^2 > 3\gamma$ P_7^- if $\frac{2}{3} < \gamma \leq 2$, $\lambda^2 > 2$	P_3^- if $\lambda < \sqrt{6}$, $\gamma \neq 2$ P_1^- if $\lambda > -\sqrt{6}$, $\gamma \neq 2$
$-1 < \mathcal{H} < 1$	P_3^+ if $\lambda > -\sqrt{6}$, $\gamma \neq 2$ P_1^+ if $\lambda < \sqrt{6}$, $\gamma \neq 2$	as above
$\mathcal{H} = 1$	as above	P_4^+ if $0 < \gamma < 2$, $\lambda^2 < 3\gamma$ or if $\frac{2}{3} < \gamma \leq 2$, $\lambda^2 < 2$ P_5^+ $\quad 0 < \gamma < \frac{2}{3}$, $\lambda^2 > 3\gamma$ P_7^+ if $\frac{2}{3} < \gamma \leq 2$, $\lambda^2 > 2$

is the same as in the previous case but there are subtle differences with respect the late time dynamics. In fact, in the invariant set $-1 < \mathcal{H} < 1$ the future attractors are P_3^- or P_1^- depending if $\lambda < \sqrt{6}$, $\gamma \neq 2$ or $\lambda > -\sqrt{6}$, $\gamma \neq 2$. If $|\lambda| < \sqrt{6}$ the system is bistable. Such solutions represent contracting stiff-fluid cosmologies. This means that a typical quinstant negatively curved cosmologies allows the collapse of matter when the time evolves. For contracting cosmologies with ($H < 0, H \to \infty, \mathcal{H} = -1$), i.e., the standard contracting model near the initial singularity, the late time dynamics is the same as in the previously described case, i.e., the collapse. However, there are subtle differences concerning the early time dynamics. The late time dynamics in $\mathcal{H} = -1$ is given by either a scalar field dominated solution ($\Omega_\phi \to 1$) represented by P_4^- provided $0 < \gamma < 2, \lambda^2 < 3\gamma$ or $\frac{2}{3} < \gamma \leq 2, \lambda^2 < 2$; or by a scaling solution ($\Omega_m/\Omega_\phi = O(1)$) represented by P_5^- for $0 < \gamma < \frac{2}{3}, \lambda^2 > 2$; or by a curvature scaling solution represented by P_7^- provided $\frac{2}{3} < \gamma \leq 2, \lambda^2 > 2$.

5.3. Quinstant Cosmology with Positive Curvature

5.3.1. Normalization, State Space, and Dynamical System

Let us consider the normalization factor $\hat{D} = \sqrt{H^2 - \Lambda/3 + \frac{1}{a^2}} > 0$ and the time variable $d\hat{\tau} = \hat{D}dt$, and the phase space variables:

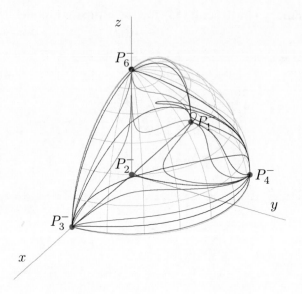

Figure 10. Some orbits of (98-101) in the invariant set $\mathcal{H} = -1$ for the values $\lambda = -\sqrt{\frac{3}{2}}$ and $\gamma = 1$.

$$\hat{x} = \frac{\dot{\phi}}{\sqrt{6}\hat{D}}, \; \hat{y} = \frac{\sqrt{V(\phi)}}{\sqrt{3}\hat{D}}, \; \hat{H} = \frac{H}{\hat{D}}, \hat{z} = \frac{1}{a\hat{D}}. \tag{108}$$

The former variables lies in the compact phase space

$$\left\{ (\hat{x}, \hat{y}, \hat{z}, \hat{H}) : \hat{x}^2 + \hat{y}^2 \leq 1, \; \hat{y} \geq 0, \; -1 \leq \hat{H} \leq 1, 0 \leq \hat{z} \leq 1 \right\}.$$

The dynamical system is given by:

$$\hat{x}' = \frac{3}{2}\hat{x}\hat{H}\left((2-\gamma)\left(1-\hat{x}^2\right)-\gamma\hat{y}^2\right)-\frac{\sqrt{6}}{2}\lambda\hat{y}^2, \tag{109}$$

$$\hat{y}' = \frac{3}{2}\hat{y}\hat{H}\left((2-\gamma)\hat{x}^2+\gamma\left(1-\hat{y}^2\right)\right)+\frac{\sqrt{6}}{2}\lambda\hat{x}\hat{y}, \tag{110}$$

$$\hat{z}' = \frac{1}{2}\hat{z}\hat{H}\left(3\left(2-\gamma\right)\hat{x}^2+3\gamma\left(1-\hat{y}^2\right)-2\right) \tag{111}$$

$$\hat{H}' = -\frac{3}{2}\left(\hat{H}^2-1\right)\left((\gamma-2)\hat{x}^2+\left(\hat{y}^2-1\right)\gamma\right)+\hat{z}^2 \tag{112}$$

where the prime denotes derivative with respect to $\hat{\tau}$. As before, we will re-express the cosmological magnitudes of interest in terms of the normalized variables.

The deceleration parameter is explicitly

$$q \equiv -\ddot{a}a/\dot{a}^2 = -1 + \frac{3}{2}\left[\frac{\hat{x}^2\left(2-\gamma\right)+\left(1-\hat{y}^2\right)\gamma}{\hat{H}^2}\right] - \frac{\hat{z}^2}{\hat{H}^2}; \tag{113}$$

the fractional energy density of the scalar field and curvature are given respectively by

$$\Omega_\phi = \frac{\hat{x}^2 + \hat{y}^2}{\hat{H}^2}; \Omega_k = \frac{\hat{z}^2}{\hat{H}^2} \tag{114}$$

and the 'effective' EoS parameter is given by

$$\omega_{eff} \equiv \frac{P_{tot}}{\rho_{tot}} \equiv \frac{\frac{1}{2}\dot{\phi}^2 - V(\phi) + (\gamma - 1)\rho_M - \Lambda}{\frac{1}{2}\dot{\phi}^2 + V(\phi) + \rho_M + \Lambda} = -1 + \frac{(2-\gamma)\tilde{x}^2 + (1 - \tilde{y}^2)\gamma}{\hat{H}^2 + \hat{z}^2}. \tag{115}$$

5.3.2. Form Invariance under Coordinate Transformations

The system (109-112) is form invariant under the coordinate transformation and time reversal

$$\left(\hat{\tau}, \hat{x}, \hat{y}, \hat{z}, \hat{H}\right) \rightarrow \left(-\hat{\tau}, -\hat{x}, \hat{y}, \hat{z}, -\hat{H}\right). \tag{116}$$

Thus, it is sufficient to discuss the behavior in one part of the phase space, the dynamics in the other part being obtained via the transformation (116). Observe that equations (109-112) are form invariant under the coordinate transformation $\hat{y} \rightarrow -\hat{y}$ and $\hat{z} \rightarrow -\hat{z}$. Then, (109-112) is form invariant under they composition with (116).

There are two obvious invariant sets under the flow of (109-112), they are $\hat{y} = 0$, $\hat{z} = 0$.

From equation (112) we can immediately see that the surfaces $\hat{H} = \pm 1$ are not invariant provided $\hat{z} \neq 0$. In fact, the surfaces $\hat{H} = \pm 1$ act as membranes (that can be crossed). This follows from the fact that $\hat{H}'|_{\hat{H}=\pm 1} = \hat{z}^2 > 0$ for $\hat{z} \neq 0$. Observe that if initially $\hat{z} > 0$, then, from equation (111), follows that the sign of $\hat{z}$ is invariant. Only if $\hat{z} = 0$, the surfaces $\hat{H} = \pm 1$ could be invariant.

5.3.3. Monotonic Functions

Let be defined in the invariant set

$$\left\{(\hat{x}, \hat{y}, \hat{z}, \hat{H}) : \hat{x}^2 + \hat{y}^2 < 1, \hat{y} > 0, 0 < \hat{z} < 1, -1 < \hat{H} < 1\right\},$$

the function

$$Z = \frac{\hat{z}^4}{(1 - \hat{x}^2 - \hat{y}^2)^2}, \quad Z' = -2(2 - 3\gamma)\hat{H}Z \tag{117}$$

This function is monotonic in the regions $\hat{H} < 0$ and $\hat{H} > 0$ for $\gamma \neq \frac{2}{3}$.

The existence of this monotonic function allows to state that there can be no periodic orbits or recurrent orbits in the interior of the phase space. Furthermore, it is possible to obtain global results. From the expression Z we can immediately see that asymptotically $\hat{z} \rightarrow 0$ or $\hat{x}^2 + \hat{y}^2 \rightarrow 1$.

Table 19. Location and existence conditions of the critical points of the system (109-111)

Label	Coordinates: $(\hat{x}, \hat{y}, \hat{z}, \hat{H})$	Existence
$\hat{P}_1^\pm$	$(-1, 0, 0, \pm 1)$	All λ
$\hat{P}_2^\pm$	$(0, 0, 0, \pm 1)$	All λ
$\hat{P}_3^\pm$	$(1, 0, 0, \pm 1)$	All λ
$\hat{P}_4^\pm$	$\left(\mp \frac{\lambda}{\sqrt{6}}, \sqrt{1 - \frac{\lambda^2}{6}}, 0, \pm 1 \right)$	$-\sqrt{6} < \lambda < \sqrt{6}$
$\hat{P}_5^\pm$	$\left(\mp \sqrt{\frac{3}{2}} \frac{\gamma}{\lambda}, \sqrt{\frac{3}{2}} \sqrt{\frac{(2-\gamma)\gamma}{\lambda^2}}, 0, \pm 1 \right)$	$\gamma = 0, \; \lambda \neq 0$ $0 < \gamma \leq 2, \; \lvert \lambda \rvert \geq \sqrt{3\gamma}$
$\hat{P}_6$	$\left(x^\star, 0, \sqrt{\frac{3}{2}\left(\gamma + (2-\gamma)x^{\star 2}\right)}, 0 \right)$	$0 \leq \gamma \leq \frac{2}{3}, \; \lvert x^\star \rvert \leq \sqrt{\frac{2-3\gamma}{3(2-\gamma)}}$
$\hat{P}_7^\pm$	$\left(\mp \frac{1}{\sqrt{3}}, \sqrt{\frac{2}{3}}, \sqrt{1 - \frac{\lambda^2}{2}}, \pm \frac{\lambda}{\sqrt{2}} \right)$	$-\sqrt{2} \leq \lambda \leq \sqrt{2}$

5.3.4. Local Analysis of Critical Points

Using (116), it is possible to infer the local stability of $\hat{P}_1^+, \hat{P}_2^+, \hat{P}_3^+, \hat{P}_4^+, \hat{P}_5^+, \hat{P}_7^+$ from the local stability of $\hat{P}_3^-, \hat{P}_2^-, \hat{P}_1^-, \hat{P}_4^-, \hat{P}_5^-, \hat{P}_7^-$. Thus, we will analyze only the critical points in the "negative" branch.

The critical points $\hat{P}_1^+$ to $\hat{P}_5^+$ have similar properties, dynamical character as unhatted ones in section 5.2. and the same physical interpretation of the unhatted ones characterized in section 5.1.. Thus, we will not comment about its stability in detail. For completeness in table 20 are summarize the existence conditions and stability properties. We will submit the interested reader to previous sections for more details.

For the positive curvature model there are new critical points: the curve of nonhy-perbolic critical points $\hat{P}_6$ represent the Einstein's static universe. Special critical points of this family are those with the choice $x^\star = 0$ (it exists provided $0 \leq \gamma \leq \frac{2}{3}$) and $x^\star = \pm \sqrt{\frac{3(2-3\gamma)}{2-\gamma}}$ (it exists provided $0 \leq \gamma < \frac{2}{3}$), and the curvature-scaling solution $\hat{P}_7^\pm$. Concerning the stability of curvature-scaling solutions we have that P_7^- is always a sad-dle point (at least one eigenvalues has negative real part, and the others are of different sign). Its stable manifold is 1D provided $-\sqrt{2} < \lambda < 0, \, 0 < \gamma < \frac{2}{3}$; or 2D provided $0 < \lambda < \sqrt{2}, \, \frac{2}{3} < \gamma < 2$ or $-\sqrt{2} < \lambda < 0, \, \frac{2}{3} < \gamma < 2$; or 3D if $0 < \lambda < \sqrt{2}, \, 0 \leq \gamma < \frac{2}{3}$.

Table 20. Eigenvalues, and dynamical character of the fixed points of the system (98-101). We use the notation $\Delta = (2 - \gamma)(24\gamma^2 + \lambda^2(2 - 9\gamma))$ and
$$\mu = \sqrt{\tfrac{3}{2}\left((x^{\star 2} - 1)\,\gamma\,(2 - 3\gamma) + 24x^{\star 2}\right)}.$$

Label	Eigenvalues	Dynamical character
$\hat{P}_1^-$	$-6, -3 - \sqrt{\tfrac{3}{2}}\lambda, -3(2 - \gamma), -2$	nonhyperbolic if $\gamma = 2$ or $\lambda = -\sqrt{6}$; stable (node) if $\lambda > -\sqrt{6}$ and $\gamma \neq 2$; saddle, otherwise.
$\hat{P}_2^-$	$\tfrac{3}{2}(2 - \gamma), -3\gamma, -\tfrac{3\gamma}{2}, 1 - \tfrac{3\gamma}{2}$	nonhyperbolic if $\gamma = 0$ or $\gamma = 2$ or $\gamma = \tfrac{2}{3}$ saddle otherwise.
$\hat{P}_3^-$	$-6, -3 + \sqrt{\tfrac{3}{2}}\lambda, -3(2 - \gamma), -2$	nonhyperbolic if $\gamma = 2$ or $\lambda = \sqrt{6}$; stable (node) if $\lambda < \sqrt{6}$ and $\gamma \neq 2$; saddle, otherwise.
$\hat{P}_4^-$	$-\lambda^2, \tfrac{1}{2}(6 - \lambda^2), -\lambda^2 + 3\gamma, 1 - \tfrac{\lambda^2}{2}$	nonhyperbolic if $\lambda = 0$ or $\lambda^2 = 3\gamma$ or $\lambda^2 = 2$; saddle otherwise.
$\hat{P}_5^-$	$-3\gamma, \tfrac{3}{4}\left(2 - \gamma \pm \tfrac{1}{\lambda}\sqrt{\Delta}\right), 1 - \tfrac{3\gamma}{2}$	nonhyperbolic if $\gamma = 0$ or $\lambda^2 = 3\gamma$ or $\gamma = \tfrac{2}{3}$; saddle, otherwise.
$\hat{P}_6$	$0, \sqrt{\tfrac{3}{2}}\lambda x^{\star}, -\mu, \mu$	nonhyperbolic
$\hat{P}_7^-$	$-\sqrt{2}\lambda, \tfrac{\lambda \pm \sqrt{8 - 3\lambda^2}}{\sqrt{2}}, \tfrac{(3\gamma - 2)\lambda}{\sqrt{2}}$	nonhyperbolic if $\lambda = 0$ or $\lambda^2 = 2$; or $\gamma = \tfrac{2}{3}$; unstable (saddle), otherwise.

5.3.5. Bifurcations

The critical points $\left(\hat{P}_4^+, \hat{P}_4^-\right)$ reduce to $\left(\hat{P}_1^+, \hat{P}_3^-\right)$ as $\lambda \to (\sqrt{6})^-$. The critical points $\left(\hat{P}_4^+, \hat{P}_4^-\right)$ reduce to $\left(\hat{P}_3^+, \hat{P}_1^-\right)$ as $\lambda \to (-\sqrt{6})^+$. The critical points $\hat{P}_5^{\pm}$ reduce to $\hat{P}_2^{\pm}$ as $\gamma \to 0^+$. On the other hand, $\hat{P}_5^{\pm}$ reduce to $\hat{P}_4^{\pm}$ as $\lambda \to (\sqrt{3\gamma})^+$ or $\lambda \to (-\sqrt{3\gamma})^-$. $(\hat{P}_7^+, \hat{P}_7^-)$ reduce to $(\hat{P}_4^+, \hat{P}_4^-)$ as $\lambda \to (\sqrt{2})^-$ and $(\hat{P}_7^+, \hat{P}_7^-)$ reduce to $(\hat{P}_4^-, \hat{P}_4^+)$ as $\lambda \to (-\sqrt{2})^+$. For these values of the parameters a bifurcation arises.

5.3.6. Typical Behavior

As a consequence that $\hat{H} = \pm 1$ are not invariant sets, the determination of past and future attractors is more simpler. If $\lambda < -\sqrt{6}, \gamma \neq 2$, then P_1^+ is the past attractor and P_3^- is the future attractor. If $-\sqrt{6} < \lambda < \sqrt{6}$, $\gamma \neq 2$ the past attractors are both $\hat{P}_3^+$ and $\hat{P}_1^+$ and the future attractors are both $\hat{P}_1^-$ and $\hat{P}_3^-$. Finally, if $\lambda > \sqrt{6}, \gamma \neq 2$ then P_3^+ is the past attractor and P_1^- is the future attractor. In any case the Universe evolves from a stiff regime to a stiff regime by crossing the value $\hat{H} = 0$, allowing the collapse of the Universe.

In table 21, where we present a summary of attractors (both past and future) for the quinstant model with $k = 1$.

Table 21. Summary of attractors for for the quinstant model with $k = 1$ (system (109-112)) for $z > 0$.

Past attractor		Future attractor	
$\hat{P}_3^+$	if $\lambda > -\sqrt{6}$, $\gamma \neq 2$	$\hat{P}_3^-$	if $\lambda < \sqrt{6}$, $\gamma \neq 2$
$\hat{P}_1^+$	if $\lambda < \sqrt{6}$, $\gamma \neq 2$	$\hat{P}_1^-$	if $\lambda > -\sqrt{6}$, $\gamma \neq 2$

6. Observational Test and Dynamical Systems: The Interplay

Dynamical systems techniques by one way and Observational test by the other are strongly enough to discriminate among the wide variety of dark energy models nowadays under investigation. The first one is more mathematical in character: they can be used to select the better behaved models, with appropriate attractors in the past and future. These tools to analyse and interpreted results has gained a lot of attention in recent years. The techniques are so powerful when we want to investigate the asymptotic (and even, intermediate) behavior of models. The other class of tools is rather physical: they can be used of astrophysical observations to crack the degeneracy of classes of dark energy models. In the interplay, both serve to constraint the free parameters of the models under consideration. In our case, quintom and quinstant dark energy, with flat and curved geometry.

Quintom Dark Energy Paradigm

The model of quintom, which is mainly favored by current SNIa only, needs to be confronted with other observations in the framework of concordance cosmology. Since SNIa offer the only direct detection of DE, this model is the most promising to be distinguished from the cosmological constant and other dynamical DE models which do not get across -1, by future SNIa projects on the low redshift (for illustrations see e.g. [44]).

From the dynamical systems viewpoint we have obtained further results in support of the previous results in [12, 13, 16]. For negative curvature models, we have devised two dynamical systems adapted to the study of expanding ($\epsilon = \text{sign } H > 0$) and contracting ($\epsilon = \text{sign } H < 0$) models. Also, we have devised another dynamical system well suited for investigating positive curvature models. We have characterized the critical points of each system and interpreted the cosmological solutions associated. By devising well defined monotonic functions we were able to get global results for ever expanding and contracting models (for both negative and positive curvature models). We have reviewed the results concerning the flat case. It is known that, for flat ever expanding models the attractor will be the matter scaling solution [16]. If matter scaling solutions do not exist, the attractor will be phantom ($w < -1$) or de Sitter ($w = -1$) like. This is a difference with respect to the results in [12] and [13]. It was proved there, that the attractor solutions are de Sitter-like, unless some trajectories cross, transiently, the $w = -1$ boundary to become even smaller before ending in a de Sitter phase.

The new results we survey here are as follows:

For negative-curvature ever-expanding models ($\epsilon = \text{sign}\, H > 0$) we have obtained the existence of scaling curvature attractors (without matter) (provided $\delta < \frac{1}{3}$). The attractor solution will be dominated by DE whenever its existence precludes the existence of scaling curvature attractors. These solutions can be: phantom-like ($w < -1$), de Sitter-like ($w = -1$), or quintessence-like. This is a difference with respect the situation in [16]. We must notice, however, that if we consider other values for γ, other than $\gamma = 1$, then the attractor of the system can be the matter scaling solution. This is the case if $0 < \gamma < \frac{2}{3}$, $\delta > \frac{\gamma}{2}$. Under the above conditions on the parameters DM mimics DE. For contracting models ($\epsilon = \text{sign}\, H < 0$), the attractor will be a MSF solution that mimics a stiff fluid. Towards the past, the typical situation is the reverse of the former described. For positive curvature (closed) models, we have obtained conditions under which there is an orbit of type $_+\hat{K} \to _-\hat{K}$. This represents a cosmological solution starting in a ending towards a singularity described by a MSF cosmology. We have obtained, also, a flat FRW solution starting in a big-bang in $_+F$ y recolapsing in a "big-crunch" in $_-F$. We have illustrated this results by means of numerical integrations of the system of ODEs describing this cosmological model. We have obtained conditions for the existence of global attractors. We have offered, here, only a simplified qualitative analysis (as a difference of the mathematical analysis in [88] pages 69-73). However, our study have relevance by its own right, and can be considered in some way as a complement of the former since we have added a phantom field in the dynamics. We must to restate, however, that our analysis is not as detailed as in that reference. But its is suffice to illustrated our goals. The qualitative analysis in multi- scalar field (coventional) cosmologies with exponential potentials (in the context of assisted inflation) was done in the same reference, section VII, and in [89, 90], particularly for two fields. They do not consider phantom field as we do here. Our monotonic functions were able to discard the existence of periodic orbits, homoclinic orbits, or recurrent orbits.

Quinstant Dark Energy Paradigm

From the stability analysis of all studied models of (exponential) quinstant dark energy ([38], [39]), the typical behavior, irrespective the curvature choice, is the evolution from an stiff-regime near the past, to a stiff-regime in the far future. Besides, from our dynamical analysis it seems that for negatively curvature and flat models, the model shows the divergence of the Hubble parameter (H) in the asymptotic regimes. There are other cosmological models of composite dark energy having stiff-matter domination as an attractor in the past, but usually they would not be global attractors, but local. This is the case of several models of quintom dark energy. From the structure formation we see that QDE makes reasonable predictions for the formation of linear large scale structure of the Universe. It reproduces reasonably well the non-linear structures from today up to redshifts a bit larger than one, but fails to reproduce the perturbations in the non-linear regime for redshifts a bit larger than one. This models are dynamically equivalent models of $f(R)$ modified gravity. It would be interesting to study how these $f(R)$ models behave concerning structure formation, and then we would have a better understanding on how these observations would crack the degeneracy dark energy-$f(R)$ modified gravity.

In summary, the new results concerning quinstant dark energy are as follows:

For the standard flat expanding cosmology near the big-bang, a typical model behaves like a massless scalar field (kinetic dominated energy density) and the late time dynamics is given by either a scalar field dominated solution ($\Omega_\phi \to 1$) or by a scaling solution ($\Omega_m/\Omega_\phi = O(1)$) represented by P_5^+ provided $\lambda^2 < 3\gamma$. This is the standard behavior for quintessence models (without Λ). For finite values of H, the early time dynamics is the same as in the previous case but there are subtle differences with respect the late time dynamics. In fact, in this invariant set the future attractors are stiff-like (contracting) solutions with $H \to \pm\infty$. This means that a typical quinstant cosmologies allows the collapse of matter when the time evolves. For the standard contracting model near the initial singularity, the late time and early time dynamics is the reverse of the previously described.

The behavior of a typical negatively curved quinstant model is similar to the flat situation, but not the same. The differences is in that in the limit $H \to \infty$ the late time dynamics given by either a scalar field dominated solution ($\Omega_\phi \to 1$) provided $0 < \gamma < 2, \lambda^2 < 3\gamma$ or $\frac{2}{3} < \gamma \le 2, \lambda^2 < 2$; or by a scaling solution ($\Omega_m/\Omega_\phi = O(1)$) for $0 < \gamma < \frac{2}{3}, \lambda^2 > 2$; or by a curvature scaling solution provided $\frac{2}{3} < \gamma \le 2, \lambda^2 > 2$. For finite values of H, the future attractors are stiff-like solutions. For positive curvature models the Universe evolves from a stiff regime to a stiff regime by crossing the value $\hat{H} = 0$, allowing the collapse of the Universe. Thus, from the dynamical view point there are not significant differences between quinstant and quintom dark energy paradigms.

Our opinion is that any dark energy model which presents a stiff-like equation of state in the past, during a long period of time, will predict abrupt peaks of structure formation, which would be the result of enhanced accretion of the forming structures, both because of gravitational and viscous forces.

References

[1] E. J. Copeland, M. Sami, and S. Tsujikawa, *Int. J. Mod. Phys.* **D15** (2006) 1753–1936.

[2] R. R. Caldwell and M. Kamionkowski, arXiv:0903.0866 [astro-ph.CO].

[3] S. M. Carroll, W. H. Press, and E. L. Turner, *Ann. Rev. Astron. Astrophys.* **30** (1992) 499–542.

[4] V. Sahni and A. A. Starobinsky, *Int. J. Mod. Phys.* **D9** (2000) 373–444.

[5] T. Padmanabhan, *Phys. Rept.* **380** (2003) 235–320.

[6] V. Sahni, *Class. Quant. Grav.* **19**, 3435 (2002) [arXiv:astro-ph/0202076].

[7] Feng B., Wang X. L. and Zhang X. M., 2005, *Phys. Lett. B* **607**, 35 [arXiv:astro-ph/0404224].

[8] Wei H. and Cai R. G., 2006, *Phys. Lett. B* **634**, 9 [arXiv:astro-ph/0512018].

[9] Sadjadi H. M. and Alimohammadi M., 2006, *Phys. Rev. D* **74**, 043506 [arXiv:gr-qc/0605143].

[10] M. R. Setare and E. N. Saridakis, *JCAP* **0809**, 026 (2008) [arXiv:0809.0114 [hep-th]].

[11] M. R. Setare and E. N. Saridakis, *Phys. Rev. D* **79**, 043005 (2009) [arXiv:0810.4775 [astro-ph]].

[12] Guo Z. K., Piao Y. S., Zhang X. M. and Zhang Y. Z., 2005, *Phys. Lett. B* **608**, 177 [arXiv:astro-ph/0410654].

[13] Zhang X. F., Li H., Piao Y. S. and Zhang X. M., 2006, *Mod. Phys. Lett. A* **21**, 231 [arXiv:astro-ph/0501652].

[14] Wei H. and Cai R. G., 2005, *Phys. Rev. D* **72**, 123507 [arXiv:astro-ph/0509328].

[15] Wei H., Cai R. G. and Zeng D. F., 2005, *Class. Quant. Grav.* **22**, 3189 [arXiv:hep-th/0501160].

[16] Lazkoz R. and Leon G., 2006, *Phys. Lett. B* **638**, 303 [arXiv:astro-ph/0602590].

[17] Zhang X., *Int. J. Mod. Phys. D* **14**, 1597 [arXiv:astro-ph/0504586].

[18] Zhang X. and Wu F. Q., *Phys. Rev. D* **72**, 043524 [arXiv:astro-ph/0506310].

[19] Zhang X., *Phys. Rev. D* **74**, 103505 [arXiv:astro-ph/0609699].

[20] E. N. Saridakis, *Phys. Lett. B* **661**, 335 (2008) [arXiv:0712.3806 [gr-qc]].

[21] M. R. Setare and E. N. Saridakis, *Phys. Lett. B* **671**, 331 (2009) [arXiv:0810.0645 [hep-th]].

[22] Y. f. Cai, M. z. Li, J. X. Lu, Y. S. Piao, T. t. Qiu and X. m. Zhang, *Phys. Lett. B* **651**, 1 (2007) [arXiv:hep-th/0701016].

[23] Zhang S. and Chen B., arXiv:0806.4435 [hep-ph].

[24] Sadeghi J., Setare M. R., Banijamali A. and Milani F., 2008, *Phys. Lett. B* **662**, 92 [arXiv:0804.0553 [hep-th]].

[25] Cai Y. F. and Wang J., 2008, *Class. Quant. Grav.* **25**, 165014 [arXiv:0806.3890 [hep-th]].

[26] E. N. Saridakis, P. F. Gonzalez-Diaz and C. L. Siguenza, *Class. Quant. Grav.* **26**, 165003 (2009) [arXiv:0901.1213 [astro-ph]].

[27] Lazkoz R., Leon G. and Quiros I., 2007, *Phys. Lett. B* **649**, 103 [arXiv:astro-ph/0701353].

[28] Alimohammadi M. and Sadjadi H. M., 2007, *Phys. Lett. B* **648**, 113 [arXiv:gr-qc/0608016].

[29] Alimohammadi M., 2008, *Gen. Rel. Grav.* **40**, 107 [arXiv:0706.1360 [gr-qc]].

[30] M. R. Setare and E. N. Saridakis, *Int. J. Mod. Phys.* D **18**, 549 (2009) [arXiv:0807.3807 [hep-th]].

[31] M. R. Setare and E. N. Saridakis, *Phys. Lett.* B **668**, 177 (2008) [arXiv:0802.2595 [hep-th]].

[32] E. Elizalde, S. Nojiri and S. D. Odintsov, *Phys. Rev.* D **70**, 043539 (2004) [arXiv:hep-th/0405034].

[33] P. S. Apostolopoulos and N. Tetradis, *Phys. Rev.* D **74**, 064021 (2006) [arXiv:hep-th/0604014].

[34] K. Bamba, S. Nojiri and S. D. Odintsov, *Phys. Rev.* D **77**, 123532 (2008) [arXiv:0803.3384 [hep-th]].

[35] K. Bamba, C. Q. Geng, S. Nojiri and S. D. Odintsov, *Phys. Rev.* D **79**, 083014 (2009) [arXiv:0810.4296 [hep-th]].

[36] M. R. Setare and E. N. Saridakis, *JCAP* **0903**, 002 (2009) [arXiv:0811.4253 [hep-th]].

[37] S. Nojiri and S. D. Odintsov, *eConf* **C0602061**, 06 (2006) [*Int. J. Geom. Meth. Mod. Phys.* **4**, 115 (2007)] [arXiv:hep-th/0601213].

[38] R. Cardenas, T. Gonzalez, Y. Leyva, O. Martin, and I. Quiros, *Phys. Rev.* **D67** (2003) 083501.

[39] V. F. Cardone, R. P. Cardenas, and Y. Leyva Nodal, *Class. Quant. Grav.* **25** (2008) 135010.

[40] J. Grande, J. Sola and H. Stefancic, *AIP Conf. Proc.* **878**, 220 (2006) [arXiv:astro-ph/0609683].

[41] J. Grande, J. Sola and H. Stefancic, *Phys. Lett.* B **645**, 236 (2007) [arXiv:gr-qc/0609083].

[42] J. Grande, J. Sola and H. Stefancic, *JCAP* **0608**, 011 (2006) [arXiv:gr-qc/0604057].

[43] J. Grande, J. Sola and H. Stefancic, *J. Phys. A* **40**, 6787 (2007) [arXiv:gr-qc/0701090].

[44] D. Huterer and A. Cooray, *Phys. Rev.* D **71**, 023506 (2005) [arXiv:astro-ph/0404062].

[45] D. Huterer and G. Starkman, *Phys. Rev. Lett.* **90**, 031301 (2003) [arXiv:astro-ph/0207517].

[46] Y. Wang and M. Tegmark, *Phys. Rev. Lett.* **92**, 241302 (2004) [arXiv:astro-ph/0403292].

[47] U. Alam, V. Sahni and A. A. Starobinsky, *JCAP* **0406**, 008 (2004) [arXiv:astro-ph/0403687].

[48] Y. Wang and P. Mukherjee, *Astrophys. J.* **606**, 654 (2004) [arXiv:astro-ph/0312192].

[49] U. Alam, V. Sahni, T. D. Saini and A. A. Starobinsky, *Mon. Not. Roy. Astron. Soc.* **354**, 275 (2004) [arXiv:astro-ph/0311364].

[50] T. Padmanabhan and T. R. Choudhury, *Mon. Not. Roy. Astron. Soc.* **344**, 823 (2003) [arXiv:astro-ph/0212573].

[51] Z. H. Zhu, M. K. Fujimoto and X. T. He, *Astron. Astrophys.* **417**, 833 (2004) [arXiv:astro-ph/0401095].

[52] M. Chevallier and D. Polarski, *Int. J. Mod. Phys. D* **10**, 213 (2001) [arXiv:gr-qc/0009008].

[53] E. V. Linder, *Phys. Rev. Lett.* **90**, 091301 (2003) [arXiv:astro-ph/0208512].

[54] B. Feng, X. L. Wang and X. M. Zhang, *Phys. Lett. B* **607**, 35 (2005) [arXiv:astro-ph/0404224].

[55] S. Hannestad and E. Mortsell, *JCAP* **0409**, 001 (2004) [arXiv:astro-ph/0407259].

[56] J. Q. Xia, B. Feng and X. M. Zhang, *Mod. Phys. Lett. A* **20**, 2409 (2005) [arXiv:astro-ph/0411501].

[57] J. Q. Xia, G. B. Zhao, B. Feng, H. Li and X. Zhang, *Phys. Rev. D* **73**, 063521 (2006) [arXiv:astro-ph/0511625].

[58] J. Q. Xia, G. B. Zhao, B. Feng and X. Zhang, *JCAP* **0609**, 015 (2006) [arXiv:astro-ph/0603393].

[59] G. B. Zhao, J. Q. Xia, B. Feng and X. Zhang, *Int. J. Mod. Phys. D* **16**, 1229 (2007) [arXiv:astro-ph/0603621].

[60] J. Q. Xia, G. B. Zhao, H. Li, B. Feng and X. Zhang, *Phys. Rev. D* **74**, 083521 (2006) [arXiv:astro-ph/0605366].

[61] J. Q. Xia, G. B. Zhao and X. Zhang, *Phys. Rev. D* **75**, 103505 (2007) [arXiv:astro-ph/0609463].

[62] G. B. Zhao, J. Q. Xia, H. Li, C. Tao, J. M. Virey, Z. H. Zhu and X. Zhang, *Phys. Lett. B* **648**, 8 (2007) [arXiv:astro-ph/0612728].

[63] Y. Wang and P. Mukherjee, *Phys. Rev. D* **76**, 103533 (2007) [arXiv:astro-ph/0703780].

[64] E. L. Wright, *Astrophys. J.* **664**, 633 (2007) [arXiv:astro-ph/0701584].

[65] H. Li, J. Q. Xia, G. B. Zhao, Z. H. Fan and X. Zhang, *Astrophys. J.* **683**, L1 (2008) [arXiv:0805.1118 [astro-ph]].

[66] J. Q. Xia, H. Li, G. B. Zhao and X. Zhang, *Phys. Rev. D* **78**, 083524 (2008) [arXiv:0807.3878 [astro-ph]].

[67] H. Li *et al.*, arXiv:0812.1672 [astro-ph].

[68] E. Komatsu *et al.* [WMAP Collaboration], *Astrophys. J. Suppl.* **180**, 330 (2009) [arXiv:0803.0547 [astro-ph]].

[69] Y. F. Cai, E. N. Saridakis, M. R. Setare and J. Q. Xia, arXiv:0909.2776 [hep-th].

[70] G. B. Zhao, J. Q. Xia, M. Li, B. Feng and X. Zhang, *Phys. Rev. D* **72**, 123515 (2005) [arXiv:astro-ph/0507482].

[71] C. P. Ma and E. Bertschinger, *Astrophys. J.* **455**, 7 (1995) [arXiv:astro-ph/9506072].

[72] D. Wands, K. A. Malik, D. H. Lyth and A. R. Liddle, *Phys. Rev. D* **62**, 043527 (2000) [arXiv:astro-ph/0003278].

[73] M. Kawasaki, T. Moroi and T. Takahashi, *Phys. Rev. D* **64**, 083009 (2001) [arXiv:astro-ph/0105161].

[74] T. Moroi and T. Takahashi, *Phys. Rev. Lett.* **92**, 091301 (2004) [arXiv:astro-ph/0308208].

[75] C. Gordon and D. Wands, *Phys. Rev. D* **71**, 123505 (2005) [arXiv:astro-ph/0504132].

[76] A. Lewis, A. Challinor and A. Lasenby, *Astrophys. J.* **538**, 473 (2000) [arXiv:astro-ph/9911177].

[77] J. Weller and A. M. Lewis, *Mon. Not. Roy. Astron. Soc.* **346**, 987 (2003) [arXiv:astro-ph/0307104].

[78] R. R. Caldwell, R. Dave and P. J. Steinhardt, *Phys. Rev. Lett.* **80**, 1582 (1998) [arXiv:astro-ph/9708069].

[79] A. G. Riess *et al.* [Supernova Search Team Collaboration], *Astrophys. J.* **607**, 665 (2004) [arXiv:astro-ph/0402512].

[80] C. R. Contaldi, M. Peloso, L. Kofman and A. Linde, *JCAP* **0307**, 002 (2003) [arXiv:astro-ph/0303636].

[81] B. Feng and X. Zhang, *Phys. Lett. B* **570**, 145 (2003) [arXiv:astro-ph/0305020].

[82] A. Kogut *et al.* [WMAP Collaboration], *Astrophys. J. Suppl.* **148**, 161 (2003) [arXiv:astro-ph/0302213].

[83] G. Hinshaw *et al.* [WMAP Collaboration], *Astrophys. J. Suppl.* **148**, 135 (2003) [arXiv:astro-ph/0302217].

[84] S. Dodelson, *Amsterdam, Netherlands: Academic Pr. (2003) 440 p.*

[85] B. Jain and A. Taylor, *Phys. Rev. Lett.* **91**, 141302 (2003) [arXiv:astro-ph/0306046].

[86] P. Zhang and U. L. Pen, *Mon. Not. Roy. Astron. Soc.* **367**, 169 (2006) [arXiv:astro-ph/0504551].

[87] P. Zhang and U. L. Pen, *Phys. Rev. Lett.* **95**, 241302 (2005) [arXiv:astro-ph/0506740].

[88] Coley A. A., "Dynamical systems and cosmology," *Kuwler Academic Publishers, (2003).*

[89] R. J. van den Hoogen and L. Filion, *Class. Quant. Grav.* **17**, 1815 (2000).

[90] A. A. Coley and R. J. van den Hoogen, *Phys. Rev. D* **62**, 023517 (2000) [arXiv:gr-qc/9911075].

[91] R. Tavakol, "Introduction to dynamical systems," *ch. 4, Part one, pp. 84–98, Cambridge University Press, Cambridge, England, 1997.*

[92] Y. Leyva Nodal, V. F. Cardone, and R. P. Cardenas, *AIP Conf. Proc.* **1083** (2008) 128–135.

[93] M. Fukugita and P.J.E Peebles, *ApJ,* **616**, 643, 2004.

[94] D. Eisenstein et al., *ApJ,* **633**, 560, 2005.

[95] M.A. Strauss et al., *AJ,* **124**, 1810, 2002.

[96] J.R. Bond, G. Efstathiou, M. Tegmark, *MNRAS,* **291**, L33, 1997.

[97] Y. Leyva, R. Cardenas, and V. Cardone, *Astrophys. Space Sci.* **323** (2009) 107.

Reviewed by Professor Alan A. Coley
Mailing Address: Dalhousie University
Dept. of Mathematics and Statistics
Halifax, Nova Scotia CANADA B3H 3J5

In: Dark Energy: Theories, Developments and Implications   ISBN 978-1-61668-271-2
Editors: K. Lefebvre and R. Garcia, pp. 215-239   © 2010 Nova Science Publishers, Inc.

Chapter 10

ON ACCRETION OF DARK ENERGY
ONTO BLACK- AND WORM-HOLES

José A. Jiménez Madrid[1],* *and Prado Martín-Moruno*[2],†
[1]Department of Applied Mathematics and Theoretical Physics,
Wilberforce Road, Cambridge, CB3 0WA, United Kingdom
[2]Colina de los Chopos, Instituto de Física Fundamental,
Consejo Superior de Investigaciones Científicas, Serrano 121, 28006 Madrid, Spain

Abstract

We review some of the possible models that are able to describe the current Universe which point out the future singularities that could appear. We show that the study of the dark energy accretion onto black- and worm-holes phenomena in these models could lead to unexpected consequences, allowing even the avoidance of the considered singularities. We also review the debate about the approach used to study the accretion phenomenon which has appeared in literature to demonstrate the advantages and drawbacks of the different points of view. We finally suggest new lines of research to resolve the shortcomings of the different accretion methods. We then discuss future directions for new possible observations that could help choose the most accurate model.

PACS 04.70.-s, 95.36.+x, 98.80.-k.

Keywords: dark energy, black holes, wormholes, accretion.

1.   Introduction

The discovery of the cosmic acceleration indicated by the observational data [1, 2, 3] has caused a break in the belief of what could be the matter content of the universe and what might be its possible future evolution. The interpretation of this data in the framework of General Relativity implies that the majority of the content in the universe should be new stuff, which has been called dark energy, possessing anti-gravitational properties, i.e. the

---

*E-mail adresses: J.Madrid@damtp.cam.ac.uk, madrid@imaff.cfmac.csic.es
†E-mail address:pra@imaff.cfmac.csic.es

equation of state parameter of the dark energy must be $w < -1/3$ ($w = p/\rho$). It even seems to be possible that the equation of state parameter is less than -1. In that case the new stuff is known as phantom energy [4] and the consideration of this fluid could lead the universe to a catastrophic end by the appearance of a future singularity. The most popular of such singularities is the so-called big rip [5], which is a possible doomsday of the universe where both of its size and its energy density become infinitely larger. However, the big rip is not the only possibility which has been suggested as the catastrophic end to the universe in the new phantom models. It has also been argued that the universe could finish its evolution at a time where its energy density becomes infinitely larger maintaining the scale factor as a finite value, known as the big freeze singularity [6, 7].

It is well known that dark energy should be accreted onto black holes in a different way that ordinary matter does, since that new fluid covers the whole space. Therefore, the study of dark energy accretion onto black holes becomes an interesting field of study which could lead to surprising effects as the possible disappearance of black holes in phantom environments. As we will show in this chapter the mentioned accretion phenomenon was originally studied by Babichev et al. [21], although a great number of works have been done to improve the method used by those authors [28, 31, 33].

On the other hand, the accretion process could also imply unexpected consequences in the case that one considers the evolution of cosmological objects even stranger than black holes, wormholes[1]. Traversable wormholes are short-cuts between two regions of the same universe or between two universes, which could be used to construct time-machines [16]. The reason for its strangeness is not related to its bridge character but to that, in order to be traversable and stable, the wormholes must be surrounded by some kind of exotic matter which do not fulfil the null energy condition. Nevertheless, the consideration of phantom models as the possible current description of the universe has caused a revival of interest in traversable wormholes, since this fluid would also violate the null energy condition. Even more, it has been shown that an inhomogeneous version of phantom energy can be the exotic stuff which supports wormholes [14]. In some cosmological models the accretion of phantom energy onto a wormhole could lead to an enormous growth of its mouth, engulfing the whole universe which would travel through it in a big trip [25], avoiding the big rip [19] or big freeze [6] singularity in the corresponding cases.

In the present chapter we show the method of how to treat the accretion phenomenon onto black- and worm- holes and its cosmological consequences in some models. In Sec. II, we review some candidates that could be responsible for the current cosmological acceleration and which pay special attention to the possible future singularities appearing in some of them. In Sec. III, the procedure of the study of the dark energy accretion onto black holes based on the Babichev et al. method is shown and its application to the models included in Sec. II is presented. The corresponding study in the case of accretion onto wormholes is shown in Sec. IV, where the possible avoidance of the future singularities is highlighted. Since the study of dark energy accretion onto black- and worm- holes is still an open issue, we refer in Sec. V some interesting works which have produced a debate about the used method and we also outline some possible lines for future research to solve the shortcomings. Finally, in Sec. VI, the results are discussed and further comments are added.

---

[1]For information about the historical development of wormholes and a deep study of their spacetime, please see Ref.[17]

2.   Brief Review of Some Candidates to Cosmic Acceleration

The origin of the current accelerating expansion of the universe is one of the most interesting challenges in cosmology. Therefore, a plethora of cosmological models have been developed in recent years in order to take into account such acceleration. Although modifications of the Lagrangian of General Relativity or considerations of more than four dimensions could explain the current phase of the universe, the acceleration can also be modelled in the framework of General Relativity theory, which has shown agreement with the observational tests up to now. Nevertheless, the consequences of using the Einstein's theory is that most part of the universe's content must be some kind of fluid with anti-gravitational properties, called dark energy.

In order to show the necessity of the inclusion of dark energy as a new component of a universe described by General Relativity, we must consider an homogeneous, isotropic and spatially flat universe, i.e. a Friedmann-Lemaître-Robertson-Walker (FLRW) model with $k = 0$. As an approximation, one can consider that this model is only filled with one fluid[2]. Throughout this chapter we shall use natural units so that G = c = 1. The Friedmann equations can be expressed in the usual way [47]

$$3 \left(\frac{\dot{a}}{a} \right)^2 = 8\pi\rho, \tag{1}$$

$$3\frac{\ddot{a}}{a} = -4\pi(\rho + 3p), \tag{2}$$

where $a(t)$ is the scale factor, which can be used to define the Hubble parameter $H = \dot{a}/a$, ρ and p are the energy density and the pressure of the fluid, respectively. Eq. (2) shows that, in order to obtain an accelerating universe, the dark energy must have an energy density and pressure such that $\rho + 3p < 0$, violating at least the strong energy condition. If one wants to minimise the strange character of the dark energy, then $w > -1$ ($w = p/\rho$) could be imposed, but it must be noted that such restriction is not based in direct observations but in theoretical wishes.

In fact, as we have already mentioned in the introduction, the observational data indicates that w must be around -1 and, therefore, values less than -1 are not excluded. In that case the fluid is called phantom energy [4] and violates even the dominant energy condition.

The limiting case, $w = -1$, is also possible and is equivalent to the introduction of a positive cosmological constant. We shall not pay much attention to this case because it is the well-known de Sitter solution and more importantly, the cosmological constant would be accreted neither by black- nor by worm-holes, as it could be expected and we show in the next sections.

In this section we present the simplest dark and phantom energy models, where the equation of state parameter is considered to be closely constant. As we shall see, such a

---

[2]It must be noted that such an approximation is justified because the contribution of the current dark energy density is around 74% of the total energy density of the universe and, from the evolution of the energy density in terms of the scale factor, it is expected that dark energy density decays slower than the ordinary matter energy density when the scale factor increases, being therefore the future dynamic of the universe governed by the dark component. Even more, in the phantom case the phantom energy density would increase with the scale factor, so the same conclusions can be, of course, recovered in this case.

phantom energy model implies the occurrence of a big rip [5]. Since it could seem that phantom energy implies the occurrence of a big rip singularity, we want to show that such a singularity is not an inherent property of that fluid. So we shall consider Phantom Generalized Chaplygin Gas (PGCG) models, in order to clarify that phantom models could present no future singularities [8] or future singularities of other kinds [6, 7].

2.1. Quintessence with a Constant Equation of State Parameter

The most popular candidate to describe dark energy, allowing a dynamic evolution for this unknown component, is the quintessence model. In this model a spatially homogeneous massless scalar field is considered, which can be interpreted as a perfect fluid with negative pressure, taking the equation of state parameter values on the range $-1 < w < -1/3$. Supposing that the equation of state parameter is approximately constant, the conservation law of the fluid in a FLRW spacetime, $\dot{\rho} + 3H(p + \rho) = 0$, can be integrated to obtain

$$p = w\rho = w\rho_0 \left[a(t)/a_0\right]^{-3(1+w)}, \tag{3}$$

which can be introduced in Eq. (1) leading

$$a(t) = a_0 \left(1 + \frac{3}{2}(1 + w)C(t - t_0)\right)^{2/[3(w+1)]}, \tag{4}$$

with $C = (8\pi\rho_0/3)^{1/2}$ and the subscript 0 denoting the value at the current time t_0. Therefore, a universe described with that model would accelerate forever, decreasing the dark energy density in the process.

2.2. Phantom Quintessence with a Constant Equation of State Parameter

Phantom energy can be considered to be a fluid with an equation of state parameter less than -1 which would, therefore, violate not only the strong energy condition but also the dominant energy condition[3]. But, since the observational data suggests that it could be responsible for the current accelerating expansion therefore, such a pathological fluid should be seriously considered as the possible dominating matter content of our universe.

Analogous to the quintessence case, one can easily obtain an expression for the scale factor in this model considering that w is approximately constant. One has

$$a(t) = a_0 \left(1 - \frac{3}{2}(|w| - 1)C(t - t_0)\right)^{-2/[3(|w|-1)]}, \tag{5}$$

with $\rho = \rho_0 \left[a(t)/a_0\right]^{3(|w|-1)}$ and C taking the already mentioned value. Therefore, the scale factor (5) increases with time even faster than the scale factor of a de Sitter universe (which has an exponential behaviour) up to

$$t_{br} = t_0 + \frac{2}{3(|w| - 1)C} > t_0. \tag{6}$$

---

[3]It must be noted that if we want to express the phantom fluid by using a scalar field like in the quintessence dark energy case, then it will possess a negative kinetic term.

At this time both the scale factor and the energy density of the fluid blow up in which is known as the big rip singularity [5]. If our Universe is described by this model, then an observer located on the Earth would see how progressively it could be ripped apart the galaxies, the stars, our solar system and finally, the atoms and nuclei, up until the moment when every component of the universe would be out of the Hubble horizon of the other components.

2.3. Phantom Generalized Chaplygin Gas

It could seem that the consideration of a universe filled with phantom energy implies the occurrence of a future big rip singularity, but this is not necessarily the case. We support this claim with the example taken from the Phantom Generalized Chaplygin Gas (PGCG) [6, 7, 8].

The Generalized Chaplygin Gas (GCG) is a fluid with an equation of state of the form [9]

$$p = -\frac{A}{\rho^\alpha},\tag{7}$$

where A is a positive constant and $\alpha > -1$ is a parameter. In the particular case $\alpha = 1$ we recover the equation of state of a Chaplygin gas. Inserting Eq. (7) in the conservation of the energy momentum tensor, one obtains

$$\rho = \left(A + \frac{B}{a^{3(1+\alpha)}}\right)^{\frac{1}{1+\alpha}},\tag{8}$$

with B a constant parameter. It can be seen that, maintaining $A > 0$, such a fluid fulfils the dominant energy condition for $B > 0$ and it is violated otherwise. The rather strange equation of state expressed through Eq. (7) has been considered firstly in cosmology because the GCG could reproduce a transition from a dust dominated universe at early time to de Sitter behaviour at late time.

On the other hand, PGCG [6, 7, 8, 13] are fluids with an equation of state of the GCG type, Eq. (7), with the parameters A, B and α taking values in intervals such that the energy density and pressure of the fluid fulfil the requirements $\rho > 0$ and $p + \rho < 0$. It can be seen that such requirements lead to four classes of PGCG with

- type I: $A > 0$, $B < 0$ and $1 + \alpha > 0$.

- type II: $A > 0$, $B < 0$ and $1 + \alpha < 0$.

- type III: $A < 0$, $B > 0$ and $(1 + \alpha)^{-1} = 2n > 0$.

- type IV: $A < 0$, $B > 0$ and $(1 + \alpha)^{-1} = 2n < 0$.

We want to emphasise that when PGCG is considered, the sign of the parameters A and B must not to be necessarily positive and also α can be bigger or less than -1.

It can be seen that for type I the scale factor is bounded from below by $a_{\min} = |B/A|^{1/[3(1+\alpha)]}$ and, therefore, it takes values in the interval $a_{\min} \leq a < \infty$, which corresponds to $0 \leq \rho < |A|^{1/(1+\alpha)}$, approaching the energy density a finite constant value when the scale factor tends to infinity. The Friedmann Eq. (1) can be analytically integrated for the energy density (8) to lead a functional dependence of the cosmic time and the scale factor in terms of a hypergeometric series [8]. Since this expression is rather complicated and can be found in the literature, [8], we consider that it is enough to comment that the resulting expression implies that when the scale factor diverges the cosmic time also blows up, that is, there is no a singularity at a finite time in the future. The future normal behaviour is also indicated by the Hubble parameter which approaches a constant finite non-vanishing value for large scale factors and, therefore, we can conclude that the late time evolution of such a model is asymptotically de Sitter. It can be seen [7] that the type III model has a similar evolution for late times than the type I, although both behaviours can differ greatly at early times[4].

Although we have just discussed that the consideration of phantom models could avoid the occurrence of a future big rip singularity, another kind of doomsday could appear in phantom models. In order to show this fact, let us consider the type II PGCG. In this case the scale factor is bound from above by $a_{max} = |B/A|^{1/[3(1+\alpha)]}$ taking, therefore, values in the interval $0 < a \leq a_{max}$ which correspond to $A^{1/(1+\alpha)} \leq \rho < \infty$. As in the type I an analytical expression for the scale factor depending on the cosmic time can be found in terms of hypergeometric series [7]. Such an expression can be approximated close to the maximum scale factor value and inverted leading to [6, 7]

$$a \simeq a_{max} \left\{ 1 - \left(\frac{8\pi}{3} \right)^{\frac{1+\alpha}{1+2\alpha}} \left[\frac{1+2\alpha}{2(1+\alpha)} \right]^{\frac{2(1+\alpha)}{1+2\alpha}} A^{\frac{1}{1+2\alpha}} |3(1+\alpha)|^{\frac{1}{1+2\alpha}} (t_{max} - t)^{\frac{2(1+\alpha)}{1+2\alpha}} \right\},$$
(9)

which implies that the cosmic time elapsed since the universe has a given scale factor a (closed to a_{max}) until it reaches its maximum value is finite, i.e., $t_{max} - t < \infty$. Therefore, this model would end at a finite singularity where its energy density blows up whereas the scale factor remains finite, called big freeze singularity [6]. The type IV model would have a similar behaviour and it exhibits a singularity of the same kind, [7]. So, both models describe a universe which would expand accelerating until it freezes its evolution at a finite time where it is infinitely full of phantom energy.

It must be pointed out that from a classical point of view, as such we are considering through this chapter, a singularity would break down the spacetime. Nevertheless, it has been argued that the consideration of quantum effects could avoid the big rip [22] and the big freeze singularity [10].

3. Dark Energy Accretion onto Black Holes

As we have pointed out in the previous section, dark energy is filling our Universe and, therefore, it could be expected that it would interact with different cosmological objects like black- and worm-holes. This section is dedicated to the study of the accretion process onto black holes by using the most accepted model dealing with this phenomenon. Nevertheless, it must be pointed out that some controversy has originated around this method and as such other approaches have been proposed, which will be discussed in Sec. V.

---

[4]The quantised α parameter in the type IV model eliminates possible past singularities that could appear in type I. A discussion about the evolution of these models at early times is out in the scope of the present chapter, so we refer the interested reader to Ref. [7].

The standard method treating the dark energy accretion onto black holes was firstly presented by Babichev et al. [21] and is based on the consideration of a black hole described by the Schwarzschild metric, surrounded by a perfect fluid which represents dark energy. In such a framework they considered the zero component of the energy-momentum conservation equation and the projection of this equation along the four-velocity, to derive the dynamical evolution of the black hole mass. We want to summarise a generalisation of this method, presented in Ref. [28], which allows an internal nonzero energy-flow component Θ_0^r by the consideration of the simplest non-static generalisation of the Schwarzschild metric, in which the black hole mass can depend generically on time. That is, the metric can be given by

$$ds^2 = \left(1 - \frac{2M(t)}{r}\right)dt^2 - \left(1 - \frac{2M(t)}{r}\right)^{-1}dr^2 - r^2\left(d\theta^2 + \sin^2 d\phi^2\right), \qquad (10)$$

where $M(t)$ is the black hole mass.

The zero component of the conservation law for energy-momentum tensor and its projection along the four-velocity can be integrated in the radial coordinate considering a surrounding perfect fluid. On the other hand, it is known that the rate of change of the black hole mass due to accretion of dark energy can be derived by integrating over the surface area the density of momentum T_0^r. Taking into account these considerations one gets the following equation [28]

$$\dot{M} = 4\pi A_M M^2\left(p + \rho\right)e^{-\int_\infty^r f(r,t)dr}, \qquad (11)$$

relating the temporal rate of change of the mass with the pressure and energy density of the perfect fluid which is considered to describe dark energy.

For the relevant physical case of an asymptotic observer, i.e. $r \to \infty$, the previous equation simplifies to

$$\dot{M} = 4\pi A_M M^2\left(p + \rho\right), \qquad (12)$$

where A_M is a positive constant of order unity. It must be pointed out that this expression is the same as that obtained by Babichev et al. [21] using the usual Schwarzschild metric, with the difference that in the current case it is only valid for asymptotic observers.

It can be noted that Eq. (12) shows that the black hole mass, and with it its size, must decrease when the black hole accretes a fluid which violates the dominant energy condition, i.e. a phantom fluid. This would increase when the dominant energy condition is preserved and it remains constant in the cosmological constant case, since a cosmological constant can be modelled by a perfect fluid with $p + \rho = 0$. It must be emphasised that Eq. (12) has been obtained for a general perfect fluid, therefore it would be valid in a great variety of dark energy models, since a load of them are based on a perfect fluid.

Since accretion of dark energy onto black holes would increase the black hole size in dark energy models fulfilling the dominant energy condition, an interesting question is whether this growth could be large enough to that the black hole might engulf the whole universe. In order to find a possible answer for this question, one can take into account the

Friedmann equations to integrate Eq. (12), obtaining the temporal evolution of the black hole mass. That is, [12]

$$M = \frac{M_0}{1 + \sqrt{\frac{8\pi}{3}} A_M M_0 \left(\rho^{1/2} - \rho_0^{1/2} \right)}. \tag{13}$$

This equation can also be expressed in terms of the Hubble parameter in the following way

$$M(t) = \frac{M_0}{1 + D M_0 \left[H(t) - H_0 \right]}. \tag{14}$$

Therefore, as mentioned in Ref. [28], a black hole capable of engulfing the whole universe as shown in a model with an equation of state parameter bigger than minus one, should have a current mass which, roughly speaking, is bigger than all the matter content of the current observable universe, making the occurrence of such a phenomenon impossible.

In the following subsections we consider the models reviewed in the previous section to study the evolution of a black hole living in a universe filled with dark energy.

3.1. Application to a Quintessence Model

We assume that dark energy is modelled by a quintessence model satisfying the equation of state (3) with $w > -1$ constant. As we have already mentioned, Eq. (12) implies that the rate of mass change is positive in this model, therefore the black hole mass grows along time due to accretion of quintessence. In order to obtain the dynamical behaviour of the black hole mass, one can introduce the equation of state (3) in Eq. (13) which leads to

$$M = \frac{M_0 \left[1 + \frac{3}{2} (1 + w) C (t - t_0) \right]}{1 + \frac{3}{2} (1 + w) C (t - t_0) - 4\pi A_M \rho_0 M_0 (1 + w) (t - t_0)}. \tag{15}$$

There are something very interesting in this expression for the black hole mass, because it allows the occurrence of a bizarre fate for our universe as we have already pointed out. That is, Eq. (15) expresses the possibility that our universe might be engulfed by a black hole, since accretion of dark energy could make the mass of the black hole increase so quickly as to yield a black hole size that would eventually exceed the size of the universe in a finite cosmic time. In fact, the time in which the black hole might reach a infinite size would be

$$t_{bs} = t_0 + \frac{1}{(1 + w) \left(4\pi A_M \rho_0 M_0 - \sqrt{6\pi \rho_0} \right)}. \tag{16}$$

This time is finite but, although a universe filled with quintessence has no future singularity, present observational data seems to imply that w is not constant and suggests values less than -1, making it unlikely that the occurrence of the considered bizarre phenomenon at any time in the far future would happen, at least in principle. However, if $\dot{w} > 0$ then the black holes would undergo a larger growth due to accretion of dark energy.

Nevertheless, if one studies in deeper detail Eq. (15), then one obtains (see Ref. [28]) that in order to have a black hole able to reach an infinite mass in an infinite time, this black hole must possess an initial mass such as $(8\pi \rho_0 / 3)^{1/2} A_M M_0 = 1$ which means, taking into account the observational data, $M \sim 10^{23} M_\odot$ (where $M_\odot$ is the Sun's mass). Therefore, in order to have an infinitely large black hole in a finite time in the future, the current black hole mass should be bigger than $10^{23} M_\odot$, which is an extremely large value even for black holes in the galaxies centres. Even more, one can estimate the current matter content of the universe assuming that the observable Universe expands at the speed of light, obtaining a total of 10^{23} starts [28]. Therefore, it seems that in order to have a black hole able to engulf the whole universe, it should have a current mass equal to the mass of all the observable universe, which would not be possible.

Finally, we want to point out that, even in the case that the accretion phenomenon of dark energy onto black holes could not produce cosmological consequences in terms of a catastrophic end, it could help in the determination of the correct dark energy model. So, if astronomers were able, in practice, to observe a growth bigger than expected of those black holes living in the centre of the galaxies, then this could be an observational measure of the effects of dark energy with $w > -1$.

3.2. Application to a Phantom Quintessence Model

As we have already pointed out, the observational data not only allows that the equation of state parameter takes a value less than -1 but even they seem to suggest it, acquiring therefore, special interest in the study of the evolution of a black hole in a universe filled with phantom energy. So, now we consider $w < -1$ and constant, consequently (12) shows that the black hole mass decreases with time. More precisely, inserting Eqs. (3) and (5) in Eq. (13), we get an accurate expression of the evolution of the black hole mass, i.e.

$$M = \frac{M_0 \left[1 - \frac{3}{2}(|w|-1)C(t-t_0)\right]}{1 - \frac{3}{2}(|w|-1)C(t-t_0) + 4\pi A_M \rho_0 M_0 (|w|-1)(t-t_0)}. \tag{17}$$

Taking into account that in a universe filled with phantom energy with w constant a big rip singularity will take place in the future, one can introduce the time of occurrence of the big rip, t_{br}, in (17) to get that the black hole mass vanishes at t_{br}, independently of the current black hole mass, M_0; that is, all black holes disappear at the big rip [21]. It can be noted, by inspection of Eq. (14), that this is not only an interesting property of this phantom model but it can be found in all models which present a singularity with a divergence of the Hubble parameter at a finite time in the future.

Finally, the decrease of black holes due to the phantom energy accretion phenomenon could provide us with a possible observational test of these models. Therefore, if future observations of black holes in the centre of galaxies (or other possible black holes) indicate a growth of those objects less than expected, then it could be associated to accretion of phantom energy, providing us with another measure able to discriminate between different dark energy models, which would complete those that come from GRB [44, 48], supernova, or other observational data.

3.3. Application to a Generalized Chaplygin Model

Let us now study the evolution of a black hole in a universe filled with a Generalized Chaplygin Gas. Since the consideration of dark energy modelled by some kinds of Phantom

Generalized Chaplygin Gas could prevent the occurrence of a big rip [8], one could expect the avoidance of the weird behaviours that appear in quintessence or phantom models in these frameworks. In order to see if that is the case, one must take into account Eq. (13).

As we have mentioned in Sec. II, a Generalized Chaplygin Gas, with $A > 0$ and $\alpha > -1$, preserves dominant energy condition when the parameter[5] $B > 0$ and violated otherwise. So when $B > 0$ the black hole mass increases with cosmic time up to a constant value, but there are a set of parameters where if α is close to -1 then it would seem that the black hole mass could eventually exceed the size of the universe at finite time in the future [24]. When dominant energy condition is violated, $B < 0$, black hole mass decreases along cosmic time, tending to a nonzero constant value, therefore black holes do not disappear in this model.

On the other hand, it can be seen [12] that performing a deeper study of the mentioned four types of PGCG (where the sign of A and the range on α is not previously fixed) at late times, where the phantom fluid would drive the dynamical evolution of the universe, the results can be summarised as follows:

- Type I and III. Black holes decrease with time, where the mass tend to a nonzero value when the time goes to infinity.

- Type II and IV. Black holes masses decrease, but now all black holes disappear at big freeze, i.e., the mass of all black holes tend to zero when the universe reaches the big freeze singularity with independence of their initial mass (as it should be expected by inspection of equation (14)).

To end this section, we consider that it would be quite interesting to explore the region of parameter space $(\alpha, A, H_0, \Omega_K, \Omega_\phi)$ allowed by current observations in order to determine whether there exists any allowed sections leading to a big freeze or a big rip. However, all available analyses [37, 38, 39, 40, 41, 42, 43, 44, 45, 46] are restricted to the physical region where no dominant energy condition is violated. Therefore, the section described by the interval implied by a PGCG necessarily is outside the analysed regions. One has to extend the investigated domains to include values of parameter $A > 1$, $A < 0$ or $\alpha < -1$ to probe the parameters space where the dominant energy condition is violated.

3.4. Consideration to Other Black Holes

Up to now, we have shown the evolution of a Schwarzschild black hole in a universe filled with dark energy. In this subsection, we study the accretion of dark energy onto charged or rotating black holes, to show whether charge or angular momentum have some influence in their evolution.

Let us continue by considering another black hole metric in order to understand better the application of the dark energy accretion mechanism. In [34], Babichev et al. apply a generalisation of the accretion formalism to a Reissner-Nordsröm black hole. In this case, the metric is given by

---

[5]B is the constant parameter that appears in the energy momentum tensor conservation law (8) for a GCG.

$$ds^2 = \left(1 - \frac{2M}{r} + \frac{e^2}{r^2}\right) dt^2 - \left(1 - \frac{2M}{r} + \frac{e^2}{r^2}\right)^{-1} dr^2 - r^2\left(d\theta^2 + \sin^2\theta d\phi^2\right), \quad (18)$$

where $m^2 > e^2$. It can be noted that if $m < |e|$, then the solution would represent a naked singularity, corresponding $m = |e|$ to the extreme case. By integrating the conservations laws for momentum-energy and its projection along four-velocity for the case for a perfect fluid, and taking into account that the rate of change of the black hole mass due to accretion of dark energy can be derived by integration over the surface area the density of momentum T_0^r, Babichev et al. get again the same expression (12) which relates the temporal rate of change of the black hole mass to the pressure and the energy density of the perfect fluid. So, also in a Reissner-Nordsröm black hole, the black hole mass increases when it is accreting dark energy holding the dominant energy condition and its mass decreases when phantom energy is considered in the accretion process.

At this point the next question arises again, if phantom energy is getting involved then black hole mass decreases, vanishing at big rip singularity. Therefore, since the electric charge e remains constant due to phantom energy accretion, then in a finite time, the black hole must reach the extreme case, transforming the black hole into a naked singularity. Nevertheless, the authors of Ref. [34] perform a more detailed study about this possible transformation in a naked singularity, concluding that there is no accretion of the perfect fluid onto the Reissner-Nordström naked singularity when $m^2 < e^2$ and that, in this situation, a static atmosphere of the fluid around the naked singularity would be formed.

It must be emphasised that, although when one is considering cases far from the extreme case, the back reaction can be neglected and the perfect fluid approximation appears to be valid, it seems that this approximation breaks down close to the extremal case, where one has to take into account the back reaction of the perfect fluid onto the background metric. Even more, the same consideration about the avoidance of transformation of a black hole into a naked singularity, can also be applied to a Kerr black hole. Nevertheless, if the back reaction does not prevent the process of phantom accretion onto a charged black hole or rotating black hole, then this process could be a way to violate the cosmic censorship conjecture [51].

4.   Dark Energy Accretion onto Wormholes

The first solution of the Einstein's equations describing a traversable wormhole was found by Morris and Thorne in their seminal work [15]. That solution, obtained under the assumption of staticity and spherical symmetry, describes a throat connecting two asymptotically flat regions of the spacetime without any horizon and can be expressed as

$$ds^2 = -e^{2\Phi(r)}dt^2 + \frac{dr^2}{1 - K(r)/r} + r^2\left(d\theta^2 + \sin^2 d\varphi^2\right), \quad (19)$$

where $\Phi(r)$ and $K(r)$ are the shift and shape functions, respectively, both tending to a constant value when the radial coordinate $r \to \infty$ in order to have asymptotic flatness. It must be noted that, in these coordinates, two coordinate patches are needed to cover the two asymptotically flat regions, each with $r_0 \leq r \leq \infty$, with r_0 the minimum radius which corresponds to the throat radius, where $K(r_0) = r_0$.

It can be seen [15] that solution (19) must fulfil some additional requirements in order to describe a traversable wormhole. In particular the outward flaring condition imposes $K'(r_0) < 1$ what, through the Einstein's equations, implies that $p_r(r_0) + \rho(r_0) < 0$ (where here p_r denotes the radial component of the pressure). Therefore, the wormhole must be surrounded by some material with unusual characteristics, called exotic matter, which could lead to the neglect of such spacetime.

Nevertheless, as we have already mentioned in the introduction, the discovery of the current accelerated expansion of the Universe and the consideration of phantom energy as a possible candidate for its origin has produced a more natural consideration of the properties of exotic matter, since it seems that phantom energy could be precisely the exotic stuff which supports wormholes. Gonzalez-Diaz [19] considered that similarly to black holes accrete dark energy, wormholes could accrete phantom energy producing a great increase of their size, in such a way that the size of a wormhole could be infinitely larger before the universe reaches the big rip singularity. Such a process would produce the moment that the size of the wormhole equals the size of the universe, the universe boards itself in a travel through the wormhole, called big trip. Even more, the notion of phantom energy has been extended to inhomogeneous spherically symmetric spacetimes showing that it can be in fact the exotic material which supports wormholes [14], which backs up the mentioned idea.

In order to study such a process one can follow a similar method to the one used by Babichev et al. for the black hole case. So, considering the non-static generalisation of Eq. (19) obtained by the consideration of an arbitrary dependence of the shape function on the time, $K(r,t)$, and an energy momentum-tensor of a perfect fluid, one can find the equivalent of Eq. (12) for the wormhole case, that is the temporal mass rate evolution as measured by an asymptotic observer, which is [30]

$$\dot{m} = -4\pi Q m^2 (p + \rho), \tag{20}$$

with Q a positive constant. That expression shows that the wormhole mass, and with it, its size must increase when the wormhole accretes phantom energy, it decreases by the accretion of dark energy, remaining constant in the cosmological constant case. It must be remarked that in the achieving of Eq. (20) no assumption about the possible dependence on the energy density or on the pressure of the fluid have been done, allowing an arbitrary dependence with the time and with the radial component, therefore, such an equation is general and take into account the possible back reaction in an asymptotically flat wormhole spacetime[6].

If one now considers as an approximation that the fluid which surrounds the wormhole is a cosmological one, i.e., an homogeneous and isotropic fluid fulfilling the Friedmann equations (1) and (2) and the conservation law, then Eq. (20) can be integrated to obtain [12, 24]

$$m(t) = \frac{m_0}{1 - Q m_0 \left[H(t) - H_0 \right]}. \tag{21}$$

---

[6]It must be emphasised that, whereas in the case of the study of the dark energy accretion onto black hole phenomenon, the solution is not able to take into account the back reaction of the spacetime, in this case we are treating with a non-vacuum solution and allowing arbitrary time dependence on the involved functions and, therefore, up to now we are taking into account any possible back reaction.

This expression shows that in phantom models, where $H(t)$ is an increasing function, the wormhole throat could become infinitely big if the Hubble parameter reaches the value $H_* = H_0 + 1/(Qm_0) < \infty$ at some time t_* in the future. It can be seen that this would be the case at least in models which show a future singularity in a finite time in the future characterised by a divergence of the Hubble parameter, because in those models one has $H_0 < H_* < H_{\text{sing}} = \infty$ and, since the Hubble parameter is a strictly increasing and continuous function before the time of the singularity, this implies $t_0 < t_* < t_{\text{sing}}$. Therefore, the size of a wormhole would be bigger than the size of the universe before the occurrence of the future singularity in all models possessing a future singularity where the Hubble parameter blows up, i.e., in such models the universe would travel through a big trip.

In this section we show the implications of the phenomenon of dark and phantom energy accretion onto wormholes in the models presented in Sec. II. That procedure lead, as it is expected, to the decrease of the wormhole size when it accretes dark energy with $w > -1$ and to a growth of the wormhole mouth in phantom cases. Even more, the big rip and big freeze singularities, in the corresponding models, can be avoided by a big trip phenomenon since, although these singularities present a different behaviour of the scale factor, at both singularities the Hubble parameter blows up.

4.1. Application to a Quintessence Model

Let us consider that the dark energy is modelled by a quintessence model, satisfying the equation of state (3) with $w > -1$ constant. Eq. (20) tells us that the rate of mass change is negative, so the wormhole mass would decrease with time due to the accretion of quintessence. Furthermore, taking into account the equation of state (3), one can solve (20) getting the following expression which relates the wormhole mass to the cosmic time [25],

$$m = \frac{m_0}{1 + \frac{4\pi Q\rho_0 m_0(1+w)(t-t_0)}{1+\frac{3}{2}(1+w)C(t-t_0)}}. \tag{22}$$

This expression shows us how a wormhole loses mass due to the accretion of quintessence. Even more, if due to any additional hypothetical process this wormhole would have a macroscopic size, then it would be subjected to chronology protection [36]; therefore, vacuum polarisation created particles would catastrophically accumulated on the chronology horizon of the wormhole, letting the corresponding normalised stress-energy tensor to diverge which, at the end of the day, would imply the disappearance of the wormhole.

4.2. Application to a Phantom Quintessence Model

Now, we are interested in study the evolution of a wormhole in a universe filled with phantom energy with $w < -1$ constant. In order to obtain the temporal evolution of the wormhole, one can introduce the equation of state of phantom energy into the r.h.s. of Eq. (20), getting [25],

$$m = \frac{m_0}{1 + \frac{4\pi Q\rho_0 m_0(|w|-1)(t-t_0)}{1+\frac{3}{2}(|w|-1)C(t-t_0)}}. \tag{23}$$

This expression implies that the exotic mass of the wormhole diverges at the time

$$t_{bt} = t_0 + \frac{t_{br} - t_0}{1 + \frac{8\pi Q m_0 a_0^{3(|w|-1)/2}}{3C}}, \tag{24}$$

where t_{br} is the finite time at which the big rip singularity takes place. Since the wormhole size diverges before that the universe reaches the big rip singularity, it would be a previous time at which the size of the wormhole would be bigger than the universe, being at this time where properly starts the travel of the universe through the wormhole.

The huge growth of the wormhole throat poses the following two problems. On the one hand, since the wormhole spacetimes are usually considered to be asymptotically flat, when the wormhole increases more than the universe it is impossible to place the wormhole on this universe. On the other hand, since the universe is travelling through the wormhole, one can ask where is the universe travelling to? The solution of these equivalent problems requires the consideration of a multiverse scenario. In such a framework the wormhole could be re-infixed in another universe where the wormhole would be asymptotically flat to, giving also a final destination to the universal travel[7].

4.3. Application to a Generalized Chaplygin Gas Model

Finally, we will study the evolution of a wormhole in the case of a universe filled with a Generalized Chaplygin Gas. Since type I and III of PGCG avoid the occurrence of a future singularity [7, 8], it is of special interest to study the possible occurrence of a big trip phenomenon in these models. Following this line of thinking, in Ref. [24] it is analysed the phantom energy accretion phenomenon onto wormholes when the phantom energy is modelled by a type I PGCG. In order to perform this study, let us temporarily fix $A > 0$ and $\alpha > -1$, therefore, solving Eq. (21) for the equation of state of a GCG, one obtains

$$m = \frac{m_0}{1 - Q m_0 \sqrt{\frac{8\pi}{3}} \left(\rho^{\frac{1}{2}} - \rho_0^{1/2} \right)}. \tag{25}$$

For the case where the dominant energy condition is violated, i.e. $B < 0$, we obtain that m increases with time and tends to a maximum, nonzero constant value. If the dominant energy condition is assumed to be hold, i.e. $B > 0$, then m decreases with time, with m tending to nonzero constant values.

It can be seen that, when the cosmic time goes to infinity, then the exotic mass of wormhole approaches to

$$m = \frac{m_0}{1 - Q m_0 \sqrt{\frac{8\pi}{3}} \left(A^{\frac{1}{2(1+\alpha)}} - \rho_0^{1/2} \right)}, \tag{26}$$

that is a generally finite value both for $B > 0$ and $B < 0$. Thus, it could be thought that the presence of a Generalized Chaplygin Gas prevents the eventual occurrence of the big trip

---

[7]It is worth noticing that in models showing one big trip, as it is the considered case, the universe would travel through the time of the arrival universe being, in this case, not a proper time travel. On the other hand, in models which present more than one big trip phenomenon the consideration of a multiverse framework would be not more necessary, since the wormhole mouth at the moment that it is bigger than the universe could be connected to the other infinitely large wormhole mouth, travelling in this case the universe along its own time from future to past. The reader interested on this topic is advised to consult Ref. [49].

phenomenon. However, such a conclusion cannot be guaranteed as the size of the wormhole throat could still exceed the size of the universe during its previous evolution. The question is whether the wormhole would grow rapidly enough or not to engulf the universe during the evolution to its final classically stationary state. To avoid a big trip one needs that the radius of the wormhole does not exceed the size of the universe. It can be checked [24] that GCG generally prevents the occurrence of a big trip when α does not reach values sufficiently close to -1, but when α is inside the interval

$$-1 < \alpha < \frac{\ln A}{\ln \left(\sqrt{\frac{3}{8\pi} \frac{1}{m_0 D}} + \rho_0^{1/2} \right)^2} - 1, \tag{27}$$

a big trip would still take place.

It is worth noticing that, when there is no big trip phenomenon, the wormhole size tends to become constant at the final stages of its evolution being a rather a macroscopic object. So, the wormhole at this stage would be subjected to chronology protection [36] and vacuum polarisation created particles would catastrophically accumulate on the chronology horizon of the wormhole making the corresponding renormalised stress-energy tensor to diverge and hence the wormhole would disappear.

On the other hand, one can study the wormholes' evolution living in a universe with phantom energy modelled by a type II or IV PGCG [12], where a future big freeze singularity is predicted. Since a big freeze singularity implies the divergence of the Hubble parameter at this singularity, as we have mentioned in the introduction of the present section, this implies that the wormhole size would blow up before the occurrence of the singularity, implying a big trip phenomenon. That can be easily proved taking into account Eqs. (1), (8) and (21) for type II PGCG, which yields

$$m(x) = \frac{m_0}{1 + \sqrt{\frac{8\pi}{3}} \frac{Qm_0}{A^{\frac{1}{2|1+\alpha|}}} \left[\frac{1}{(1-x_0^{3|1+\alpha|})^{\frac{1}{2|1+\alpha|}}} - \frac{1}{(1-x^{3|1+\alpha|})^{\frac{1}{2|1+\alpha|}}} \right]}, \tag{28}$$

where $x = a/a_{\max}$ ($0 \leq x \leq 1$) and a similar expression can be obtained for type IV replacing A with $|A|$. In order to study the behaviour of the wormhole mass, one can define the function $F(x) = m_0/m(x)$ which is continuous in the interval $[x_0, 1)$. This function takes a value $F(x_0) = 1 > 0$ and tends to minus infinity when x goes to 1 (which corresponds to $a \to a_{\max}$), which implies that $F(x)$ vanishes at some x_* with $x_0 < x_* < 1$. Therefore, $m(x)$ blows up at x_* being the throat size infinitely large before the universe reaches the big freeze singularity (at $x = 1$). So the whole universe will travel through the wormhole before the occurrence of the doomsday.

The results can be summarised as follows

- Types I and III. The evolution of the wormhole is the same for GCG.

- Types II and IV. A big trip phenomenon would prevent the expected cosmological doomsday, i.e., the big freeze.

5. Debate and New Lines of Research

In the present chapter we have used methods based in the Babichev et al. one, in order to consider the dark and phantom energy accretion onto black- and worm-holes. Our intention is not to claim that the study of these processes is a closed issue, on the contrary, it remains open up to now and a lot of discussion has been originated in this way.

In order to point out the shortcomings of the current available methods which deal with the mentioned accretion phenomenon and suggest possible new lines of research. In this section we also include in chronological order some comments which have appeared in the literature supporting, improving or criticising the Babichev et al. method [11]. We alternate works regarding the accretion onto black holes with others dealing with the accretion onto wormholes, because both phenomena can be studied following the same procedure. Nevertheless, as it has been and will be pointed out, the wormhole case is free of some shortcomings which affect the black hole one, since the first one can never be considered as a vacuum solution (at least if one restricts oneself to the traversable wormhole case).

The first work about accretion of dark energy onto black holes was due to E. Babichev, V. Dokuchaev and Yu. Eroshenko [11]. They considered the spherically symmetric accretion of dark energy onto black holes adjusting the analytic relativistic accretion solution onto the Schwarzschild black hole developed by Michel [18], eliminating from the equations the particle number density. So they obtain the expression for the black hole temporal mass rate

$$\dot{M} = 4\pi A_M M^2 \left[\rho_\infty + p\left(\rho_\infty\right)\right], \qquad (29)$$

showing that the black hole mass could decrease by the accretion phenomenon. The authors pointed out that such a decrease of the black hole size is due to the violation of the dominant energy condition, since this condition is assumed to be fulfilled in the derivation of the black hole non-decrease area theorem. On the other hand, by integrating (29) in a phantom Friedmann universe, they found that the masses of all black holes tend towards zero when the universe approaches the big rip, independently of their initial masses.

Soon after, P. F. Gonzalez-Diaz [19] considered the spherically symmetric accretion of dark and phantom energy onto Morris-Thorne wormholes. He assumed that, since the mass of the spherical thin shell of the exotic matter in a Morris-Thorne wormhole, $\mu = -\pi b_0/2$ (where b_0 is the radius of the spherical wormhole throat), is approximately just the negative of the amount of the mass required to produce a Schwarzschild wormhole, then the rate of change of the wormhole throat radius should be similar to that obtained by Babichev et al. [11] for the black hole mass but with a minus sign, i.e.

$$\dot{b_0} = -2\pi^2 Q b_0^2 \left(1 + w\right)\rho, \qquad (30)$$

with $Q \simeq A_M$. He concluded, therefore, that the wormhole throat should increase by the accretion of phantom energy. Even more, he showed, by integrating Eq. (30) in a phantom model with constant equation of state parameter, that the wormhole increases even faster than the universe itself, engulfing the whole universe before it reaches the big rip singularity. Therefore, the universe would embarks itself in a big trip.

Later on, P. F. Gonzalez-Diaz and C. L. Siguenza, [20], obtained that the phantom energy accretion onto black holes leads to the disappearance of the black holes at the big rip even when Eq. (29) is integrated in more complicated models and Babichev et al. recovered their previous result in a more detailed work [21] where they also included a deeper study of two dark energy models admitting analytical solutions. On the other hand, works containing relevant implications of the result of Babichev et al. were also published during that time, like the influence of the accretion phenomenon on the black hole and phantom thermodynamics, Ref. [20], and the possible survival of black holes at the big rip due to the same phenomenon which could smooth the big rip singularity when quantum effects are taken into account, Ref. [22].

But the previous mentioned works did not had the final say about this topic. In 2005 V. Faraoni and W. Israel [23] considered the time evolution of a wormhole in a phantom Friedmann universe finding no big trip phenomenon. In that work they commented that the way in which Gonzalez-Diaz applied the Babichev et al. method to the wormhole case in Ref. [19] could be wrong, since $\mu = -\pi b_0/2$ must not to be necessarily valid for a time-dependent wormhole embedded in a FLRW universe and, therefore, the wormhole mass time rate due to the accretion phenomenon would not be simply the analogous negative of the black hole mass time rate.

Soon after, Gonzalez-Diaz [25] followed a similar procedure as it has been done by Babichev et al. [11], adjusting the Michel theory to the case of Morris-Thorne wormholes, in order to study the dark energy accretion onto wormholes. He obtained Eq. (20) for the case of an asymptotic observer (which is equivalent to Eq. (30) with a re-definition of the constants). He also claimed that the results obtained by Faraoni and Israel [23] just take into account the inflationary effects of the accelerated expansion of the universe on the wormhole size (also considered by himself years ago in another work [26]) and do not include the superposed effects due to the accretion phenomenon, which existence is clarified in that work [25].

On the other hand, a number of difficulties related to the big trip process were treated also by Gonzalez-Diaz in Ref. [27]. First, he showed how the corrections appearing in the expressions of the study of a wormhole metric with a non-static shape function applying the Babichev et al. method [11] should disappear on the asymptotic limit, coinciding with those corresponding calculated expressions in the static case in that regime. Even more, he considered explicitly a metric able to describe a wormhole in a Friedmann universe, arguing that it would ultimately imply the occurrence of a big trip phenomenon. Second, since wormhole spacetimes are usually considered to be asymptotically flat, then when the wormhole increases more than the universe, this object can neither be placed on it nor be asymptotically flat to it. He proposed that in such a situation a multiverse context must be considered, what would allow to re-insert the wormhole in another universe, recovering the meaning of the asymptotic regime where the accretion process has been calculated. In the third place, he considered the possible instability of wormholes due to the quantum creation of vacuum particles on the chronology horizon when the wormhole throat grows at a rate smaller than or nearly the same as the speed of light. However, in a phantom model the accreting wormhole would clearly grow at a rate which exceeds the speed of light asymptotically and so the vacuum particles would never reach the chronology horizon where they have being created, keeping the wormhole stability. Moreover, although it was known that quantum effects could affect the big rip singularity [22], Gonzalez-Diaz showed that those effects have no influence in the big trip, which would take place before that singularity. In the fourth place, he argued that the big trip phenomenon would not imply any contradiction with the holographic bound, since wormholes able to connect regions after and before the big rip extend the evolution of the universe up to infinite time.

Following that line of thinking, the authors of [28] applied the Babichev et al. method to the simplest non-static generalisation of the Schwarzschild metric, in order to study the dark energy accretion onto black holes with arbitrary accretion rates. Although they are still using a test fluid approach and, therefore, the validity of their result on arbitrary accretion rates is debatable, the non-static metric is enough to take into account internal non-zero energy flow Θ_0^r. As it was suggested by Gonzalez-Diaz in the case of wormholes [27], a study of the accretion phenomenon using a non-static metric recovers the result obtained in the static case, Eq. (29), for asymptotic observers.

Later on, Faraoni published "No "big trips" for the universe", [29], where a sceptical attitude about the big trip phenomenon is adopted, based on some shortcomings of the works of Gonzalez-Diaz [19, 25, 27] in particular and the method of Babichev et al. [11] in general. His principal objection regarding the mentioned works of Gonzalez-Diaz was that the use of a static metric can never produce a non-zero radial energy flow onto the hole, i.e. static metrics always imply $\Theta_0^r = 0$. Even more, the solution of Gonzalez-Diaz (and the corresponding of Babichev et al. in the black hole case) cannot be adjusted to satisfy the Einstein's equations, as the used conservation laws would only strictly correspond to vacuum solutions. On the other hand, he also showed that if the phantom fluid is modelled by a perfect fluid, as it is done in the Babichev et al. method and its application to wormholes, the proper radial velocity of the fluid is $v \sim a^{3(1+w)/2}$ which vanishes at the big rip stopping the accretion phenomenon.

Soon after, Gonzalez-Diaz et al., [30], applied the method of Babichev et al. to a nonstatic generalisation of the Morris-Thorne metric, introducing a shape function with an arbitrary dependence on time, $K(r, t)$. They recovered again Eq. (20) for the temporal mass rate of a wormhole in the asymptotic limit. It must be noted that in their derivation of such expression they allowed an arbitrary dependence of the energy density, pressure and the four velocity of the fluid on both the time and radial coordinates. Therefore, the wormhole mass rate expression Eq. (20) must be valid in general for asymptotically flat wormholes. It must be emphasised that a wormhole is a non-vacuum solution and that the consideration of a time dependence in the shape function leads also to a non zero Θ_0^r, which take into account the non-zero energy flow onto the hole, therefore taking into account the back reaction. Nevertheless, the authors of Ref. [30] considered that the most crucial argument against the big trip included in the paper of Faraoni [29] is the vanishing of the proper radial velocity at the big rip and its quickly decrease close to it, since it is in the point of introducing an explicitly equations of state for the fluid in order to integrate Eq. (20) where they were considering an approximation. First of all, the authors noted that, besides the fact that the time where the universe is engulfed by a wormhole is not only before the big rip but even before the divergence of the wormhole mouth, the important quantity which refers to an accretion process is the proper radial flow, which is approximately $\rho v \sim a^{-3(1+w)/2}$ which increases with time for $w < -1$ and diverges at the big rip, what guarantees the process would not be stopped. In the second place, they point out that the accretion of dark and phantom energy onto astronomical objects differs from the accretion of usual energy concentrated in a given region of space onto those objects, because in the first case the energy pervades the whole space being, therefore, a phenomenon not based on any fluid motion, but on increasing more and more space filled with such kind of energy inside the boundary of the considered object.

Later on, C. Gao, X. Chen, V. Faraoni and Y. G. Shen, [31], emphasising that the method of Babichev et al. applied to black holes is not taking into account the backreaction of the fluid on the background, claimed that the results obtained by using that method can only be valid in a low matter density background. In that spirit, they used a generalised McVittie metric and inserted a radial heat flux term in the energy-momentum tensor, in order to show that a cosmological black hole (non-asymptotically flat) should increase by the accretion of phantom energy. The authors also pointed out any difficulties to compare their solution with the corresponding of Babichev et al. [11], arguing that this fact could be due to the simplifications taken in both cases.

The work of Gao et al. originated some interesting comments. First, in Ref. [32], X. Zhang pointed out the shortcomings of the Babichev et al. method, in particular the use of a non-cosmological metric and the exclusion of the backreaction of the phantom fluid on the black hole metric. Nevertheless, Zhang found the results of Ref. [31] highly speculative, giving the example of the use of a hypothesised metric. Therefore, he decided to use for the moment the method of Babichev et al. in order to extract at least some tentative conclusions, lacking a complete method to the study of the dark energy accretion phenomenon. Second, in [33] where a first attempt to include cosmological effects in the study of the accretion of dark energy onto black holes was done by the consideration of an Schwarzschild-de Sitter spacetime, it was noted that the results achieved in [31] were obtained under the assumption of a premise (contradictory with all studies of this problem in the literature) in which their desired result is contained, making circular their whole argument, and their result invalid. Third, in [34], Babichev et al. included some comments expressing their doubts regarding the conclusions of [31]. They claimed that the heat flux term is introduced in an unnatural way in the solution of Gao et al. to support their configuration, because the perfect fluid is not accreted in the mentioned solution and that such an introduction could may be lead to instabilities to small perturbations. They also pointed out that the temperature of the fluid blows up at the event horizon.

In summary, regarding the phenomenon of dark and phantom energy accretion onto black holes the method developed by Babichev et al. [11] has been improved to take into account the backreaction on the black hole size [28] in an asymptotically flat space and also in a cosmological one [33] by using a non-static generalisation of the Schwarzschild and Schwarzschild-de Sitter metric, respectively. Nevertheless, the study still lacks the consideration of the complete backreaction of the dark or phantom fluid in an asymptotically dark or phantom universe.

Although the method used to study the dark and phantom energy accretion onto worm-holes is similar to the one treating the above mentioned phenomenon, in the case of the phantom energy accretion onto wormholes the backreaction originated by the consideration of the phantom fluid is automatically taken into account by using a non-static generalisation of the Morris-Thorne metric [30]. In this case there are no studies, up to our knowledge, considering rigorously the wormhole accretion phenomenon in a cosmological spacetime.

We want however to point out, that the increase (decrease) of the black hole size by the accretion of dark (phantom) energy, and the contrary in the case of wormholes, seems considerably well supported. Such affirmation can be also understood taken into account a different method. It is well known that the formalism developed by Hayward for spherically symmetric spacetimes [35] implies that a dynamical black hole (characterised by a future outer trapping horizon) would increase if it is considering in an environment fulfilling $p + \rho > 0$ and decrease if $p + \rho < 0$, phenomenon which is due to a flow of such surrounding material into the hole. Therefore, this totally independent study confirms the results presented in this chapter about black holes at least in a qualitative way. The question would be, how large are the quantitative differences which could appear by the consideration of the backreaction and the cosmological space?

On the other hand, regarding wormholes it has been shown [50] that in order to recover using the Hayward formalism the results obtained by the accretion method where the backreaction is included and by the very basis of wormhole physics, a wormhole must be characterised by a past outer trapping horizon. Since it seems that there would be no reason to change the local characterisation of an astronomical object because of the consideration of such an object in a space with a different asymptotically behaviour, the qualitative increase (decrease) of wormholes by the accretion of phantom (dark) energy should be recovered by considering cosmological wormholes spacetimes. Nevertheless, whether a wormhole would suffer a so huge increase to include the whole universe, occurring a big trip, is still an open question.

We want to point out that there are several interesting opened questions concerning accretion onto black holes. More improvement in the accretion theory is needed to take into account the backreaction of the space-time and study the situation where a perfect fluid approximation is not valid, what would clarify whether a black hole can become a naked singularity? A more detailed study about the spin or charge super-radiance is also needed, in order to show whether these processes could concur in such a way that finally the cosmic censorship conjecture would keep hold. Preliminary results indicate that this is the case, but the possibility of violation of the cosmic censorship conjecture produced by the process of dark energy accretion onto a charged or rotating black holes is still open.

Finally, it must be worth noticing that recent papers, [52, 53], consider the Babichev et al. method to propose a new observational dark energy test. The main idea is based on the black hole mass change induced by the dark energy accretion process which, as we have shown in this chapter, is proportional to $1 + w$. Although a direct observation of this change is beyond our current detection possibilities, because the time scale to produce a change in the black hole mass measurable with our present devices would be too long, it would cause observable modifications in the orbital radius of the supermassive black hole binaries, since the black hole binaries would either merge in a more accelerated way than expected if $1 + w > 0$ or the merging would be progressively stopped if the dominant energy condition is violated, being also possible that the binaries would rip apart in the second case[8]. At this moment there are two candidates, Galaxy 0402+379 and Radio Galaxy OJ287, for the observation of the mentioned phenomenon what could provide us with more information about the nature of dark energy, and probably more interesting candidates can be expected in the future.

---

[8]The interested reader can look up in [52, 53] for more details.

6. Conclusion

In the present chapter we have shown that if the current accelerating expansion of the universe is explained in the framework of General Relativity, then the consideration of a dynamical dark energy fluid would produce other effects besides the modelling of that acceleration. In this sense, the consideration of dark energy would not be simply the consideration of an ether covering the whole space, since footprints of such a fluid could appear in our Universe by observing the evolution of black- and worm-holes. Whether such effects are measurable in practice, is a question related to the accuracy of the observational data.

The effects in question regarding the dynamical evolution of black- and worm-holes, would be an additional increase (decrease) of the black hole size in the case that the dark energy fulfils (violates) the dominant energy condition, and the contrary in the wormhole case. Even more, these effects would produce changes in the orbital radius of black hole binaries which could be large enough to be detected in practice, helping us to get new constrains to dark energy equation of state parameter.

If one considers the used test fluid approximation to hold in phantom models possessing a singularity at a finite time in the future, then the black holes would tend to disappear at that singularity, which would never be reached since a big trip phenomenon may take place before. On the other hand, in dark energy models with $w > -1$ black holes would not engulf the whole universe, since the current mass of a hole able to exceed the universe size in a finite time should be so huge that it would be bigger than the mass of the observable Universe.

Although, by the arguments presented in this chapter, the qualitative evolution of the considered astronomical objects by the accretion of dark energy seems to be a solid result, the quantitative dynamical behaviour could differ from the mentioned results, since at the final step we are considering the approximation that the surrounding fluid is a cosmological one. In the black hole case this approximation is even stronger, since the non-static generalisation of the Schwarzschild metric, though taking into account the radial flow, lacks of a consideration of the backreaction.

Finally, whether or not the above features studied in this chapter can be taken to imply that certain dark energy models are more consistent than others is a matter that will depend on both the intrinsic consistency of the different models and the current and future observational data.

Acknowledgments

We acknowledge La Casa de Aragón to provide us a relaxing and inspiring place where this chapter was designed and discussed. JAJM thanks the financial support provided by Fundación Ramón Areces and the DGICYT Research Project MTM2008-03754. P. M. M. gratefully acknowledges the financial support provided by the I3P framework of CSIC and the European Social Fund and by a Spanish MEC Research Project No.FIS2008-06332/FIS. Special thanks to Jodie Holdway for English revision of this chapter.

References

[1] D. J. Mortlock and R. L. Webster, "The statistics of wide-separation lensed quasars," *Mon. Not. Roy. Astron. Soc.* **319**, 872 (2000) [arXiv:astro-ph/0008081].

[2] A. G. Riess *et al.* [Supernova Search Team Collaboration], "Observational Evidence from Supernovae for an Accelerating Universe and a Cosmological Constant," *Astron. J.* **116**, 1009 (1998) [arXiv:astro-ph/9805201].

S. Perlmutter *et al.* [Supernova Cosmology Project Collaboration], "Measurements of Omega and Lambda from 42 High-Redshift Supernovae," *Astrophys. J.* **517**, 565 (1999) [arXiv:astro-ph/9812133].

J. L. Tonry et al. [Supernova Search Team Collaboration], "Cosmological Results from High-z Supernovae," *Astrophys. J.* **594**, 1 (2003) [arXiv:astro-ph/0305008].

[3] D. N. Spergel et al. [WMAP Collaboration], "First Year Wilkinson Microwave Anisotropy Probe (WMAP) Observations: Determination of Cosmological Parameters," *Astrophys. J. Suppl.* **148**, 175 (2003) [arXiv:astro-ph/0302209].

C. L. Bennett et al., "First Year Wilkinson Microwave Anisotropy Probe (WMAP) Observations: Preliminary Maps and Basic Results," *Astrophys. J. Suppl.* **148**, 1 (2003) [arXiv:astro-ph/0302207].

M. Tegmark et al. [SDSS Collaboration], "Cosmological parameters from SDSS and WMAP," *Phys. Rev. D* **69**, 103501 (2004) [arXiv:astro-ph/0310723].

[4] R. R. Caldwell, *Phys. Lett. B* **545** (2002) 23; S. M. Carroll, M. Hoffman and M. Trodden, *Phys. Rev. D* **68** (2003) 023509.

[5] R. R. Caldwell, M. Kamionkowski and N. N. Weinberg, "Phantom Energy and Cosmic Doomsday," *Phys. Rev. Lett.* **91**, 071301 (2003);

P. F. Gonzalez-Diaz, "K-essential phantom energy: Doomsday around the corner?," *Phys. Lett. B* **586**, 1 (2004);

P. F. Gonzalez-Diaz, "Axion phantom energy," *Phys. Rev. D* **69**, 063522 (2004);

S. Nojiri and S. D. Odintsov, "The final state and thermodynamics of dark energy universe," *Phys. Rev. D* **70**, 103522 (2004).

[6] M. Bouhmadi-Lopez, P. F. Gonzalez-Diaz and P. Martin-Moruno, *Phys. Lett. B* **659** (2008) 1 [arXiv:gr-qc/0612135].

[7] M. Bouhmadi-Lopez, P. F. Gonzalez-Diaz and P. Martin-Moruno, *Int. J. Mod. Phys. D* **17** (2008) 2269 [arXiv:0707.2390 [gr-qc]].

[8] M. Bouhmadi-López and J. A. Jiménez Madrid, *JCAP* **0505**, (2005) 005 [arXiv:astro-ph/0404540].

[9] A. Y. Kamenshchik, U. Moschella and V. Pasquier, *Phys. Lett. B* **511**, (2001) 265 [arXiv:gr-qc/0103004];

N. Bilić, G. B. Tupper and R. D. Viollier, *Phys. Lett. B* **535**, (2002) 17 [arXiv:astro-ph/0111325];

M. C. Bento, O. Bertolami and A. A. Sen, *Phys. Rev. D* **66**, (2002) 043507 [arXiv:gr-qc/0202064].

[10] M. Bouhmadi-Lopez, C. Kiefer, B. Sandhofer and P. V. Moniz, *Phys. Rev. D* **79**, (2009) 124035 [arXiv:0905.2421[gr-qc]].

[11] E. Babichev, V. Dokuchaev and Yu. Eroshenko, *Phys. Rev. Lett.* **93**, 021102 (2004) [arXiv:gr-qc/0402089].

[12] P. Martin-Moruno, *Phys. Lett. B* **659**, (2008) 40.

[13] I. M. Khalatnikov, *Phys. Lett. B* **563**, (2003) 123.

[14] S. V. Sushkov, *Phys. Rev. D* **71** 043520 (2005). F. S. N. Lobo, *Phys. Rev. D* **71** 084011 (2005).

[15] M. S. Morris and K. S. Thorne, *Am. J. Phys.* **56** 395 (1988).

[16] M. S. Morris, K. S. Thorne and U. Yurtsever, *Phys. Rev. Lett.* **61**, 1446 (1988).

[17] M. Visser, *Lorentzian Wormholes: from Einstein to Hawking*, AIP Press (1996).

[18] F. C. Michel, *Astrophys. Sp. Sc.* **15**, 153 (1972).

[19] P. F. Gonzalez-Diaz, *Phys. Rev. Lett.* **93**, 071301 (2004) [arXiv:astro-ph/0404045].

[20] P. F. Gonzalez-Diaz and C. L. Siguenza, *Nucl. Phys. B* **697**, 363 (2004) [arXiv:astro-ph/0407421].

[21] E. Babichev, V. Dokuchaev and Y. Eroshenko, *J. Exp. Theor. Phys.* **100**, 528 (2005) [*Zh. Eksp. Teor. Fiz.* **127**, 597 (2005)] [arXiv:astro-ph/0505618].

[22] S. Nojiri and S. D. Odintsov, *Phys. Rev. D* **70**, 103522 (2004) [arXiv:hep-th/0408170].

[23] V. Faraoni and W. Israel, *Phys. Rev. D* **71**, 064017 (2005) [arXiv:gr-qc/0503005].

[24] J. A. Jiménez Madrid, *Phys. Lett. B* **634** (2006) 106 [arXiv:astro-ph/0512117].

[25] P. F. Gonzalez-Diaz, *Phys. Lett. B* **632**, 159 (2006) [arXiv:astro-ph/0510771].

[26] P. F. Gonzalez-Diaz, *Phys. Rev. D* **68**, 084016 (2003) [arXiv:astro-ph/0308382].

[27] P. F. Gonzalez-Diaz, *Phys. Lett. B* **635**, 1 (2006) [arXiv:hep-th/0607137].

[28] P. Martin-Moruno, J. A. J. Madrid and P. F. Gonzalez-Diaz, *Phys. Lett. B* **640**, 117 (2006) [arXiv:astro-ph/0603761].

[29] V. Faraoni, *Phys. Lett. B* **647**, 309 (2007) [arXiv:gr-qc/0702143].

[30] P. F. González-Díaz and P. Martín-Moruno, *Proceedings of the eleventhMarcel Grossmann Meeting on General Relativity*, Editors: H. Kleinert, R. T. Jantzen and R. Ruffini, World Scientific, New Jersey 2190-2192 (2008).

[31] C. Gao, X. Chen, V. Faraoni and Y. G. Shen, *Phys. Rev. D* **78**, 024008 (2008) [arXiv:0802.1298 [gr-qc]].

[32] X. Zhang, *Eur. Phys. J. C* **60**, 661 (2009) [arXiv:0708.1408 [gr-qc]].

[33] P. Martin-Moruno, A. E. Marrakchi, S. Robles-Perez and P. F. Gonzalez-Diaz, *Gen. Rel. Grav.* **41** 2797 (2009) [arXiv:0803.2005 [gr-qc]].

[34] E. Babichev, S. Chernov, V. Dokuchaev and Yu. Eroshenko, arXiv:0806.0916 [gr-qc].

[35] S. A. Hayward, *Phys. Rev. D* **70**, 104027 (2004).

[36] S. W. Hawking, *Phys. Rev. D* **46**, 603 (1992).

[37] O. Bertolami, A. A. Sen, S. Sen and P. T. Silva, *Mon. Not. Roy. Astron. Soc.* **353**, 329 (2004) [arXiv:astro-ph/0402387].

[38] M. Biesiada, W. Godlowski and M. Szydlowski, *Astrophys. J.* **622**, 28 (2005) [arXiv:astro-ph/0403305].

[39] Z. H. Zhu, *Astron. Astrophys.* **423**, 421 (2004) [arXiv:astro-ph/0411039].

[40] R. J. Colistete and J. C. Fabris, *Class. Quant. Grav.* **22**, 2813 (2005) [arXiv:astro-ph/0501519].

[41] J. Lu, Y. Gui and L. X. Xu,

[42] S. del Campo and J. Villanueva, arXiv:0909.5258 [astro-ph.CO].

[43] Z. Li, P. Wu and H. W. Yu, *JCAP* **0909**, 017 (2009) [arXiv:0908.3415 [astro-ph.CO]].

[44] F. Y. Wang, Z. G. Dai and S. Qi, *Res. Astron. Astrophys.* **9**, 547 (2009).

[45] P. Wu and H. W. Yu, *Astrophys. J.* **658**, 663 (2007).

[46] J. Lu, L. Xu, J. Li, B. Chang, Y. Gui and H. Liu, *Phys. Lett. B* **662**, 87 (2008).

[47] R. M. Wald, "General Relativity," The University of Chicago,1984

[48] G. Ghirlanda, G. Ghisellini and C. Firmani, *New J. Phys.* **8**, 123 (2006) [arXiv:astro-ph/0610248].

[49] A. V. Yurov, P. Martin Moruno and P. F. Gonzalez-Diaz, *Nucl. Phys. B* **759** (2006) 320 [arXiv:astro-ph/0606529].

[50] P. Martin-Moruno and P. F. Gonzalez-Diaz, *Phys. Rev. D* **80** (2009) 024007 [arXiv:0907.4055 [gr-qc]].

[51] R. Penrose, *Gravitional Collapse : The Role of General Relativity*, Riv. Nuovo Cimento 1, special number, (1969), pp.252-276.

[52] L. Mersini-Houghton and A. Kelleher, arXiv:0808.3419 [gr-qc].

[53] L. Mersini-Houghton and A. Kelleher, *Nucl. Phys. Proc. Suppl.* **194**, 272 (2009) [arXiv:0906.1563 [gr-qc]].

Reviewed by Prof. V. I. Dokuchaev

In: Dark Energy: Theories, Developments and Implications   ISBN 978-1-61668-271-2
Editors: K. Lefebvre and R. Garcia, pp. 241-294   © 2010 Nova Science Publishers, Inc.

Chapter 11

ANALYTIC APPROACHES TO THE STRUCTURE FORMATION IN THE ACCELERATING UNIVERSE

*Takayuki Tatekawa[1] and Shuntaro Mizuno[2],**
[1]Center for Computational Science and e-Systems, Japan Atomic Energy Agency,
6-9-3 Higashi-Ueno, Taito, Tokyo, 110-0015, Japan
[2]School of Physics and Astronomy, University of Nottingham,
University Park, Nottingham NG7 2RD, United Kingdom

Abstract

The exixtence of dark energy is a serious problem in modern cosmology. For the origin of the dark energy, many models including a cosmological constant have been proposed. Although these models can explain the present acceleration of the Universe, some of the models would not be able to explain the observed large-scale structure of the universe. Therefore, in order to constrain the models of the dark energy, we should consider the structure formation in the universe. From primordial density fluctuation, the large-scale structure is formed via its own self-gravitational instability. Even though numerical simulations are necessary to follow the full history of the structure formation, in order to understand the physics behind the structure formation, analytic approaches play important roles. In this review, we summarise various analytic approaches to the evolution of the density fluctuation in Newtonian cosmology and show they can be helpful to distinguish models when applied to the quasi-nonlinear region. We also mention several applications of the analytic approaches including the initial condition problems for cosmological N-body simulations, higher-order Lagrangian perturbation theory.

1. Introduction

How the structure observed in the Universe has been formed is one of the most important problems in modern cosmology. Because the large-scale structure of the Universe reflects the whole evolution history of the Universe, which cosmological model is the actual one would be decided by observations of the structure of the Universe.

---

*E-mail address: shuntaro.mizuno@nottingham.ac.uk

The structure of the Universe is formed from the primordial density fluctuation via its own self-gravitational instability [21, 57, 76, 77, 79, 82, 114]. While the self-gravity make matter cluster, the matter move away due to the cosmic expansion. Since the rate of the cosmic expansion depends on the model of the dark energy, the structure formation is affected strongly by the detail of the model of the dark energy. Therefore, it is possible to constrain the dark energy model by observations of large-scale structure of the Universe.

For the discussion of the plausible dark energy model, not only observations but also theoretical predictions are important. By comparisons between observations and theoretical predictions, we can decide the feature of the dark energy. As one of the explanations of the dark energy, although the cosmological constant has been considered, it has several scerious problems such as a fine-tuning problem [113]. At present, instead of the cosmological constant, many dynamical dark energy models have been also proposed abundantly (For a review, see [24]). For the constraint to dark energy models, the large-scale structure is regarded as one of the useful observational evidences.

In early era, spacial two-point correlation function was derived by the observations [78, 111]. The function was fitted by the power-law function of the separation distance. For the explanation of the origin of this power-law behaviour, self-similar solutions in BBGKY equation have been discussed [31, 75, 116, 117]. Although this approaches are partly successful, quantitatively precise time evolution is not obtained in this approach. Therefore we should consider dynamical evolution.

For the dynamical evolution of the large-scale structure, analytic approaches have been used for a long time. Regarding the primordial fluctuation as small quantity, the evolution of the fluctuations is solved with perturbative equations [6, 52, 82, 101]. If the scale of the primordial fluctuation is less than cosmological horizon scale, the fluctuation can be handled with Newtonian cosmology. If the dark energy does not couple with the matter, the dark energy affects the motion of the matter through the modification of the cosmic expansion. In Newtonian cosmology, the Eulerian approach in which the density fluctuation is regarded as the perturbative quantity was adopted from the old days. Although this approach seems easy to handle, the approximation breaks down when the density fluctuation approaches to unity. In other words, the Eulerian approach is valid only around the linear region.

Zel'dovich [122] proposed a new approach in which the displacement from the homogeneous distribution is regarded as the perturbative quantity. Because the relation between the perturbative quantity and the density fluctuation is nonlinear, this approach can describe the evolution in the quasi-nonlinear regime better. This approach is called as the Lagrangian approach.

After that, the nonlinear perturbations for both Eulerian and Lagrangian approaches have been proposed. Furthermore, several modified or improved methods have been considered. In generic cases, the matter is regarded as dust fluid. However, when we based on collisionless Boltzmann's equation (Vlasov's equation) [7], the velocity dispersion term which is equivalent with the effective pressure appears. Following this result, the Lagrangian approach for non-dust fluid has been proposed.

Even if analytic approaches are improved, these approaches have critical issues. Because these approaches are based on fluid dynamics, they are no longer valid when the stream lines cross, i.e., the density fluctuation diverges. Furthermore, because these approaches are based on perturbation theories, the description of the evolution seems well only until the quasi-nonlinear regime. Therefore, in the strongly nonlinear regime, cosmo-logical N-body simulations need to be carried out [30, 47, 69, 91].

However, recently, analytic approaches have been noticed again. As we mentioned be-fore, analytic approaches are valid until quasi-nonlinear regime. Therefore they cannot be applied for the present structure. In the near future, high-z observations for the galaxies are scheduled. Because the large-scale structure at high-z still be in the quasi-nonlinear regime, the analytic approach seems valid. Furthermore, recently the initial condition prob-lem for cosmological N-body simulations have been discussed [25, 49, 106]. The initial condition for N-body simulations are set up around redshift $z \sim 30$ with linear Lagrangian perturbation theory for a long time. However, if we set the initial condition with nonlinear Lagrangian perturbation theories, the evolution of the large-scale structure changes. Al-though the differences are small, it is necessary in precise cosmology to establish what initial condition for cosmological N-body simulations is most reliable. For these purposes, analytic approaches for the structure formation seems to play important roles.

This paper is organized as follows. In Sec. 2., we present basic equations for the cos-mological fluid in Newtonian cosmology. After that first, we derive the basic equations for Eulerian perturbation theories in Sec. 3.. In this section, we mention both linear pertur-bations (Sec. 3.1.) and nonlinear perturbations (Sec. 3.2.). Then we introduce Lagrangian description. The explanation of Lagrangian description is devided to two sections: basic and advanced. In the basic section (Sec. 4.), we introduce Lagrangian perturbations. Then both linear (Sec. 4.2.) and nonlinear perturbations (Sec. 4.3.) in standard approach are shown. In many cases, we consider only the longitudinal mode. However, even if we consider only the longitudinal mode at linear order, the transverse mode is shown to appear at third-order (Sec. 4.4.). In the advanced section (Sec. 5.), we mention modified methods, improvements, renormalization group approaches, and wave mechanical approach. As generalized cases, we also mention non-dust model and multi-component model.

In Sec. 6., we discuss a couple of applications of analytic approaches. In many cases, the primordial fluctuations are regarded as Gaussian. However, during nonlinear evolution, non-Gaussianity of the density fluctuation would appear. In Sec. 6.1., we discuss non-Gaussianity in weakly nonlinear region. Recently, baryon acoustic oscillations (BAO) in the two-point correlation function has drawn much interests [37]. In Sec. 6.2., we mention recent progress for the analysis of BAO with analytic approaches. The initial condition problem for cosmological N-body simulations has been also highlighted in the context of the precision cosmology. We have analyzed about what is the reasonable initial condition for the simulations. Our results are shown in Sec. 6.3.. In Sec. 7., we summarize the success and failure of analytic approaches for the structure formation and future perspective.

We notice the procedure about connection between before and after shell-crossing mak-ing use of a simple toy model so-called one-dimensional sheet model where we can obtain the analytic exact solutions in Appendix A.. In Appendix B., in order to catch the structure of the system which can not be seen by perturbative approaches, we show the derivation of basic equations from Vlasov's equation. In addition to past work, we discuss the ef-fect of anisotropic velocity dispersion. In Appendix C., we provide tables summarizing the physical quantities used in this paper.

2. Basic Equations

In this section, we show basic equations for the cosmological fluid. Here we assume that the motion of the fluid is described by Newtonian dynamics. The background cosmic expansion is given by the Friedmann equation. For example, if the dark matter behaves as relativistic fluid, or it couples directly with dark energy, this description is no longer valid.

In this case, the motion of the fluid is described by a continuous equation, Euler's equation, and Poisson's equation [21, 79, 82].

$$\left(\frac{\partial \rho}{\partial t}\right)_r + \nabla_r \cdot (\rho \boldsymbol{u}) = 0, \tag{2.1}$$

$$\left(\frac{\partial \boldsymbol{u}}{\partial t}\right)_r + (\boldsymbol{u} \cdot \nabla)_r \boldsymbol{u} = -\frac{1}{\rho}\nabla_r P + \boldsymbol{g}, \tag{2.2}$$

$$\boldsymbol{g} = -\nabla_r \Phi, \qquad \nabla \cdot \boldsymbol{g} = 4\pi G \rho. \tag{2.3}$$

The effect of the cosmic expansion is given by scale factor $a(t)$, where t is the cosmic time. We adopt the coordinate transformation from the physical coordinates to the comoving coordinates.

$$\boldsymbol{x} = \frac{\boldsymbol{r}}{a(t)}, \quad a(t) : \text{scale factor}, \tag{2.4}$$

where $\boldsymbol{r}$ and $\boldsymbol{x}$ are variables in the physical coordinates and comoving coordinates, respectively. The evolution of scale factor is given by the Friedmann equations:

$$\left(\frac{\dot{a}}{a}\right)^2 = H^2 = \frac{8\pi G}{3}\rho_b - \frac{\mathcal{K}}{a^2} + \frac{\Lambda}{3}, \tag{2.5}$$

$$\frac{\ddot{a}}{a} = -\frac{4\pi G}{3}\rho_b + \frac{\Lambda}{3}, \tag{2.6}$$

$$H \equiv \frac{\dot{a}}{a}, \tag{2.7}$$

with a curvature constant $\mathcal{K}$ and a cosmological constant Λ. $H = \dot{a}/a$ and ρ_b are Hubble parameter and background density, respectively. Here we define density parameter Ω_M:

$$\Omega_M \equiv \frac{8\pi G}{3H^2}\rho_b. \tag{2.8}$$

When we consider modified gravity models such as scalar-tensor gravity, higher dimensional model, instead of the Friedmann equations, we must adopt modified equations to obtain the evolution of the scale factor.

Under the transformation (2.4), the velocity and partial differential operator are transformed as

$$\boldsymbol{u} = \dot{\boldsymbol{r}} = \dot{a}\boldsymbol{x} + \boldsymbol{v}(\boldsymbol{x}, t), \quad (\boldsymbol{v} \equiv a\dot{\boldsymbol{x}}), \tag{2.9}$$

$$\nabla_x = a\nabla_r, \tag{2.10}$$

$$\left(\frac{\partial f(\boldsymbol{x} = \boldsymbol{r}/a, t)}{\partial t}\right)_r = \left(\frac{\partial f}{\partial t}\right)_x - \frac{\dot{a}}{a}(\boldsymbol{x} \cdot \nabla_x)f, \tag{2.11}$$

where v denotes the peculiar velocity. Here we define the density fluctuation δ as follows:

$$\rho = \rho_b(t)\{1 + \delta(x, t)\}, \quad \rho_b \propto a^{-3}. \tag{2.12}$$

In the comoving coordinates, the basic equations for the cosmological fluid are described as

$$\frac{\partial \delta}{\partial t} + \frac{1}{a}\nabla_x \cdot \{v(1 + \delta)\} = 0, \tag{2.13}$$

$$\frac{\partial v}{\partial t} + \frac{1}{a}(v \cdot \nabla_x)v + \frac{\dot{a}}{a}v = \frac{1}{a}\tilde{g} - \frac{1}{a\rho}\nabla_x P, \tag{2.14}$$

$$\nabla_x \times \tilde{g} = 0, \tag{2.15}$$

$$\nabla_x \cdot \tilde{g} = 4\pi G \rho_b a \delta. \tag{2.16}$$

3. Eulerian Perturbations

First, we consider Eulerian perturbations. In Eulerian perturbations, we regard the density fluctuation δ as a perturbative quantity. Therefore, when the density fluctuation evolves to $\delta \simeq 1$, the perturbation breaks down. Furthermore, in Eulerian perturbations, negative mass region ($\delta < -1$) will appear. We should use Eulerian perturbations paying attention to these points.

3.1. Linear Perturbations

Keeping only the linear terms in (2.13), we can obtain the relation between the density perturbation and the peculiar velocity

$$\frac{\partial \delta}{\partial t} + \frac{1}{a}\nabla_x \cdot v = 0. \tag{3.1}$$

Comoving Euler's equation (2.14) is linearized as

$$\frac{\partial v}{\partial t} + \frac{\dot{a}}{a}v = \frac{1}{a}\tilde{g} - \frac{1}{a\rho}\nabla_x P. \tag{3.2}$$

Therefore we obtain from (2.16) and (3.2),

$$\nabla_x \cdot \left(\frac{\partial v}{\partial t} + \frac{\dot{a}}{a}v\right) = -4\pi G \rho_b \delta - \nabla_x \left(\frac{1}{a\rho}\nabla_x P\right). \tag{3.3}$$

Since most of the component in the Universe except dark enregy is considered as collision-less cold dark matter (CDM) [92], in generic cases, we ignore the pressure term. When we subtitute (3.1) to the above equation, we obtain the Eulerian linear perturbative equation:

$$\frac{\partial^2 \delta}{\partial t^2} + 2\frac{\dot{a}}{a}\frac{\partial \delta}{\partial t} = -4\pi G \rho_b \delta. \tag{3.4}$$

The effect of dark energy is exerted on the evolution of the density perturbation through the modification of the cosmic expansion.

For the Einstein-de Sitter (E-dS) Universe ($\Omega_M = 1, \Lambda = 0$), the scale factor is given by

$$a(t) \propto t^{2/3} \,. \tag{3.5}$$

Therefore, in the E-dS Universe, the solutions of the Eulerian linear perturbation theory are described as

$$
\begin{align}
\delta(t, \boldsymbol{x}) &= g_{1+}(t)\delta_+(\boldsymbol{x}) + g_{1-}(t)\delta_-(\boldsymbol{x}) \,, \tag{3.6} \\
g_{1+}(t) &\propto t^{2/3} \,, \tag{3.7} \\
g_{1-}(t) &\propto t^{-1} \,, \tag{3.8}
\end{align}
$$

i.e., the growth factor of the linear perturbation theory is proportional to the scale factor.

For the Λ-flat Universe ($\Omega_M < 1, \Lambda = 0, \mathcal{K} = 0$), $g_{1\pm}$ is written as

$$
\begin{align}
g_{1+}(t) &= h \int_h^\infty \frac{\mathrm{d}\chi}{\chi^2(\chi^2 - 1)} = \frac{h}{2} B_{1/h^2}\left(\frac{5}{6}, \frac{2}{3}\right) \,, \tag{3.9} \\
g_{1-}(t) &= h \,, \tag{3.10} \\
h &= \frac{H(t)}{\sqrt{\Lambda/3}} \,, \tag{3.11}
\end{align}
$$

where B_{1/h^2} is incomplete Beta function:

$$B_z(\mu, \nu) \equiv \int_0^z y^{\mu-1}(1 - y)^{\nu-1}\mathrm{d}y \,. \tag{3.12}$$

In de Sitter (dS) Universe ($\Omega_M = 0, \Lambda \neq 0$, flat), the perturbation hardly grows, because the scale factor in dS Universe is described as the exponential function of the time, the matter cannot cluster.

From (2.15) and (3.2), we obtain the evolution equation for vorticity.

$$
\begin{align}
\frac{\partial^2 \boldsymbol{\omega}}{\partial t^2} + \frac{\dot{a}}{a}\frac{\partial \boldsymbol{\omega}}{\partial t} &= 0 \,, \tag{3.13} \\
\boldsymbol{\omega} &\equiv \nabla_x \times \boldsymbol{v} \,. \tag{3.14}
\end{align}
$$

In the E-dS Universe, the solution of (3.13) is described as

$$\boldsymbol{\omega} \propto t^0 \,, \ t^{-1/3} \,. \tag{3.15}$$

Therefore, in linear perturbations, the vorticity does not have growing mode.

3.2. Non-linear Perturbations

For improvements of the solutions obtained by linear perturbations, we would consider higher-order perturbations. For example, we notice second-order perturbations. The perturbative quantities are described as

$$
\begin{align}
\boldsymbol{v} &= \boldsymbol{v}^{(1)} + \boldsymbol{v}^{(2)} \,, \tag{3.16} \\
\delta &= \delta^{(1)} + \delta^{(2)} \,, \tag{3.17}
\end{align}
$$

where the superscript means the order of the perturbations.

We substitute above perturbations to (2.13), (2.14), and (2.16). Although it is a little complicated, we can obtain the second-order perturbative equations. The generic discussion for second-order perturbations is discussed by Hunter [48] and Tomita [109, 110].

For higher-order perturbations, we can consider the perturbative solutions in Eulerian perturbations in the Fourier space. Here we define Fourier transform of a field $F(\boldsymbol{x}, t)$ as

$$\widehat{F}(\boldsymbol{k}, t) = \int \frac{\mathrm{d}^3\boldsymbol{x}}{(2\pi)^3} \exp(-i\boldsymbol{k} \cdot \boldsymbol{x}) F(\boldsymbol{x}, t) \,. \tag{3.18}$$

When we apply Fourier transform for (2.13) and the divergence of (2.14), we obtain

$$\frac{\partial \widehat{\delta}}{\partial t} + \widehat{\theta} = \int \mathrm{d}^3\boldsymbol{k}_1 \mathrm{d}^3\boldsymbol{k}_2 \delta_D(\boldsymbol{k} - \boldsymbol{k}_{12}) \cdot \alpha_1(\boldsymbol{k}_1, \boldsymbol{k}_2) \widehat{\theta}(\boldsymbol{k}_1, t) \widehat{\delta}(\boldsymbol{k}_2, t) \,, \tag{3.19}$$

$$\frac{\partial \widehat{\theta}}{\partial t} + \frac{\dot{a}}{a}\widehat{\theta} + 4\pi G\rho_b a^2 \widehat{\delta} = -\int \mathrm{d}^3\boldsymbol{k}_1 \mathrm{d}^3\boldsymbol{k}_2 \delta_D(\boldsymbol{k} - \boldsymbol{k}_{12}) \cdot \alpha_2(\boldsymbol{k}_1, \boldsymbol{k}_2) \widehat{\theta}(\boldsymbol{k}_1, t) \widehat{\theta}(\boldsymbol{k}_2, t) \tag{3.20}$$

$$\theta \equiv \nabla_x \cdot \boldsymbol{v} \,, \tag{3.21}$$

$$\boldsymbol{k}_{12} \equiv \boldsymbol{k}_1 + \boldsymbol{k}_2 \,, \tag{3.22}$$

where δ_D means Dirac's delta function. α_1 and α_2 is given by

$$\alpha_1(\boldsymbol{k}_1, \boldsymbol{k}_2) \equiv \frac{\boldsymbol{k}_{12} \cdot \boldsymbol{k}_1}{k_1^2} \,, \quad \alpha_2(\boldsymbol{k}_1, \boldsymbol{k}_2) \equiv \frac{k_{12}^2(\boldsymbol{k}_1 \cdot \boldsymbol{k}_2)}{2k_1^2 k_2^2} \,. \tag{3.23}$$

The right-hand sides of equations (3.19) and (3.20) show mode-coupling of perturbations. In higher-order perturbations, the evolution of $\widehat{\delta}$ and $\widehat{\theta}$ is affected by mode-coupling of the density and velocity fields at all pairs of wave vectors $\boldsymbol{k}_1$ and $\boldsymbol{k}_2$.

By solving (3.19) and (3.20), we obtain higher-order perturbative solutions. For example, we show the case of E-dS Universe. In this case, the solutions are described as

$$\widehat{\delta}(\boldsymbol{k}, t) = \sum_{n=1}^{\infty} a^n(t) \widehat{\delta}^{(n)}(\boldsymbol{k}) \,, \tag{3.24}$$

$$\widehat{\theta}(\boldsymbol{k}, t) = -\frac{\dot{a}}{a} \sum_{n=1}^{\infty} a^n(t) \widehat{\theta}^{(n)}(\boldsymbol{k}) \,. \tag{3.25}$$

Here we consider only the strongest growing mode. In linear perturbations, the relation between the density fluctuation and the divergence of the peculiar velocity is given by (3.1). Therefore we can describe all perturbative solutions by the linear density fluctuation $\delta^{(1)}$.

From (3.19) and (3.20), the nonlinear perturbative solutions are written as,

$$\widehat{\delta}^{(n)}(\boldsymbol{k}) = \int \mathrm{d}^3\boldsymbol{k}_1' \cdots \int \mathrm{d}^3\boldsymbol{k}_n' \delta_D(\boldsymbol{k} - \boldsymbol{k}_{1\cdots n}') \mathcal{F}_n(\boldsymbol{k}_1', \cdots, \boldsymbol{k}_n') \delta^{(1)}(\boldsymbol{k}_1) \cdots \delta^{(1)}(\boldsymbol{k}_n) \tag{3.26}$$

$$\widehat{\theta}^{(n)}(\boldsymbol{k}) = \int \mathrm{d}^3\boldsymbol{k}_1' \cdots \int \mathrm{d}^3\boldsymbol{k}_n' \delta_D(\boldsymbol{k} - \boldsymbol{k}_{1\cdots n}') \mathcal{G}_n(\boldsymbol{k}_1', \cdots, \boldsymbol{k}_n') \delta^{(1)}(\boldsymbol{k}_1) \cdots \delta^{(1)}(\boldsymbol{k}_n) \tag{3.27}$$

where $\mathcal{F}$ and $\mathcal{G}$ are functions of wavevectors $(\boldsymbol{k}'_1, \cdots, \boldsymbol{k}'_n)$. They are constructed from α_1, α_2 with the recursive relations [41].

$$\mathcal{F}_n(\boldsymbol{k}'_1, \cdots, \boldsymbol{k}'_n) = \sum_{m=1}^{n-1} \frac{\mathcal{G}_m(\boldsymbol{k}'_1, \cdots, \boldsymbol{k}'_m)}{(2n+3)(n-1)} \left[(2n+1)\alpha_1(\boldsymbol{k}_1, \boldsymbol{k}_2)\mathcal{F}_{n-m}(\boldsymbol{k}'_{m+1}, \cdots, \boldsymbol{k}'_n) \right.$$
$$\left. + 2\alpha_2(\boldsymbol{k}_1, \boldsymbol{k}_2)\mathcal{G}_{n-m}(\boldsymbol{k}'_{m+1}, \cdots, \boldsymbol{k}'_n) \right], \qquad (3.28)$$

$$\mathcal{G}_n(\boldsymbol{k}'_1, \cdots, \boldsymbol{k}'_n) = \sum_{m=1}^{n-1} \frac{\mathcal{G}_m(\boldsymbol{k}'_1, \cdots, \boldsymbol{k}'_m)}{(2n+3)(n-1)} \left[3\alpha_1(\boldsymbol{k}_1, \boldsymbol{k}_2)\mathcal{F}_{n-m}(\boldsymbol{k}'_{m+1}, \cdots, \boldsymbol{k}'_n) \right.$$
$$\left. + 2n\alpha_2(\boldsymbol{k}_1, \boldsymbol{k}_2)\mathcal{G}_{n-m}(\boldsymbol{k}'_{m+1}, \cdots, \boldsymbol{k}'_n) \right], \qquad (3.29)$$

$$\mathcal{F}_1 = 1, \ \mathcal{G}_1 = 1, \qquad (3.30)$$
$$\boldsymbol{k}_1 \equiv \boldsymbol{k}'_1 + \cdots + \boldsymbol{k}'_m, \ \ \boldsymbol{k}_2 \equiv \boldsymbol{k}'_{m+1} + \cdots + \boldsymbol{k}'_n,$$
$$\boldsymbol{k} = \boldsymbol{k}_1 + \boldsymbol{k}_2.$$

For example, in the case of $n = 2$, we obtain

$$\mathcal{F}_2(\boldsymbol{k}'_1, \boldsymbol{k}'_2) = \frac{5}{7} + \frac{1}{2} \frac{\boldsymbol{k}'_1 \cdot \boldsymbol{k}'_2}{k'_1 k'_2} \left(\frac{k'_1}{k'_2} + \frac{k'_2}{k'_1} \right) + \frac{2}{7} \frac{(\boldsymbol{k}'_1 \cdot \boldsymbol{k}'_2)}{k'^2_1 k'^2_2}, \qquad (3.31)$$

$$\mathcal{G}_2(\boldsymbol{k}'_1, \boldsymbol{k}'_2) = \frac{3}{7} + \frac{1}{2} \frac{\boldsymbol{k}'_1 \cdot \boldsymbol{k}'_2}{k'_1 k'_2} \left(\frac{k'_1}{k'_2} + \frac{k'_2}{k'_1} \right) + \frac{4}{7} \frac{(\boldsymbol{k}'_1 \cdot \boldsymbol{k}'_2)}{k'^2_1 k'^2_2}. \qquad (3.32)$$

The expansion of higher order perturbations depends on the cosmological parameters. For generic Friedmann Universes, the perturbative solutions are obtained, too. Furthermore, in non-dust matter models, dark energy models, or modified gravity models such as Brans-Dicke theory, the nonlinear perturbative solutions are derived [39].

Because the ratio of growth factor between first- and second-order perturbations depends on the cosmological model, in weakly nonlinear regime, Eulerian nonlinear perturbations can describe the dependency of cosmological models for the evolution of density fluctuation. Therefore, for example, the evolution of the power spectrum in the quasi-nonlinear regime is discussed with Eulerian nonlinear perturbations. However, as we mentioned before, when the density fluctuation approaches to unity, Eulerian perturbation would break down.

4.   Lagrangian Perturbations I – Basic

4.1.   Lagrangian Description

Instead of the density fluctuation itself, Zel'dovich found it more convenient to regard the displacement from the homogeneous distribution as perturbative quantity [122]. In the method he proposed, the motion of the matter fluid is given by Lagrangian description. In the Lagrangian description for hydrodynamics, the coordinates $\boldsymbol{x}$ of the fluid elements are represented in terms of the Lagrangian coordinates $\boldsymbol{q}$ as

$$\boldsymbol{x} = \boldsymbol{q} + \boldsymbol{s}(\boldsymbol{q}, t), \qquad (4.1)$$

where q is defined as initial values of x, and s denotes the Lagrangian displacement vector due to the presence of inhomogeneities. The peculiar velocity is given by

$$v = a\dot{s} . \tag{4.2}$$

Then we introduce the Lagrangian time derivative:

$$\frac{\mathrm{d}}{\mathrm{d}t} \equiv \frac{\partial}{\partial t} + \frac{1}{a}v \cdot \nabla_x . \tag{4.3}$$

Using the Lagrangian derivative, the nonlinear term of the peculiar velocity in comoving Euler's equation (2.14) disappears. Using the Lagrangian time derivative, divergence and rotation of comoving Euler's equation are written as

$$\nabla_x \times \left(\frac{\mathrm{d}v}{\mathrm{d}t} + \frac{\dot{a}}{a}v \right) = 0 , \tag{4.4}$$

$$\nabla_x \cdot \left(\frac{\mathrm{d}v}{\mathrm{d}t} + \frac{\dot{a}}{a}v \right) = -4\pi G\rho_b\delta - \nabla_x \cdot \left(\frac{1}{\rho}\nabla_x P \right) . \tag{4.5}$$

The exact form of the energy density in the Lagrangian space is obtained from Eq. (2.13) as

$$\rho = \rho_b J^{-1} , \tag{4.6}$$

where $J \equiv \det(\partial x_i/\partial q_j) = \det(\delta_{ij} + \partial s_i/\partial q_j)$ is the Jacobian of the coordinate transformation from x to q. J is described by expansion of the derivative of Lagrangian perturbation as follows:

$$J = 1 + \nabla_q \cdot s + \frac{1}{2}\left((\nabla_q \cdot s)^2 - \frac{\partial s_i}{\partial q_j}\frac{\partial s_j}{\partial q_i} \right) + \det\left(\frac{\partial s_i}{\partial q_j} \right) . \tag{4.7}$$

Next, we consider the description of the vorticity in the Lagrangian space. We define the vorticity as

$$\omega \equiv \frac{1}{a}\nabla_x \times v . \tag{4.8}$$

From Eq. (4.4), we obtain the vorticity equation.

$$\frac{\mathrm{d}\omega}{\mathrm{d}t} + 2\frac{\dot{a}}{a}\omega + \frac{\omega}{a}(\nabla_x \cdot v) = (\omega \cdot \nabla_x \dot{x}) . \tag{4.9}$$

In the Lagrangian description, Eq. (4.9) is solved exactly.

$$\omega = \frac{(\omega_0(q) \cdot \nabla_q)\,x(q,t)}{a^2 J} , \tag{4.10}$$

where $\omega_0(q)$ is the primordial vorticity.

From Eqs. (4.2) and (4.5), the peculiar gravitational field is written as

$$\tilde{g} = a\left(\ddot{s} + 2\frac{\dot{a}}{a}\dot{s} - \frac{1}{a^2}\frac{\mathrm{d}P}{\mathrm{d}\rho}(\rho)\,J^{-1}\nabla_x J \right) , \tag{4.11}$$

where an overdot ($\dot{}$) denotes $\mathrm{d}/\mathrm{d}t$. Hence, from Eqs. (2.15) and (2.16), we obtain the following equations for s:

$$\nabla_x \times \left(\ddot{s} + 2\frac{\dot{a}}{a}\dot{s}\right) = 0, \tag{4.12}$$

$$\nabla_x \cdot \left(\ddot{s} + 2\frac{\dot{a}}{a}\dot{s} - \frac{1}{a^2}\frac{\mathrm{d}P}{\mathrm{d}\rho}J^{-1}\nabla_x J\right) = -4\pi G\rho_b(J^{-1} - 1). \tag{4.13}$$

If we find solutions of Eqs. (4.12) and (4.13) for s, the dynamics of the system considered is completely determined. Since these equations are highly nonlinear and hard to solve exactly, we will advance a perturbative approach. Remark that, in solving the equations for s in the Lagrangian coordinates q, the operator ∇_x will be transformed into ∇_q by the following rule:

$$\frac{\partial}{\partial q_i} = \frac{\partial x_j}{\partial q_i}\frac{\partial}{\partial x_j} = \frac{\partial}{\partial x_i} + \frac{\partial s_j}{\partial q_i}\frac{\partial}{\partial x_j}. \tag{4.14}$$

We decompose s into the longitudinal and transverse modes as $s = \nabla_q S + S^{\mathrm{T}}$ with $\nabla_q \cdot S^{\mathrm{T}} = 0$.

4.2. Linear Perturbations (Zel'dovich Approximation)

Zel'dovich derived a first-order Lagrangian solution of the longitudinal mode for dust fluid ($P = 0$) [122]. This first-order approximation is called Zel'dovich approximation (ZA). Especially when we consider the plane-symmetric case, ZA gives exact solutions [1, 34]. The evolution equation at linear-order is written as

$$\ddot{S}^{(1)} + 2\frac{\dot{a}}{a}\dot{S}^{(1)} - 4\pi G\rho_b S^{(1)} = 0. \tag{4.15}$$

The form of the evolution equation (4.15) is same as that of evolution equation in the Eulerian linear perturbation theory (3.4). Therefore the time components of linear perturbation coincides. The first-order solutions are written as follows:

$$S^{(1)}(q, t) = g_{1+}(t)\psi_+^{(1)}(q) + g_{1-}(t)\psi_-^{(1)}(q). \tag{4.16}$$

For the E-dS Universe ($\Omega_M = 1$), $g_{1\pm}$ are written as

$$g_{1+}(t) = t^{2/3} \propto a(t), \tag{4.17}$$

$$g_{1-}(t) = t^{-1} \propto a(t)^{-3/2}. \tag{4.18}$$

For the Λ-flat Universe, the solutions are given by (3.9).

Using Eqs. (4.6) and (4.16), we can describe density fluctuation. Here we define the deformation tensor $X_{\alpha\beta}$:

$$X_{\alpha\beta} \equiv \frac{\partial x_\alpha}{\partial q_\beta} = \delta_{\alpha\beta} + \frac{\partial s_\alpha}{\partial q_\beta}. \tag{4.19}$$

The eigenvalues of the deformation tensor (4.19) are written by

$$w_i = -g_1(t)\lambda_i^0(q), \tag{4.20}$$

where λ_i^0 is the eigenvalue of $\partial\psi_{,\alpha}/\partial q_\beta$. Using these eigenvalues, the density fluctuation is described by

$$(1+\delta)^{-1} = \prod_{i=1}^{3}(1+w_i)\,. \tag{4.21}$$

When w_i becomes -1, the caustic ("Zel'dovich's pancake") is formed and the density fluctuation diverges. After that, the perturbative solution does not have physical meaning. In order to avoid the formation of caustics, several modified methods have been proposed. We will discuss these models in Sec. 5.2.. By generalizing the eigenvalues of the deformation tensor (Eq. (4.20)), more exact approximations can be proposed. These models will be expressed in Sec. 5.3.2.. In the plane-symmetric case, we can describe time evolution beyond shell-crossing. For detail, we will mention in Appendix A..

4.3.　Non-linear Perturbations

ZA solutions are known as perturbative solutions, which describe the structure well in the quasi-nonlinear regime. To improve approximation, higher-order perturbative solutions of Lagrangian displacement were derived. Irrotational second-order solutions (2LPT) were derived by Bouchet et al. [9] and Buchert and Ehlers [16], and third-order solutions (3LPT) were obtained by Buchert [13], Bouchet et al. [10], and Catelan [18].

Hereafter we consider only growing-mode g_{1+} in first-order Lagrangian perturbation. First, we show the evolution equation for second- and third-order perturbative equations for the longitudinal mode. Here we decompose the Lagrangian perturbation to a time and a spacial components.

$$S^{(2)}(t,\boldsymbol{q}) = g_2(t)\psi^{(2)}(\boldsymbol{q})\,, \tag{4.22}$$
$$S^{(3)}(t,\boldsymbol{q}) = g_3(t)\psi^{(3)}(\boldsymbol{q})\,, \tag{4.23}$$

where the superscript $S^{(n)}$ means n-th order Lagrangian perturbations. The second-order perturbative equation is written as

$$\left(\ddot{g}_2 + 2\frac{\dot{a}}{a}\dot{g}_2 - 4\pi G\rho_b g_2\right)\psi^{(2)}_{,ii}$$
$$= -2\pi G\rho_b g_1^2\left[(\psi^{(1)}_{,ii})^2 - \psi^{(1)}_{,ij}\psi^{(1)}_{,ji}\right]\,. \tag{4.24}$$

The third-order perturbative equation becomes a little complicated.

$$\left(\ddot{g}_3 + 2\frac{\dot{a}}{a}\dot{g}_3 - 4\pi G\rho_b g_3\right)\psi^{(3)}_{,ii}$$
$$= -8\pi G\rho_b\left[g_1(g_2 - g_1^2)\left(\psi^{(1)}_{,ii}\psi^{(2)}_{,jj} - \psi^{(1)}_{,ij}\psi^{(2)}_{,ji}\right) + g_1^3\det(\psi^{(1)}_{,ij})\right]\,. \tag{4.25}$$

The spacial component of the second-order perturbation is written as follows:

$$\psi^{(2)}_{,ii} = (\psi^{(1)}_{,jj})^2 - \psi^{(1)}_{,jk}\psi^{(1)}_{,kj}\,. \tag{4.26}$$

The time component of second-order perturbation, g_2 obeys the evolution equation.

$$\ddot{g}_2 + 2\frac{\dot{a}}{a}\dot{g}_2 - 4\pi G\rho_b g_2 = -2\pi G\rho_b g_1^2\,. \tag{4.27}$$

For third-order perturbations, we divide to two components of which one is derived from the combination $\psi^{(1)}$ and $\psi^{(2)}$ and the other is derived from $(\psi^{(1)})^3$.

$$S^{(3)} = g_{3a}(t)\psi_a^{(3)}(\boldsymbol{q}) + g_{3b}(t)\psi_b^{(3)}(\boldsymbol{q}), \tag{4.28}$$

$$\psi_{a,ii}^{(3)} = \det\left(\psi_{,jk}^{(1)}\right), \tag{4.29}$$

$$\psi_{b,ii}^{(3)} = \psi_{,jj}^{(1)}\psi_{,kk}^{(2)} - \psi_{,jk}^{(1)}\psi_{,kj}^{(2)}, \tag{4.30}$$

$$\ddot{g}_{3a} + 2\frac{\dot{a}}{a}\dot{g}_{3a} - 4\pi G\rho_b g_{3a} = -8\pi G\rho_b g_1^3, \tag{4.31}$$

$$\ddot{g}_{3b} + 2\frac{\dot{a}}{a}\dot{g}_{3b} - 4\pi G\rho_b g_{3b} = -8\pi G\rho_b g_1(g_2 - g_1^2). \tag{4.32}$$

For the E-dS Universe, g_2 and g_3 are described with analytic forms.

$$g_2(t) = -\frac{3}{7}t^{4/3} \propto g_1(t)^2, \tag{4.33}$$

$$g_{3a}(t) = -\frac{1}{3}t^2 \propto g_1(t)^3, \tag{4.34}$$

$$g_{3b}(t) = \frac{10}{21}t^2 \propto g_1(t)^3. \tag{4.35}$$

For the Λ-flat Universe, unfortunately, we need to solve evolution equations numerically [10].

$$3(h^2 - 1)\ddot{g}_2 + 2h\dot{g}_2 = 2g_2 - 2g_1^2, \tag{4.36}$$

$$3(h^2 - 1)\ddot{g}_{3a} + 2h\dot{g}_{3a} = 2g_{3a} - 2g_1^3, \tag{4.37}$$

$$3(h^2 - 1)\ddot{g}_{3b} + 2h\dot{g}_{3b} = 2g_{3b} + 2g_1^3\left(1 - \frac{g_2}{g_1^2}\right). \tag{4.38}$$

Instead of performing numerical calculations for nonlinear growth factors, fitting formula has been also obtained for specific FRW Universes [10].

Here we define the logarithmic derivative of the growth factor as

$$f_n \equiv \frac{a}{g_n}\frac{dg_n}{da}. \tag{4.39}$$

For the Λ-flat Universe, the fitting formula is given by

$$g_2 \simeq -\frac{3}{7}\Omega_M^{-1/143}g_1^2 \quad (\Omega_M \simeq 1), \tag{4.40}$$

$$g_{3a} \simeq -\frac{1}{3}\Omega_M^{-4/275}g_1^3, \; g_{3b} \simeq \frac{10}{21}\Omega_M^{-269/17875}g_1^3 \quad (\Omega_M \simeq 1), \tag{4.41}$$

$$f_1 \simeq \Omega_M^{5/9} \quad (0.1 \leq \Omega_M \leq 1), \tag{4.42}$$

$$f_2 \simeq 2\Omega_M^{6/11} \quad (0.1 \leq \Omega_M \leq 1), \tag{4.43}$$

$$f_{3a} \simeq 3\Omega_M^{13/24}, \; f_{3b} \simeq 3\Omega_M^{13/24} \quad (0.1 \leq \Omega_M \leq 1). \tag{4.44}$$

Here we show another procedure for the derivation of higher-order perturbations [18]. We have derived higher-order perturbative solutions from the first-order scalar function so far. Instead of the scalar function, here we derive the higher-order solutions from the first-order displacement vector. This derivation is sometimes more convenient.

Using the displacement vector, second- and third-order perturbative solutions in the longitudinal mode are given by

$$s_i^{(2)} = \frac{1}{2}\left[s_i^{(1)} s_{j,j}^{(1)} - s_j^{(1)} s_{i,j}^{(1)} \right] + R_i^{(2)}, \tag{4.45}$$

$$s_{a\,i}^{(3)} = \frac{1}{3}\mathcal{S}_{ij}^{(1)C} s_j^{(1)} + R_{a\,i}^{(3)}, \tag{4.46}$$

$$s_{b\,i}^{(3)} = \frac{1}{4}\left[s_i^{(1)} s_{j,j}^{(2)} - s_j^{(1)} s_{i,j}^{(2)} s_i^{(2)} s_{j,j}^{(1)} - s_j^{(2)} s_{i,j}^{(1)} \right] + R_{b\,i}^{(3)}, \tag{4.47}$$

$$\mathcal{S}_{ij} \equiv s_{i,j} \tag{4.48}$$

where $\mathcal{S}^C$ means the cofactor matrix of $\mathcal{S}$. $R_i^{(2)}, R_{a\,i}^{(3)}, R_{b\,i}^{(3)}$ are divergence-free vectors such that

$$\nabla \times s^{(2)} = 0, \ \nabla \times s_a^{(3)} = 0, \ \nabla \times s_b^{(3)} = 0. \tag{4.49}$$

Because the divergence-free vectors would be much smaller than the perturbative vector, in generic cases, we ignore these vectors.

In this description, time components of the higher-order perturbative solutions are also given by Eqs. (4.27), (4.31), (4.32).

4.4. Transverse Mode

For the transverse mode, solutions until third-order have been obtained [3, 12, 84]. The basic equation for the transverse mode is given by

$$\nabla_x \times \left(\ddot{s}^T + 2\frac{\dot{a}}{a}\dot{s}^T \right) = 0. \tag{4.50}$$

From (4.50), the evolution equation for the first-order perturbation is given as

$$\ddot{s}^T + 2\frac{\dot{a}}{a}\dot{s}^T = 0. \tag{4.51}$$

For the E-dS Universe, the first-order solution is written as

$$s^T(\boldsymbol{q}, t) = s_a^T(\boldsymbol{q}) + t^{-1/3} s_b^T(\boldsymbol{q}). \tag{4.52}$$

Therefore the transverse mode does not have a growing solution in the first-order perturbation. Hereafter, we consider only the longitudinal mode for the first order.

For the dust model, if we consider only the longitudinal mode for the first order, the second-order transverse mode does not appear. When the matter affects polytropic pressure, the second-order transverse mode is generated from the first-order longitudinal mode.

Even if we consider the dust model, the third-order transverse mode is generated by the triplet of the first-order longitudinal mode, and it grows. Here we show the evolution equation for the second- and third-order perturbation in the dust model.

For the second-order perturbations, the evolution equation is described as

$$\varepsilon_{ijk}\left(\ddot{s}_{k,j}^{(2)T} + 2\frac{\dot{a}}{a}\dot{s}_{k,j}^{(2)T}\right) - \varepsilon_{ijk}S_{,jl}^{(1)}\left(\ddot{S}_{,kl}^{(1)} + 2\frac{\dot{a}}{a}\dot{S}_{,kl}^{(1)}\right) = 0\,. \tag{4.53}$$

Because the first-order perturbations are consisted from the longitudinal mode, the evolution equation is changed to

$$\varepsilon_{ijk}\left(\ddot{s}_{k,j}^{(2)T} + 2\frac{\dot{a}}{a}\dot{s}_{k,j}^{(2)T}\right) = 0\,. \tag{4.54}$$

Therefore the evolution equation for second-order perturbations coincided with that for linear perturbations (4.51) and we can ignore the transverse mode in second-order perturbations.

For third-order perturbations, the evolution equation is derived as

$$\begin{aligned}
&\varepsilon_{ijk}\left(\ddot{s}_{k,j}^{(3)T} + 2\frac{\dot{a}}{a}\dot{s}_{k,j}^{(3)T}\right) \\
&= \varepsilon_{ijk}\left[S_{,jl}^{(1)}\left(\ddot{S}_{,kl}^{(2)} + 2\frac{\dot{a}}{a}\dot{S}_{,kl}^{(2)}\right) + S_{l,j}^{(2)}\left(\ddot{S}_{,kl}^{(1)} + 2\frac{\dot{a}}{a}\dot{S}_{,kl}^{(1)}\right) - S_{,jl}^{(1)}S_{,ml}^{(1)}\left(\ddot{S}_{,km}^{(1)} + 2\frac{\dot{a}}{a}\dot{S}_{,km}^{(1)}\right)\right] \\
&= -2\pi G g_1 g_2 \rho_b \varepsilon_{ijk}\psi_{,jl}^{(1)}\psi_{,kl}^{(2)}\,.
\end{aligned} \tag{4.55}$$

For Eq. (4.55), we take a rotation,

$$-\nabla^2\left(\ddot{s}_i^{(3)T} + 2\frac{\dot{a}}{a}\dot{s}_i^{(3)T}\right) = 2\pi G\rho_b g_1 g_2\left[\left(\psi_{,il}^{(1)}\psi_{,kl}^{(2)}\right)_{,k} - \left(\psi_{,jl}^{(1)}\psi_{,il}^{(2)}\right)_{,j}\right]\,.$$

Then we divide the perturbation to the spacial and the time components.

$$s_i^{(3)T} = g_{3T}\zeta_i^T\,, \tag{4.56}$$

$$\ddot{g}_{3T} + 2\frac{\dot{a}}{a}\dot{g}_{3T} = 2\pi G\rho_b g_1 g_2\,, \tag{4.57}$$

$$-\nabla^2\zeta_i^T = \left(\psi_{,il}^{(1)}\psi_{,kl}^{(2)}\right)_{,k} - \left(\psi_{,jl}^{(1)}\psi_{,il}^{(2)}\right)_{,j}\,. \tag{4.58}$$

In the E-dS Universe, the time components are described as

$$g_{3T} = -\frac{3}{98}t^2\,. \tag{4.59}$$

For the Λ-flat Universe, we need to solve evolution equations numerically. The evolution equation is described as

$$3(h^2 - 1)\ddot{g}_{3T} + 2h\dot{g}_{3T} = g_1 g_2\,. \tag{4.60}$$

Although the third-order solution has the growing mode, it can be shown that the vorticity does not have the growing mode.

5. Lagrangian Perturbations II – Advanced

5.1. Overview

In the last section, we showed the solutions based on Lagrangian perturbation theory until third-order. Even if we derive the solutions valid to higher-order perturbations, when the perturbative quantity approaches to unity, the accuracy becomes worse. Furthermore, the perturbative solutions have no longer physical meaning after shell-crossing.

In this section, we describe improved approximations for Lagrangian perturbations. In order to avoid shell-crossing, several modified methods such as adhesion approximation or truncated Zel'dovich approximation have been proposed. For the accurate description of the nonlinear evolution, several improvements have been proposed. For examples, local approximation, Padé approximation, Shanks approximation, and so on. Recently, the renormalization group method has been proposed. This method describes accurate quasi-nonlinear evolution. For another approach, wave mechanical approach has been considered. Using nonlinear transformation, both the density and velocity fields can be described with a "wave function". This method also describes the quasi-nonlinear evolution well. Especially this method can describes the evolution of non-Gaussianity well.

In generic cases, the cosmic fluid is given by pressureless dust fluid. If the dark matter affects short-range interaction, the description for structure formation with the dust fluid is insufficient. If the short-range interaction is equivalent with the effective pressure, the interaction can be treated as the pressure gradient in Euler's equation. As a concrete example, we notice Euler-Jeans-Newton model, in which the fluid affects polytropic pressure. Furthermore, we also consider two-fluid model, which consists of baryonic matter and CDM.

5.2. Modified Methods

Even if the Lagrangian description gives the evolution of quasi-nonlinear structure more accurately than the Eulerian picture, when a caustic is formed, the solutions are no longer physical. Furthermore, after the formation of caustics, the nonlinear structure diffuses. On the other hand, cosmological N-body simulations show that pancakes, skeletons, and clumps remain during evolution. In order to avoid shell-crossing, several modifications to Lagrangian perturbation theory have been proposed.

5.2.1. Adhesion Approximation

Adhesion approximation (AA) [43] has been proposed from a consideration based on Burgers' equation. This method is derived by the addition of an artificial viscous term to ZA. AA with small viscosity deals with the skeleton of the structure, which at an arbitrary time is found directly without a long numerical calculation.

We briefly describe the adhesion approximation. In ZA, the equation for "peculiar velocity" in the E-dS Universe is written as follows:

$$\frac{\partial u}{\partial a} + (u \cdot \nabla_x)u = 0, \tag{5.1}$$

$$u \equiv \frac{\partial x}{\partial a} = \frac{\dot{x}}{\dot{a}}, \tag{5.2}$$

where $a(\propto t^{2/3})$ is scale factor. To go beyond ZA, we add the artificial viscosity term as a source of the equation.

$$\frac{\partial \boldsymbol{u}}{\partial a} + (\boldsymbol{u} \cdot \nabla_x)\boldsymbol{u} = \nu \nabla_x^2 \boldsymbol{u}. \tag{5.3}$$

We consider the case when the viscosity coefficient $\nu \to +0$ ($\nu \neq 0$). In this case, the viscosity term especially affects the high-density region. Within the limits of a small ν, the analytic solution of Eq.(5.3) is given by

$$\boldsymbol{u}(\boldsymbol{x}, t) = \sum_\alpha \left(\frac{\boldsymbol{x} - \boldsymbol{q}_\alpha}{a} \right) j_\alpha \exp\left(-\frac{I_\alpha}{2\nu} \right) \Big/ \sum_\alpha j_\alpha \exp\left(-\frac{I_\alpha}{2\nu} \right), \tag{5.4}$$

where $\boldsymbol{q}_\alpha$ is the Lagrangian points that minimize the action

$$I_\alpha \equiv I(\boldsymbol{x}, a; \boldsymbol{q}_\alpha) = S_0(\boldsymbol{q}_\alpha) + \frac{(\boldsymbol{x} - \boldsymbol{q}_\alpha)^2}{2a} = \text{min.}, \tag{5.5}$$

$$j_\alpha \equiv \left[\det\left(\delta_{ij} + \frac{\partial^2 S_0}{\partial q_i \partial q_j} \right) \right]^{-1/2} \Bigg|_{\boldsymbol{q}=\boldsymbol{q}_\alpha}, \tag{5.6}$$

$$S_0 = S(\boldsymbol{q}, t_0), \tag{5.7}$$

considered as a function of $\boldsymbol{q}$ for fixed $\boldsymbol{x}$ [56]. In AA, because of the viscosity term, the caustic does not appear and a stable nonlinear structure can exist. However, when the evolution is advanced too much, the Universe becomes covered with a high-density structure called "skeleton."

Domínguez [32, 33] proposed the origin of the viscosity term with small-scale expansion. He considered the peculiar acceleration and the velocity dispersion in clusters. The peculiar acceleration and the velocity dispersion are expanded by a typical scale of the clusters L. If L has a finite small value, a surplus term in basic equation corresponds to the viscosity term in AA. Buchert and Domínguez [15] generalized the assumption for the expansion with L. Then they lead AA and apply both the Eulerian and Lagrangian perturbative expansions to new model. Furthermore they discussed some non-perturbative results from the starting points about nonlinear structure formation in the multi-stream regime.

5.2.2. Truncated Zel'dovich Approximation

As another modified method, truncated (or Optimized) Zel'dovich approximation (TZA) has been used well [22, 68]. During the evolution, the small scale structure contracts and forms caustics. Therefore, if we introduce some cutoff in the small scale, we will be able to avoid the formation of caustics [22, 68].

In TZA, for the avoidance of caustics, we introduce a cutoff in the initial density spectrum. Various cutoff methods were considered, and the Gaussian cutoff was shown to be the most suitable from the comparison with N-body simulations. The Gaussian cutoff is introduced to the initial density spectrum as follows:

$$\mathcal{P}(k, t_{\text{in}}) \to \mathcal{P}(k, t_{\text{in}}) \exp\left(-k^2/k_{NL} \right), \tag{5.8}$$

where k_{NL} is the "nonlinear wavenumber", defined by

$$1 = g_{1+}(t)^2 \int_{k_0}^{k_{NL}} \mathcal{P}(sk, t_{\text{in}})dk \,, \qquad (5.9)$$

where $g(t)_{1+}$ is the growth factor in ZA (Eq. (4.16)). From Eq. (5.9), it is shown that the "nonlinear wavenumber" decreases during the evolution. Therefore, when the evolution is advanced too much, small structure will vanished.

5.3. Improvements

In the last subsection, we mentioned modified methods for Lagrangian perturbation. In AA, artificial parameter for the viscosity is required. Although the viscosity term is appeared by small-scale expansion, its physical origin is not clarified. TZA seems rather good approximation than ZA. However because of the cutoff in the initial spectrum, the small-scale structure disappears.

In this subsection, we consider improvements for Lagrangian perturbations with mathematical procedures. For the improvements, artificial parameters are no longer required.

5.3.1. Padé, Shanks Approximation

Yoshisato *et al.* [120] introduced the Padé approximation for the evolution of the density fluctuation. The Padé approximation is not a simple polynomial expansion with a small parameter like the Taylor expansion but a rational polynomial expansion. For a given function $f(x)$, the Padé approximation is given as:

$$f(x) \simeq \frac{\sum_{k=0}^{M} a_k x^k}{1 + \sum_{k=1}^{N} b_k x^k} \,, \qquad (5.10)$$

where a_k and b_k are constant coefficients. Assume we already know the coefficient c_l $(0 \leq l \leq M + N)$ of the Taylor expansion around $x = 0$.

$$f(x) = \sum_{l=0}^{M+N} c_l x^l + o(x^{M+N+1})\,. \qquad (5.11)$$

Comparing the coefficents a_k, b_k, and c_k, we can determine a_k and b_k as follows:

$$a_0 = c_0\,, \qquad (5.12)$$

$$a_k = \sum_{m=1}^{N} b_m c_{k-m} \ (k = 1, \cdots, N)\,, \qquad (5.13)$$

$$\sum_{m=1}^{N} b_m c_{N-m+k} = -c_{N+k} \ (k = 1, \cdots, N)\,. \qquad (5.14)$$

The advantage of the Padé approximation is that even if we consider the expansion at same-order, the Padé approximation describes original function rather better than the one based on the Taylor expansion.

First, Yoshisato *et al.* [120] showed that the Padé approximation in Eulerian scheme improves the accuracy of the perturbations. Then Matsubara *et al.* [64] proposed the application of the Padé approximation for Lagrangian description, which improved 2LPT and 3LPT.

As another approach for the improvement of Lagrangian perturbations, the application of Shanks transformation [90] has been proposed [102]. Shanks transformation is known as a good way to promote the convergence of a slowly converging series. In other words, using this transformation, we can obtain more accurate perturbative solutions from perturbations at finite order.

In the following we explain Shanks transformation making use of a simple example. Suppose the n-th term in the sequence takes the following form:

$$A_n = A + \alpha\varepsilon^n \quad (|\varepsilon| < 1). \tag{5.15}$$

The sequence converges $A_n \to A$ as $n \to \infty$. To obtain the limit of a sequence A, we solve algebraic equations with A_{n-1}, A_n, and A_{n+1}.

$$A = \frac{A_{n+1}A_{n-1} - A_n^2}{A_{n+1} + A_{n-1} - 2A_n}. \tag{5.16}$$

This formula is exact only if the sequence A_n is described by the form in (5.15). For the generic case, we consider the nth term in the sequence takes the form:

$$A_n = A(n) + \alpha q^n, \tag{5.17}$$

where for large n, $A(n)$ is a more slowly varying function of n than A_n. Let us suppose that $A(n)$ varies sufficiently slowly so that $A(n-1), A(n)$, and $A(n+1)$ are all approximately equal. Then the above discussion motivates the nonlinear transformation

$$S(A_n) = \frac{A_{n+1}A_{n-1} - A_n^2}{A_{n+1} + A_{n-1} - 2A_n}. \tag{5.18}$$

This transformation is called Shanks transformation, creating a new sequence $S(A_n)$ which often converges more rapidly than the old sequence A_n. The sequence $S^2(A_n) = S[S(A_n)]$ and $S^3(A_n) = S[S[S(A_n)]]$ may be even more rapidly convergent.

We apply Shanks transformation for the third-order Lagrangian approximation. The Lagrangian displacement is described as

$$s_i(t, \boldsymbol{q}) = g_1(t)s_i^{(1)}(\boldsymbol{q}) + g_2(t)s_i^{(2)}(\boldsymbol{q}) + g_{3a}(t)s_i^{(3a)}(\boldsymbol{q}) + g_{3b}(t)s_i^{(3b)}(\boldsymbol{q}) + g_{3T}(t)s_i^{(3T)}(\boldsymbol{q}). \tag{5.19}$$

Applying Shanks transformation (5.18), the perturbative solution is transformed to

$$\widetilde{s}_i(t, \boldsymbol{q}) = \frac{g_1 s_i^{(1)} \zeta_i^{(3)} - g_2^2 \left\{ s_i^{(2)} \right\}^2}{g_1 s_i^{(1)} + \zeta_i^{(3)} - 2g_2 s_i^{(2)}}, \tag{5.20}$$

$$\zeta_i^{(3)}(t, \boldsymbol{q}) \equiv g_{3a}(t)s_i^{(3a)}(\boldsymbol{q}) + g_{3b}(t)s_i^{(3b)}(\boldsymbol{q}) + g_{3T}(t)s_i^{(3T)}(\boldsymbol{q}). \tag{5.21}$$

If we consider special case such as spherical collapse or evolution of spherical void, because higher-order perturbative solutions are obtained, we can improve all the more. We showed the improvement of spherical symmetric case making use of the perturbative solution at eleventh-order [102].

Now we are in a position to compare the Padé approximation with Shanks transformation. In the Padé approximation, although we can solve algebraic equations and find unique solutions, the equations are quite complicated. Furthermore, it is quite important to choose the numbers of terms M and N in (5.10). When N is sufficiently different from M, the approximation is not improved well. In the case of evolution of the spherical void, when N is sufficiently different from M, not only the approximation is not improved well, but also the solution will diverges. On the other hand, Shanks transformation does not provide the corresponding divergence.

However, in several points, Shanks transformation shows its weakness. To apply the Padé approximation, the perturbative solutions valid upto second-order are enough. On the other hand, when we consider Shanks transformation, we must obtain the perturbative solutions valid at least third-order. Using Shanks transformation, we obtain a new perturbative solution $\widetilde{R}_n$ from R_{n-1}, R_n, R_{n+1}. Then, to repeatedly apply Shanks transformation, we require $R_{n-2}, \cdots, R_{n+2}$. In general, when we apply n times transformation, at least we must know $2n + 1$-th order perturbative solutions, which means that to improve the perturbation well, we must repeat the transformation several times.

From the viewpoint of algebraic procedures, Shanks transformation has an advantage. In the Padé approximation, we must solve nonlinear equations simultaneously. Then the solution is extremely complicated in a higher-order case. For example, when we improve an eleventh-order solution with $(M, N) = (5, 6)$, we derived about fifty-digit coefficients.

A serious problem is existed in both the Padé approximation and Shanks transformation. The continued fraction expansion is applied in both methods. When the denominator becomes zero, not the density fluctuation but Lagrangian displacement diverges. In such a situation, not only the approximations fail at all but also they are unphysical.

5.3.2. Local Approximation

Another approach we mention here is known as "local" approximations. In ZA, the growing factor D is independent of the position. On the other hand, in local approximations, this factor depends on the position.

First, we derive an evolution equation for the eigenvalue of the deformation tensor (Eq. (4.19)). Here we use time variable τ.

$$d\tau \propto a^{-2}dt. \tag{5.22}$$

Differentiating Eq. (4.11) with respect to x_j and summing up the trace part, we obtain

$$X_{ij}^{-1}\frac{d^2 X_{ji}}{d\tau^2} = -4\pi G a^4 \rho_b (J^{-1} - 1). \tag{5.23}$$

If X_{ij} is diagonal,

$$X_{ij} = (1 + w_i)\delta_{ij}, \tag{5.24}$$

Eq. (5.23) is written as

$$\sum_{i=1}^{3} \frac{\ddot{w}_i}{(1 + w_i)} = -4\pi G a^4 \rho_b \left[\frac{1}{(1 + w_1)(1 + w_2)(1 + w_3)} - 1 \right]. \tag{5.25}$$

For a spherical case we have $w_1 = w_2 = w_3$; for a cylindrical case $w_1 = w_2$ and $w_3 = 0$; and for a planar case $w_2 = w_3 = 0$. ZA is a linearized form for w_i of Eq. (5.25).

Reiseneger and Miralda-Escudé [81] proposed changing the eigenvalue (Eq. (4.20)) as

$$w_i = -D(t, \lambda_i^0)\lambda_i^0(\boldsymbol{q}). \tag{5.26}$$

This is then substituted in Eq. (5.25). They named this model the modified Zel'dovich approximation (MZA) or extension of ZA (EZA). The MZA is exact for planar, spherical, and cylindrical symmetric cases. However, they also reported that the MZA may not work for underdense regions.

Audit and Alimi [2] introduced another ansatz where Eq. (5.25) can be written in the following form:

$$\sum_{i=1}^{3} \left[(1 + w_j + w_k + w_j w_k)\ddot{w}_i - 4\pi G a^4 \rho_b \left(1 + \frac{w_j + w_k}{2} + \frac{w_j w_k}{3} \right) w_i \right] = 0. \tag{5.27}$$

They split this equation into three equations for each w_i:

$$(1 + w_j + w_k + w_j w_k)\ddot{w}_i = 4\pi G a^4 \rho_b \left(1 + \frac{w_j + w_k}{2} + \frac{w_j w_k}{3} \right) w_i. \tag{5.28}$$

This approximation is called the Deformation Tensor Approximation (DTA) [2]. Although this is also exact for planar, spherical, and cylindrical symmetric cases, the splitting of Eq. (5.27) is not unique, and we could add more local terms in Eq. (5.28).

Betancort-Rijo and López-Corredoida [4] proposed another generalization. In terms of the linear solution $\lambda_i = D(t)\lambda_i^0(\boldsymbol{q})$ in ZA, they generalized Eq. (4.20) to

$$w_i = -r_i(\lambda_i, \lambda_j, \lambda_k)\lambda_i^0(\boldsymbol{q}), \tag{5.29}$$

where $r_i(\lambda_i, \lambda_j, \lambda_k)$ is the power series of λ_i.

$$r_i(\lambda_i, \lambda_j, \lambda_k) = 1 + \sum_{l,m,n=0}^{\infty} C_{l,m,n}^{p}(\lambda_j + \lambda_k)^l (\lambda_j - \lambda_k)^{2n}\lambda_i^m, \tag{5.30}$$

where $C_{l,m,n}^{p}$ are the coefficients of the p-th order terms ($p = l + 2n + m$). The ZA corresponds to $r_i = 1$. They named this generalized model the "Complete Zel'dovich approximation" (CZA). They calculated explicitly the coefficients $C_{l,m,n}^{p}$ up to fourth order of λ in the E-dS Universe model [4]. This model is also exact for planar, spherical, and cylindrical symmetric cases. Furthermore, the CZA describes the evolution of ellipsoid dust better than the MZA, the DTA, and ZA [60].

5.4. Renormalization Group Appoarches

In the previous sections, we summarized modifications and improvements for the Lagrangian perturbation theory. Regardless of the efforts, since there are several problems in these modifications and improvements, we should propose other approaches which has fewer problems and need not rely on artificial modifications.

Crocce and Scoccimarro [26] proposed another approach. They applied renormalization group for the nonlinear evolution of large-scale structure. They start with ZA. The density fluctuation in Fourier space is given by

$$\widehat{\delta}(k) \equiv \int \frac{d^3 x}{(2\pi)^3} \delta(x) e^{ik\cdot x} = \int \frac{d^3 q}{(2\pi)^3} e^{ik\cdot q} \left[e^{ik\cdot s} - 1 \right]. \tag{5.31}$$

Here we apply mass conservation, i.e., $[1 + \delta(x)] d^3 x = d^3 q$. The power spectrum of the density fluctuation is described as

$$P(k) = \int \frac{d^3 \widetilde{q}}{(2\pi)^3} e^{ik\cdot(\widetilde{q}')} \left[\left\langle e^{ik\cdot(s(q)-s(q'))} \right\rangle - 1 \right], \tag{5.32}$$

where $\widetilde{q} \equiv q - q'$. In ZA, from the relation between the Lagrangian displacement and the density fluctuation, we obtain

$$\widehat{s}^{(1)} = -i \frac{k}{k^2} \widehat{\delta}^{(1)}. \tag{5.33}$$

If the primordial density field is random Gaussian field, the Lagrangian displacement is a Gaussian field. Therefore

$$\left\langle e^{ik\cdot(s(q)-s(q'))} \right\rangle = \exp \left[\frac{1}{2} \left\langle (k\cdot(s(q)-s(q')))^2 \right\rangle \right]. \tag{5.34}$$

The power spectrum (5.32) becomes

$$P(k) = \int \frac{d^3 \widetilde{q}}{(2\pi)^3} e^{ik\cdot\widetilde{q}} \left[e^{-[k^2 \sigma_v^2 - I(k,\widetilde{q})]} - 1 \right], \tag{5.35}$$

$$I(k, \widetilde{q}) \equiv \int d^3 q \, (k\cdot q)^2 \cos(q\cdot\widetilde{q}) \frac{P_L(q)}{q^4}, \tag{5.36}$$

$$\sigma_v^2 = \frac{1}{k^2} I(k, 0) \quad \text{(the variance of the displacement field)}, \tag{5.37}$$

where $P_L(q)$ means the power spectrum in the linear approximation. In perturbative approaches, the power spectrum (5.35) is expanded by the amplitude of the linear power spectrum:

$$P(k) = \int \frac{d^3 \widetilde{q}}{(2\pi)^3} e^{ik\cdot\widetilde{q}} \sum_{n=1}^{\infty} \frac{(-1)^n}{n!} \left[k^2 \sigma_v^2 - I(k, \widetilde{q}) \right]^n \equiv \sum_{\ell=0}^{\infty} P_{PT}^{(\ell)}(k). \tag{5.38}$$

From the view point of Feynman diagram, the term with $n = 1$ gives tree-level or zero loop mode ($\ell = 0$, linear power spectrum). $n = 2$ gives one-loop ($\ell = 1$) correction.

The difference between nonlinear perturbation and loop correction appears especially in small scale. In standard perturbative expansions, the spectrum oscillates such as acoustic modes. On the other hand, in loop correction, the spectrum does not oscillates. Furthermore, the contribution of loop corrections are shown to be the same order of the one from the tree level contribution.

In the renormalization group approach, one attempt to sum an infinite classes of diagram for the propagator.

$$P(k) = \int \frac{\mathrm{d}^3\widetilde{q}}{(2\pi)^3} e^{ik\cdot\widetilde{q}} e^{-k^2\sigma_v^2} \sum_{n=1}^{\infty} \frac{[I(k,\widetilde{q})]}{n!} \equiv \sum_{\ell=0}^{\infty} P_{RPT}^{(\ell)}(k) \,. \qquad (5.39)$$

Eq. (5.39) can be rewritten as

$$P(k) = e^{-k^2\sigma_v^2} \sum_{n=1}^{\infty} n! \int \delta_D \left(k - q_{1\cdots n}\right) \left[\mathcal{F}_n\left(q_1\,,\cdots,q_n\right)\right]^2 P_L(q_1)\mathrm{d}^3 q_1 \cdots P_L(q_n)\mathrm{d}^3 q_n \,,$$
$$(5.40)$$

where $q_{1\cdots n} = q_1 + \cdots + q_n$, $\mathcal{F}_n$ represents the perturbation theory (PT) kernels, defined by Eq. (3.28), and $\delta_1(k) = \theta_1(k) = \delta_{\mathrm{ini}}(k)$.

Here we remember the basic equations in Fourier space (Eqs. (3.19) and (3.20)). The equations can be rewritten by a two-component "vector":

$$\Psi_i(k,\eta) \equiv (\delta(k,\eta), -\theta(k,\eta)/H) \,, \qquad (5.41)$$

where $i = 1, 2$. η is defined as

$$\eta \equiv \log a(t) \,. \qquad (5.42)$$

Using Ψ_i, the basic equations in Fourier space are rewritten as

$$\frac{\partial \Psi_i}{\partial \eta} + \Omega_{ij} \Psi_j = \gamma_{ijk}^{(s)}(k, k_1, k_2) \Psi_j(k_1, \eta) \Psi_k(k_2, \eta) \,, \qquad (5.43)$$

where

$$\Omega_{ij} \equiv \begin{bmatrix} 0 & -1 \\ -3/2 & 1/2 \end{bmatrix} \,. \qquad (5.44)$$

$\gamma_{ijk}^{(s)}$ is the symmetrized vertex matrix:

$$\gamma_{121}^{(s)}(k, k_1, k_2) = \delta_D(k - k_{12}) \frac{\alpha(k_1, k_2)}{2} \,, \qquad (5.45)$$

$$\gamma_{112}^{(s)}(k, k_1, k_2) = \delta_D(k - k_{12}) \frac{\alpha(k_2, k_1)}{2} \,, \qquad (5.46)$$

$$\gamma_{222}^{(s)}(k, k_1, k_2) = \delta_D(k - k_{12}) \beta(k_1, k_2) \,. \qquad (5.47)$$

The other components of γ_{ijk} are zero. An integral solution of (5.43) is given by Laplace transform with respect to η:

$$\sigma_{ij}^{-1}(\omega)\Psi_j(k,\omega) = \Psi_{i0}(k) + \gamma_{ijk}^{(s)}(k, k_1, k_2) \oint \frac{\mathrm{d}\omega_1}{2\pi i} \Psi_j(k_1, \omega_1) \Psi_k(k_2, \omega - \omega_1) \,, (5.48)$$

$$\Psi_{i0} \equiv \Psi_i(k, \eta = 0) \,, \qquad (5.49)$$

$$\sigma_{ij}^{-1} \equiv \omega\delta_{ij} + \Omega_{ij} \,, \qquad (5.50)$$

where we set the scale factor $a = 1$ at the initial condition. Therefore,

$$\sigma_{ij}(\omega) = (\omega\delta_{ij} + \Omega_{ij})^{-1} = \frac{1}{(2\omega - 3)(\omega - 1)} \begin{bmatrix} 2\omega + 1 & 2 \\ 3 & 2\omega \end{bmatrix} \,. \qquad (5.51)$$

By inverse Laplace transformation of (5.48), we obtain the solution for (5.43):

$$\Psi_i(\boldsymbol{k}, \eta) = g_{ij}(\eta)\psi_j(\boldsymbol{k}) + \int_0^\eta d\eta' g_{ij}(\eta - \eta')\gamma_{jkl}^{(s)}(\boldsymbol{k}, \boldsymbol{k}_1, \boldsymbol{k}_2)\Psi_k(\boldsymbol{k}_1, \eta')\Psi_l(\boldsymbol{k}_2, \eta'), \tag{5.52}$$

$$g_{ij}(\eta) \equiv \int_{C-i\infty}^{C+i\infty} \frac{d\omega}{2\pi i}\sigma_{ij}(\omega)e^{\omega\eta} = \frac{e^\eta}{5}\begin{bmatrix} 3 & 2 \\ 3 & 2 \end{bmatrix} - \frac{e^{-3\eta/2}}{5}\begin{bmatrix} -2 & 2 \\ 3 & -3 \end{bmatrix} \quad (\eta \geq 0). \tag{5.53}$$

Because of the causality, the liniar propagator $g_{ij}(\eta) = 0$ for $\eta < 0$. The linear propagator is the Green's function of the linearized equation for (5.43). It describes linear evolution of the density and velocity fields from any configuration of the initial conditions. By the linearized continuous equation (3.1), we can describe

$$\phi_i(\boldsymbol{k}) = u_i\delta_0(\boldsymbol{k}), \tag{5.54}$$

where u_i is a two component vector. The growing mode and the decaying mode is given by $u = (1, 1)$ and $u = (2/3, -1)$, respectively.

In Eq. (5.53), the first term corresponds to the linear evolution and the second term corresponds to the nonlinear evolution which contains mode-coupling. The interaction is described by γ_{jkl}. Then the time evolution of the interaction is given by $g_{ij}(\eta - \eta')$.

We expands $\Psi_i(\boldsymbol{k}, \eta)$ to

$$\Psi_i(\boldsymbol{k}, \eta) = \sum_{n=1}^\infty \Psi_i^{(n)}(\boldsymbol{k}, \eta), \tag{5.55}$$

$$\Psi_i^{(n)}(\boldsymbol{k}, \eta) = \int \delta_D(\boldsymbol{k} - \boldsymbol{k}_{1\cdots n})\mathcal{F}_i^{(n)}(\boldsymbol{k}_1, \cdots, \boldsymbol{k}_n; \eta)\delta_0(\boldsymbol{k}_1)\cdots\delta_0(\boldsymbol{k}_n), \tag{5.56}$$

$$\boldsymbol{k}_{1\cdots n} = \boldsymbol{k}_1 + \cdots \boldsymbol{k}_n. \tag{5.57}$$

Substituting (5.55) and (5.56) into (5.52), we obtain the recursive relation for $\mathcal{F}$.

$$\mathcal{F}_i^{(n)}(\boldsymbol{k}_1, \cdots, \boldsymbol{k}_n; \eta)\delta_D(\boldsymbol{k} - \boldsymbol{k}_{1\cdots n})$$
$$= \left[\sum_{m=1}^n \int_0^\eta g_{ij}(\eta - s)\gamma_{jkl}^{(s)}(\boldsymbol{k}, \boldsymbol{k}_{1\cdots m}, \boldsymbol{k}_{m+1\cdots n})\mathcal{F}_k^{(m)}(\boldsymbol{k}_{1\cdots m}; s)\mathcal{F}_l^{(n-m)}(\boldsymbol{k}_{m+1\cdots n}; s)\right]_{\text{symmetrized}}, \tag{5.58}$$

where the r.h.s. has to be symmetrized under the interchange of any two wave vectors. For $n = 1$,

$$\mathcal{F}_i^{(1)}(\eta) = g_{ij}(\eta)u_j. \tag{5.59}$$

It is worth noting that (5.58) reproduces the standard recursive relations (Eqs. (3.28) and (3.29)) in the limit that the initial conditions are set at infinite past ($s \to -\infty$). Otherwise (5.58) gives the full time dependence of perturbative solutions, including all transients from the initial conditions [86].

When we calculate (5.56), we obtain n-th order solution. Although the computation procedure seems systematic, the calculation becomes soon complicated with increasing n. In order to carry out systematic calculation, a diagrammatic representation of the basic objects has been introduced. The objects compose Feynman diagram. The objects of the diagram correspond to the structure formation (Table 1).

Table 1. The Correspondence between components of the Feynman diagram and structure formation in the Universe.

Diagram	Structure formation
initial condition	Primordial spectrum
vertex	nonlinearity
propagator	linear evolution

When we derive n-th order term, we draw all topologically different trees with $n-1$ vertices and n initial conditions. Then we calculate all nonlinear interactions which correspond to the trees. For example, we show explicit form for $\Psi_i^{(2)}$ and $\Psi_i^{(3)}$.

$$\Psi_i^{(2)}(\boldsymbol{k},\eta) = \int d^3k_1 \int d^3k_2 \int_0^\eta ds\, g_{ij}(\eta-s)\gamma_{jkl}^{(s)}(\boldsymbol{k},\boldsymbol{k}_1,\boldsymbol{k}_2)g_{km}(s)\phi_m(\boldsymbol{k}_1)g_{ln}(s)\phi_n(\boldsymbol{k}_2) \quad (5.60)$$

$$\Psi_i^{(3)}(\boldsymbol{k},\eta) = 2\int d^3k_1 \int d^3k_2 \int_0^\eta ds\, g_{ij}(\eta-s)\gamma_{jkl}^{(s)}(\boldsymbol{k},\boldsymbol{k}_1,\boldsymbol{k}_2)g_{km}(s)\phi_m(\boldsymbol{k}_1)\Psi_l^{(2)}(\boldsymbol{k}_2,s) \quad (5.61)$$

As one of applications to cosmology, statistical quantities such as two-point correlation function or power spectrum has been calculated. In this approach, the power spectrum is given by

$$\langle \Psi_i(\boldsymbol{k},\eta)\Psi_j(\boldsymbol{k},\eta)\rangle = \delta_D(\boldsymbol{k}+\boldsymbol{k}')P_{ij}(\boldsymbol{k},\eta). \quad (5.62)$$

Under the assumption that initial condition is Gaussian, substituting (5.55) to (5.62), we obtain the power spectrum with ℓ-loop correction.

$$P_{ij}(\boldsymbol{k},\eta) = \sum_{\ell=0}^\infty P_{ij}^{(\ell)}(\boldsymbol{k},\eta), \quad (5.63)$$

$$\delta_D(\boldsymbol{k}+\boldsymbol{k}')P_{ij}^{(\ell)}(\boldsymbol{k},\eta) = \sum_{m=1}^{2\ell+1} \left\langle \Psi_i^{(m)}(\boldsymbol{k},\eta)\Psi_j^{(2\ell+2-m)}(\boldsymbol{k}',\eta)\right\rangle. \quad (5.64)$$

Therefore, when we derive the power spectrum with ℓ-loop correction, we need to compute until $2\ell+1$-th order term for Ψ.

For the computation of the spectrum at each order, we do not mention in this paper. Following the diagrams which correspond to the ℓ-loop correction, the power spectrum is computed (For detail, see [26]).

For the summation of the perturbations, other statistical methods have been proposed. In the statistical theory of turbulence, the energy spectrum has been calculated with closure approximation for fluid. Taruya and Hiramatsu applied the closure theory for the nonlinear evolution of cosmological power spectra [95].

As resummation of the perturbative expansions, one-loop approximation has been considered [62]. In this approach, we can analyze the structure formation in not only the real space but also the redshift space.

Here we consider the relation between the position in the real space x and the redshift space x_z. The comoving distance-redshift relation in FRW Universe is given by

$$x(z) = \int_0^z \frac{cz'}{H(z')} \mathrm{d}z',$$ (5.65)

where the redshift causes by the relative velocity between the Earth and galaxies. For galaxy redshifts, the observed redshift z_{obs} includes the effect of the peculiar velocity. Therefore, the distance derived by the observed redshift z_{obs} does not coincide to actual comoving distance. The comoving distance in redshift space is defined as

$$x_z = x(z_{obs}).$$ (5.66)

When the peculiar velocity is non-relativistic, the relation between x and x_z is described as

$$x_z = x + \frac{v_z}{aH},$$ (5.67)

$$v_z \equiv \boldsymbol{v} \cdot \widehat{\boldsymbol{z}},$$ (5.68)

where v_z means the line-of-sight component of the peculiar velocity. Under the assumption that "plane-parallel" approximation is valid, the line-of-sight direction $\widehat{\boldsymbol{z}}$ is fixed. The relation between the position in the real space x and the redshift space x_z is given by

$$\boldsymbol{x}_z = \boldsymbol{x} + \frac{\boldsymbol{v} \cdot \widehat{\boldsymbol{z}}}{aH}\widehat{\boldsymbol{z}}.$$ (5.69)

Using Eq. (4.1), the displacement field in the redshift space is given by

$$\boldsymbol{s}_z = \boldsymbol{s} + \frac{\dot{\boldsymbol{s}} \cdot \widehat{\boldsymbol{z}}}{H}\widehat{\boldsymbol{z}}.$$ (5.70)

The time derivative of the displacement vector is described as

$$\dot{\boldsymbol{s}}^{(n)} = nHf_n \boldsymbol{s}^{(n)},$$ (5.71)

where f_n is defined in Eq. (4.39). Therefore, the displacement vector for each order in the redshift space is described as

$$\boldsymbol{s}_z^{(n)} = \boldsymbol{s}^{(n)} + nf\left(\dot{\boldsymbol{s}}^{(n)} \cdot \widehat{\boldsymbol{z}}\right)\boldsymbol{z}.$$ (5.72)

This linear transformation is characterized by a redshift space distortion tensor $R_{ij}^{(n)}$ for n-th order perturbation.

$$R_{ij}^{(n)} \equiv \delta_{ij} + nf\widehat{z}_i\widehat{z}_j.$$ (5.73)

Eq. (5.72) is rewritten as

$$\boldsymbol{s}_z^{(n)} = R^{(n)}\boldsymbol{s}^{(n)}.$$ (5.74)

Therefore, the perturbative kernel in the redshift space is given by

$$\mathcal{F}_z^{(n)} = R^{(n)}\mathcal{F}^{(n)}.$$ (5.75)

Using the distortion tensor, the calculation in real space is generalized to that in redshift space.

These approximations are applied to the caluclation of the power spectrum. Especially, Baryon Acoustic Oscillations (BAO) in present time is derived. Because BAO in the power spectrum is sensitive to the nonlinear evolution, the highly accurate nonlinear description is required. For more detail about BAO, we will mention in Sec. 6.2..

5.5. Wave Mechanical Approach

Recently, another approach for structute formation has been noticed. One candidate of CDM is the axion, which is proposed to explain the lack of CP violation in the strong interaction. Because the axion is predicted as quite light particle, if the axion is the most part of CDM, classical description may no longer be correct. Widrow and Kaiser [115] considered quantum mechanical description for structure formation. They proposed a new approach where Schrödiger's equation and Poisson's equation are solved simultaneously.

$$i\hbar\frac{\partial\varphi}{\partial t} = -\frac{\hbar^2}{2ma^2}\nabla^2\varphi + mV(r)\varphi, \tag{5.76}$$

$$\nabla^2 V = \frac{4\pi G}{a}\left(\varphi\varphi^* - \langle\varphi\varphi^*\rangle\right). \tag{5.77}$$

Of course they considered the cosmic expansion and the generation of weak gravitational field by axion field. After that, the development of this approach has been considered [20, 23, 51, 88, 89].

Here we derive the basic equations in the wave mechanical approach. We start with Newtonian equations in comoving coordinates (2.13)-(2.16).

Instead of t, we regard the linear growth factor $g_1(t)$ as the time variable. The time derivative is rewritten as

$$\frac{\partial}{\partial t} = \dot{g}_1\frac{\partial}{\partial g_1}. \tag{5.78}$$

Following the change of the time variable, the peculiar velocity is re-defined.

$$U \equiv \frac{\partial x}{\partial g_1} = \frac{1}{a\dot{g}_1}v. \tag{5.79}$$

Under this change, the continuous equation (2.13) is rewritten as

$$\frac{\partial\delta}{\partial g_1} + \nabla_x\cdot\{U(1+\delta)\} = 0. \tag{5.80}$$

Euler's equation (2.14) is rewritten as

$$\frac{\partial U}{\partial g_1} + \frac{3\Omega_M}{2f_1^2 g_1}U + (U\cdot\nabla_x)U + \nabla_x\Theta = 0, \tag{5.81}$$

where f_1 is defined by Eq. (4.39). If we assume the E-dS Universe, f_1 becomes 1. Here we have introduced a new potential term Θ. Θ is defined by Poisson's equation.

$$\nabla_x^2\Theta = \frac{3\Omega_M}{2f^2 g_1^2}\delta. \tag{5.82}$$

In this subsection, we consider only the irrotational flow for the cosmic fluid. Under this assumption, the peculiar velocity is derived from velocity potential.

$$U = -\nabla_x\tilde{\phi}, \tag{5.83}$$

where $\widetilde{\phi}$ is related to the gravitational potential Θ via

$$\widetilde{\phi} = \frac{2f^2 g_1}{3\Omega_M}\Theta .$$

(5.84)

When we consider a cosmological constant or some kind of dark energy, we have to modify the right-handed term of (5.82).

For the wave mechanical approach, we rewrite Euler equation with the velocity potential $\widetilde{\phi}$ (Eq. (5.83)), and integrate over the spatial coordinates.

$$\frac{\partial\widetilde{\phi}}{\partial g_1} - \frac{1}{2}\left|\nabla_x\widetilde{\phi}\right| - \mathcal{V} = 0 .$$

(5.85)

Here $\mathcal{V}$ means effective potential which is defined as

$$\mathcal{V} = \Theta - \frac{3\Omega_M}{2f^2 g_1}\widetilde{\phi} .$$

(5.86)

For the description of structure formation with a wave function, we adopt Madelung transformation:

$$\varphi = \sqrt{1+\delta}\,\exp\left(-\frac{i\widetilde{\phi}}{\nu}\right) .$$

(5.87)

Here we have introduced new parameter ν. The value of ν affects the behavior of the structure formation.

The relations between φ and other quantities become

$$\nabla_x^2\left[\mathcal{V} + \frac{3\Omega_M}{2f^2 g_1}\nu\,\arg(\varphi)\right] = \nabla_x^2\Theta ,$$

(5.88)

$$\frac{3\Omega_{M0}}{2a^2}\left(|\varphi|^2 - 1\right) = \frac{3\Omega_{M0}}{2a^2}\delta .$$

(5.89)

Using above relations, Poisson's equation is rewritten as

$$\nabla_x^2\left[\mathcal{V} + \frac{3\Omega_M}{2f^2 g_1}\nu\,\arg(\varphi)\right] - \frac{3\Omega_{M0}}{2a^2}\left(|\varphi|^2 - 1\right) = 0 ,$$

(5.90)

We also rewrite Euler's equation to

$$i\nu\frac{\partial\varphi}{\partial g_1} = \left(-\frac{\nu^2}{2}\nabla_x^2 + \mathcal{V} + \mathcal{P}\right)\varphi ,$$

(5.91)

The system described by (5.90) and (5.91) is called Schrödinger-Poisson system. $\mathcal{P}$, which is named as *quantum pressure*, is described as

$$\mathcal{P}\varphi = \frac{\nu^2}{2}\frac{\nabla_x^2|\varphi|}{|\varphi|}$$

$$= -\frac{\nu^2}{8}(1+\delta)^{-2}(\nabla\delta)^2 + \frac{\nu^2}{4}(1+\delta)^{-1}\nabla^2\delta .$$

(5.92)

In Schrödinger-Poisson system, φ can describe both the density and velocity field. $|\varphi|$ and $\arg(\varphi)$ correspond to the density and velocity field, respectively:

$$\delta = |\varphi|^2 - 1, \tag{5.93}$$

$$\tilde{\phi} = i\nu \arg(\varphi). \tag{5.94}$$

If we ignore $\mathcal{P}$ in (5.91), the evolution equation for φ can be linearized:

$$i\nu\frac{\partial\varphi}{\partial g_1} = \left(-\frac{\nu^2}{2}\nabla_x^2 + \mathcal{V}\right)\varphi. \tag{5.95}$$

On the other hand, because we ignore $\mathcal{P}$, an additional term appears in (5.85).

$$\frac{\partial\tilde{\phi}}{\partial g_1} - \frac{1}{2}|\nabla_x\tilde{\phi}|^2 - \mathcal{V} + \mathcal{P} = 0. \tag{5.96}$$

Because $\mathcal{P}$ is proportional to ν, when we take a limit $\nu \to 0$, (5.91) recovers (5.85).

For simplicity, we consider this model in the E-dS Universe. In E-dS Universe, the effective potential (5.86) vanishes. Further more the linear growth factor becomes $g_1 = a$. Therefore (5.95) becomes

$$i\nu\frac{\partial\varphi}{\partial a} = -\frac{\nu^2}{2}\nabla_x^2\varphi. \tag{5.97}$$

Applying Fourier transformation, we can solve this equation with analytic method.

$$\hat{\varphi}(a, \boldsymbol{k}) = \hat{\varphi}(a_0, \boldsymbol{k})e^{-i\nu k^2 a}, \tag{5.98}$$

$$\hat{\varphi}(a, \boldsymbol{k}) \equiv \frac{1}{(2\pi)^{3/2}}\int \varphi(a, \boldsymbol{x})e^{-i\boldsymbol{k}\cdot\boldsymbol{x}}\, \mathrm{d}^3\boldsymbol{x}. \tag{5.99}$$

Then we can obtain the solution in comoving coordinates:

$$\varphi(a, \boldsymbol{x}) = \frac{1}{(2\pi)^{3/2}}\int \hat{\varphi}(a, \boldsymbol{k})e^{i\boldsymbol{k}\cdot\boldsymbol{x}}\, \mathrm{d}^3\boldsymbol{k}. \tag{5.100}$$

Some applications of the wave mechanical approach for cosmology have been proposed. According to comparison between cosmological N-body simulations and several analytic methods, the description of nonlinear structure with the wave mechanical approach is shown to be better than that with ZA [88, 89].

Even if we apply nonlinear transformation such as Madelung transformation (Eq. (5.87)), we cannot describe the evolution of large-scale structure beyond shell-crossing. When shell-crossing occurs, the function φ diverges. After shell-crossing, even if φ has finite value, the function would not continue from past function. Because the function originally bases on fluid description, streamlines do not cross each other. At caustics, the streamlines cross. Therefore after the caustic formation, i.e., shell-crossing, the fluid description fails and φ has no longer physical meaning.

For the plane-symmetric model, by matching solutions before and after shell-crossing, the solutions have physical meaning even after shell-crossing. If such a special method for the matching of the solutions exists, the wave mechanical approach still be valid after shell-crossing.

5.6. Non-dust Model, Multi-component Model

If the velocity dispersion cannot be ignored, the perturbative solutions depend on the spacial scale. Euler-Jeans-Newton model (EJN) adopts the effect of the velocity dispersion. If the velocity dispersion is isotropic, the velocity dispersion corresponds to the effective pressure. The evolution equation with the Lagrangian description is given by (4.13). The pressure affects only the longitudinal mode in the linear approximation.

The evolution equation of the linear perturbation is described as

$$\ddot{S}^{(1)} + 2\frac{\dot{a}}{a}\dot{S}^{(1)} - 4\pi G\rho_b S^{(1)} - \frac{1}{a^2}\frac{dP}{d\rho}\bigg|_{\rho=\rho_b}\nabla^2 S^{(1)} = 0. \tag{5.101}$$

If the equation of state is given by polytropic form, i.e.,

$$P = \kappa\rho^\gamma, \tag{5.102}$$

the evolution equation becomes

$$\ddot{S}^{(1)} + 2\frac{\dot{a}}{a}\dot{S}^{(1)} - 4\pi G\rho_b S^{(1)} - \frac{\kappa}{a^2}\rho_b^{\gamma-1}\nabla^2 S^{(1)} = 0. \tag{5.103}$$

We derive perturbative solutions in the Lagrangian Fourier space. We adopt Fourier transformation in the Lagrangian space for Eq. (5.103),

$$\ddot{\widehat{S}}^{(1)} + 2\frac{\dot{a}}{a}\dot{\widehat{S}}^{(1)} - 4\pi G\rho_b \widehat{S}^{(1)} + \frac{\kappa\gamma\rho_b^{\gamma-1}}{a^2}|\boldsymbol{K}|^2\,\widehat{S}^{(1)} = 0, \tag{5.104}$$

where $\boldsymbol{K}$ is a Lagrangian wavenumber.

In the E-dS Universe, linear perturbation solutions are described with the analytic form [70]. For $\gamma \neq 4/3$,

$$\widehat{S}^{(1)}(\boldsymbol{K}, a) \propto a^{-1/4}\,\mathcal{J}_{\pm 5/(8-6\gamma)}\left(\sqrt{\frac{2C_2}{C_1}}\frac{|\boldsymbol{K}|}{|4-3\gamma|}a^{(4-3\gamma)/2}\right), \tag{5.105}$$

where $\mathcal{J}_\nu$ denotes the Bessel function of order ν, and for $\gamma = 4/3$,

$$\widehat{S}^{(1)}(\boldsymbol{K}, a) \propto a^{-1/4\pm\sqrt{25/16-C_2|\boldsymbol{K}|^2/2C_1}}, \tag{5.106}$$

where $C_1 \equiv 4\pi G\rho_b(a_{in})\,a_{in}^3/3$ and $C_2 \equiv \kappa\gamma\rho_b(a_{in})^{\gamma-1}a_{in}^{3(\gamma-1)}$. a_{in} is scale factor when an initial condition is given.

For other FRW Universes, if γ takes a special value, we obtain an analytic solution with a hypergeometric function [107]: For the Λ-flat Universe with $\gamma = 1/3, 4/3$.

$$\widehat{S}^{(1)}(\boldsymbol{K}, a) \propto a^\beta \mathcal{F}\left(\alpha_1, \alpha_2, \alpha_3; -\frac{\Lambda a^3}{6C_1}\right), \tag{5.107}$$

where

$$(\alpha_1, \alpha_2, \alpha_3, \beta) = \left(-\frac{1}{6} + \sqrt{\frac{1}{9} + \frac{C_2 |\mathbf{K}|^2}{3\Lambda}}, -\frac{1}{6} - \sqrt{\frac{1}{9} + \frac{C_2 |\mathbf{K}|^2}{3\Lambda}}, \frac{1}{6}, -\frac{3}{2} \right),$$

$$\left(\frac{2}{3} + \sqrt{\frac{1}{9} + \frac{C_2 |\mathbf{K}|^2}{3\Lambda}}, \frac{2}{3} - \sqrt{\frac{1}{9} + \frac{C_2 |\mathbf{K}|^2}{3\Lambda}}, \frac{11}{6}, 1 \right) \quad \text{for } \gamma = 1/3 , (5.108)$$

$$(\alpha_1, \alpha_2, \alpha_3, \beta) = \left(\frac{7}{12} \pm \sqrt{\frac{25}{144} - \frac{C_2 |\mathbf{K}|^2}{18 C_1}}, -\frac{1}{12} \pm \sqrt{\frac{25}{144} - \frac{C_2 |\mathbf{K}|^2}{18 C_1}}, \right.$$

$$\left. 1 \pm \sqrt{\frac{25}{36} - \frac{2 C_2 |\mathbf{K}|^2}{9 C_1}}, -\frac{1}{4} \pm \sqrt{\frac{25}{16} - \frac{C_2 |\mathbf{K}|^2}{2 C_1}} \right) \quad \text{for } \gamma = 4/3 . (5.109)$$

In these models, the behavior of the solutions strongly depends on the relation between the scale of fluctuation and the Jeans scale. Here we define the Jeans wavenumber as

$$K_{\mathrm{J}} \equiv \left(\frac{4\pi G \rho_{\mathrm{b}} a^2}{\mathrm{d}P/\mathrm{d}\rho(\rho_{\mathrm{b}})} \right)^{1/2} . \tag{5.110}$$

The Jeans wavenumber, which gives a criterion for whether a density perturbation with a wavenumber will grow or decay with oscillation, depends on time in general. If the polytropic index γ is smaller than $4/3$, all modes become decaying modes and the fluctuation will disappear. On the other hand, if $\gamma > 4/3$, all density perturbations will grow to collapse. In the case where $\gamma = 4/3$, the growing and decaying modes coexist at all times.

The second- and third-order perturbative solutions for special cases are already derived [70, 98, 99, 107]. Because of mode-coupling in the Lagrangian Fourier space, the perturbative solution is complicated. The behavior of higher-order perturbative solutions in late-time almost coincides to that in dust model for the case of $\gamma > 4/3$ [107]. On the other hand, the behavior of higher-order perturbative solutions in late-time is not known for the case of $\gamma < 4/3$.

In dust model, when we consider only the longitudinal mode for the linear perturbation, the transverse mode arises in the third-order perturbation. On the other hand, in EJN model, when we consider only the longitudinal mode for the linear perturbation, the transverse mode arises even in the second-order perturbation [70, 107].

The relation between EJN model and Adhesion approximation has been investigated [94, 97]. In dense regions, the peculiar velocity in EJN model oscillates. On the other hand, that in adhesion approximation is monotonically decelerated due to the viscosity without oscillation. Therefore it is hard to make the correspondence between the viscosity in adhesion approximation and the pressure gradient in EJN model.

EJN models are considered for a single component fluid. According to the modern cosmology, there are a couple of components in the Universe, for example, baryonic matter and dark matter. Matarrese and Mohayaee [61] proposed the Lagrangian perturbation model for two component fluids. One is dark matter which affects only gravity. Another is baryonic matter which affects both gravity and pressure.

When we consider more than one species of matter, the evolution equation for the density fluctuation is modified [76]. We consider coupled equations for multi-component. We assume a polytropic equation of state for the baryonic matter.

$$P \propto \rho^{\gamma} . \tag{5.111}$$

Taking the divergence of Euler equation for the baryonic matter, we obtain the evolution equation for the baryonic matter density fluctuation with the Lagrangian perturbation.

$$\nabla_x \cdot s_B'' = -\frac{3}{2a} \left[\nabla_x \cdot s_B' + \frac{\delta_{DM}}{a} + \frac{1}{(\gamma - 1)ak_J^2} \nabla_x^2 (1 + \delta_B)^{\gamma - 1} \right] , \tag{5.112}$$

where B and DM mean the baryonic matter and dark matter, respectively. k_J is the Jeans wavenumber of the baryonic matter. ($'$) denotes the derivative with respect to the scale factor ($' = \partial/\partial a$). The density fluctuation of the dark matter affects to that of the baryon. Here we assume that the amount of the baryonic matter is much less than that of the dark matter. Therefore, we ignore the gravitational force of the baryonic matter. Substituting Eq. (4.7) into (5.112), we obtain

$$\nabla_x \cdot s_B'' + \frac{3}{2a} \nabla_x \cdot s_B' = -\frac{3}{2a^2} \frac{1 - J_{DM}}{J_{DM}} - \frac{3}{2(\gamma - 1)a^2 k_J^2} \nabla^2 \left(\frac{1 - J_B}{J_B} \right)^{\gamma - 1} . \tag{5.113}$$

Next, we consider only the linear order terms for the Lagrangian displacement s.

$$\nabla_q \cdot s_B^{(1)''} + \frac{3}{2a} \nabla_q \cdot s_B^{(1)'} - \frac{3}{2a^2} \nabla_q \cdot s_{DM}^{(1)} = \frac{3}{2a^2 k_J^2} \nabla_q^2 \nabla_q \cdot s_B^{(1)} . \tag{5.114}$$

We note that the same Eulerian position $x(t)$ is generally obtained by the two components from different Lagrangian positions q_B and q_{DM}. Here we ignore this difference and set simply $q_B = q_{DM}$, because we consider the approximation valid at leading order.

We assume that the perturbations have only the longitudinal mode solutions ($s^{(1)} = \nabla_q S^{(1)}$). Under this assumption, we obtain the evolution equation of the Lagrangian perturbation for the baryonic matter.

$$S_B^{(1)''} + \frac{3}{2a} S_B^{(1)'} - \frac{3}{2a^2} \frac{1}{k_J^2} \nabla_q^2 S_B^{(1)} = \frac{3}{2a} S_{DM}^{(1)} . \tag{5.115}$$

This equation can be solved in the Lagrangian Fourier space.

6.   Applications

6.1.   Non-gaussianity

At the recombination era, the density fluctuations are almost regarded as Gaussian. During the nonlinear evolution, non-Gaussianity of the density fluctuations would appear. Although the non-Gaussianity of the density fluctuations is analyzed with N-body simulation in the strongly nonlinear region, we can estimate that with analytic method in the weakly nonlinear region.

Even if the relation between the density fluctuation and the perturbative quantity in ZA is nonlinear, the description of the evolution for the non-Gaussianity is insufficient [55,

104]. The non-Gaussianity is mainly generated by the nonlinear evolution. The pressure gradient for the structure formation merely affects the evolution of the non-Gaussianity [100]. Using second-order perturbative equations, we can discuss the non-Gaussiaity.

The probability distribution function (PDF) of the density fluctuations deviates from the initial Gaussian shape during the nonlinear evolution. In the standard CDM (SCDM) model, the PDF is fairly well approximated by a lognormal distribution in a weakly non-linear regime [55], while the lognormal PDF does not fit well in a highly non-linear regime [112]. After this, Plionis et al. [80] and Borgani et al. [8] analysed the non-Gaussianity of the cluster distribution for several dark matter models including CDM in the Λ-flat Universe (ΛCDM model) by which the difference of non-Gaussianity between SCDM and ΛCDM models was clearly shown. The comparison among them by N-body simulations for the quasi-nonlinear stage was also done in [54].) We investigated the dependence of PDF to the equation of state for the dark energy ($p = w\rho$) [104, 105]. For the comparison, both skewness and kurtosis are applied, which will be mentioned in this subsection. We showed that the relative difference between the non-Gaussianity of the $w = -0.8$ model and that of the $w = -1.0$ model is several percents for smoothed scale $R = 2h^{-1}$[Mpc] [105].

In this subsection, we derive the statistical quantities for the non-Gaussianity with the second-order perturbation. Here we start with Vlasov's equation for one-point distribution function of the spacial coordinates x and momentum p [79]. We integrate Vlasov's equation over the momentum space, we obtain

$$a^3 \rho_b \delta + \frac{1}{a^2} \partial_i \int p_i f \, \mathrm{d}^3 p = 0 \,. \tag{6.1}$$

Using the averaged momentum over the spacial coordinates $\bar{p}$, we derive

$$\rho_b \dot{\delta} + \frac{1}{a} \partial_i (\rho \bar{p}_i) = 0 \,. \tag{6.2}$$

In this section, we regard p_i as $\bar{p}_i$.

Next, we multiply p_j to Vlasov's equation and integrate over the momentum space.

$$\frac{\partial}{\partial t} \int p_i f \, \mathrm{d}^3 p + \frac{1}{ma^2} \partial_j \int p_i p_j f \, \mathrm{d}^3 p + a^3 \rho \phi_{,i} = 0 \,. \tag{6.3}$$

From Eq. (6.2), we obtain

$$a^2 \rho_{b0} \dot{\delta} + \partial_i (\rho \bar{p}_i) = 0 \,. \tag{6.4}$$

Then we take the time derivative, i.e.,

$$2 a \dot{a} \rho_{b0} \dot{\delta} + a^2 \rho_{b0} \ddot{\delta} + \frac{\partial}{\partial t} (\partial_i (\rho \bar{p}_i)) = 0 \,. \tag{6.5}$$

Using the above equation, we take the derivative with respect to Eq. (6.3) by x_i, we obtain

$$-a^5 \rho_b \ddot{\delta} - 2a^4 \dot{a} \rho_b \dot{\delta} + \frac{1}{ma^2} \partial_i \partial_j \int p_i p_j f \, \mathrm{d}^3 p + a^3 \partial_i (\rho \phi_{,i}) = 0 \,, \tag{6.6}$$

where we have defined

$$\overline{p_i p_j} \equiv \frac{\int f p_i p_j \, \mathrm{d}^3 p}{m^2 a^2 \int f \, \mathrm{d}^3 p} \,. \tag{6.7}$$

Finally we obtain the nonlinear evolution equation.

$$\ddot{\delta} + 2\frac{\dot{a}}{a}\dot{\delta} = \frac{1}{a^2}\partial_i\left[(1+\delta)\phi_{,i}\right] + \frac{1}{a^2}\partial_i\partial_j\left[(1+\delta)\overline{p_i p_j}\right],\tag{6.8}$$

Using this equation, we analyze the effect of second-order perturbation. Eq. (6.8) is simplified to

$$\ddot{\delta} + 2\frac{\dot{a}}{a}\dot{\delta} = 4\pi G\rho_b\delta(1+\delta) + \frac{1}{a^2}\delta_{,i}\phi_{,i} + \frac{1}{a^2}\partial_i\partial_j\left[(1+\delta)\overline{p_i p_j}\right].\tag{6.9}$$

We expand the density fluctuation until the second-order.

$$\delta = g_1(t)\delta_0(\boldsymbol{x})\left[1 + \epsilon(\boldsymbol{x},t)\right],\tag{6.10}$$

where δ_0 means primordial density fluctuation. Here we assume $\delta_0 \ll 1, \epsilon \ll 1$.

$$\dot{\delta} = \dot{g}_1\delta_0\epsilon + g_1\delta_0\dot{\epsilon},\tag{6.11}$$
$$\ddot{\delta} = \ddot{g}_1\delta_0\epsilon + 2\dot{g}_1\delta_0\dot{\epsilon} + g_1\delta_0\ddot{\epsilon}.\tag{6.12}$$

Eq. (6.9) is rewritten as

$$\left(\frac{\ddot{g}_1}{g_1}\epsilon + 2\frac{\dot{g}_1}{g_1}\dot{\epsilon} + \ddot{\epsilon}\right)\delta_0 + 2\frac{\dot{a}}{a}\left(\frac{\dot{g}_1}{g_1}\epsilon + \dot{\epsilon}\right)\delta_0$$
$$= 4\pi G\rho_b\delta_0\epsilon + 4\pi G\rho_b\delta_0^2 + \frac{1}{a^2}(-G\rho_b a^2\Delta_{,i})\delta_{0,i} + \frac{1}{16\pi^2 g_1^2}\dot{g}_1^2\left[\Delta_{,i}\Delta_{,j}\right]_{,ij},\tag{6.13}$$

where Δ means velocity potential, given by

$$\phi = -G\rho_b a^2\Delta(\boldsymbol{x}),\tag{6.14}$$
$$\Delta(\boldsymbol{x}) = \int\frac{\delta(\boldsymbol{x}')}{|\boldsymbol{x}-\boldsymbol{x}'|}\,\mathrm{d}^3\boldsymbol{x}',\tag{6.15}$$
$$v_i = \frac{a}{4\pi}\frac{\dot{g}_1}{g_1}\Delta_{,i}.\tag{6.16}$$

Using the results at linear order, we obtain evolution equation for ϵ.

$$\ddot{\epsilon} + 2\left(\frac{\dot{g}_1}{g_1} + \frac{\dot{a}}{a}\right)\dot{\epsilon} = 4\pi G\rho_b\delta_0 - G\rho_b\frac{\delta_{0,i}}{\delta_0}\Delta_{,i} + \frac{\dot{g}_1^2}{16\pi^2 g_1^2\delta_0}\left[\Delta_{,i}\Delta_{,j}\right]_{,ij}.\tag{6.17}$$

For the estimation of the non-Gaussianity in the quasi-nonlinear region, we will solve the second-order perturvative equation (6.17).

In generic case, we cannot obtain analytic solutions for Eq. (6.17). Here we show one example where we can derive analytic solutions. In the E-dS Universe, the linear growth rate is given by

$$g_1 \propto t^{2/3}, \quad 6\pi G\rho_b t^2 = 1.\tag{6.18}$$

Therefore,

$$\left[\Delta_{,i}\Delta_{,j}\right]_{,ij} = \Delta_{,ii}\Delta_{,jj} + 2\Delta_{,iij}\Delta_{,j} + \Delta_{,ij}\Delta_{,ij}$$
$$= 16\pi^2\delta^2 - 8\pi\delta_{,i}\Delta_{,i} + \Delta_{,ij}\Delta_{,ij}.\tag{6.19}$$

Then, Eq. (6.17) is written as

$$\ddot{\epsilon} + \frac{8}{3t}\dot{\epsilon} = \frac{10}{9t^2}\delta_0(t_0) - \frac{7}{18\pi t^2}\frac{\delta_{0,i}(t_0)}{\delta_0}\Delta_{,i}(t_0) + \frac{1}{36\pi^2\delta_0(t_0)t^2}\Delta_{,ij}(t_0)\Delta_{,ij}(t_0). \quad (6.20)$$

The second-order perturbative solutions are given by

$$\epsilon = \frac{5}{7}\delta_0 - \frac{1}{4\pi}\delta_{0,i}\Delta_{,i} + \frac{1}{56\pi^2}\Delta_{,ij}\Delta_{,ij}. \quad (6.21)$$

We consider the generation of non-Gaussianity from the fluctuations which is initially Gaussian. If we assume that primordial density fluctuation obeys Gaussian statistics, the n-point correlation functions of the density fluctuation become

$$\langle\delta(\boldsymbol{x})\rangle \;=\; 0\,, \quad (6.22)$$
$$\langle\delta(\boldsymbol{x}_1)\delta(\boldsymbol{x}_2)\rangle \;=\; \xi\left(|\boldsymbol{x}_1 - \boldsymbol{x}_2|\right), \quad (6.23)$$
$$\langle\delta(\boldsymbol{x}_1)\delta(\boldsymbol{x}_2)\delta(\boldsymbol{x}_3)\rangle \;=\; 0\,, \quad (6.24)$$

$$\langle\delta(\boldsymbol{x}_1)\delta(\boldsymbol{x}_2)\delta(\boldsymbol{x}_3)\delta(\boldsymbol{x}_4)\rangle$$
$$= \;\; \xi\left(|\boldsymbol{x}_1 - \boldsymbol{x}_2|\right)\xi\left(|\boldsymbol{x}_3 - \boldsymbol{x}_4|\right) + \xi\left(|\boldsymbol{x}_1 - \boldsymbol{x}_3|\right)\xi\left(|\boldsymbol{x}_2 - \boldsymbol{x}_4|\right)$$
$$+\xi\left(|\boldsymbol{x}_1 - \boldsymbol{x}_3|\right)\xi\left(|\boldsymbol{x}_2 - \boldsymbol{x}_4|\right), \quad (6.25)$$

where $\xi(r)$ means two-point spacial correlation function. If we assume Gaussian distribution, all multi-point correlations or higher-order correlations are written by the two-point correlation.

Because of the definition of the density fluctuation, even if we consider nonlinear perturbation,

$$\langle\delta(\mathbf{x})\rangle = 0\,. \quad (6.26)$$

However, higher-order correlations are changed.

$$\langle\delta_{0,i}\Delta_{,i}\rangle \;=\; -\langle\delta_0\Delta_{,ii}\rangle = 4\pi\left\langle\delta_0^2\right\rangle = 4\pi\xi(0)\,, \quad (6.27)$$
$$\langle\Delta_{,ij}\Delta_{,ij}\rangle \;=\; \langle\Delta_{,ii}\Delta_{,jj}\rangle = 16\pi^2\xi(0)\,. \quad (6.28)$$

Because the average of linear-order density fluctuation satisfies

$$\langle\delta_0\rangle = 0\,, \quad (6.29)$$

the following equation has to be satisfied,

$$\langle\epsilon\delta_0\rangle = 0\,. \quad (6.30)$$

For the effect of non-Gaussianity, we notice that the skewness is expressed as,

$$\langle\delta^3\rangle = \left\langle\delta_0^3\left(1 + 3\epsilon + 3\epsilon^2 + \epsilon^3\right)\right\rangle \simeq 3\left\langle\delta_0^3\epsilon\right\rangle. \quad (6.31)$$

From Eq. (6.25), we obtain

$$\left\langle\delta_0^4\right\rangle = 3\left\langle\delta_0^2\right\rangle^2 = 3\xi(0)^2\,. \quad (6.32)$$

The second term of the right-handed side of Eq. (6.21) is rewritten as

$$
\begin{aligned}
\langle \delta_0^2 \delta_{0,i} \Delta_{,i} \rangle &= \frac{1}{3} \left\langle (\delta_0)_{,i}^3 \, \Delta_i \right\rangle \\
&= -\frac{1}{3} \langle \delta_0^3 \Delta_{,ii} \rangle = 4\pi \xi(0)^2 \, .
\end{aligned}
\tag{6.33}
$$

The last term of the right-handed side of Eq. (6.21) is divided to the products of second-order moment.

$$
\langle \delta_0^2 \Delta_{,ij} \Delta_{,ij} \rangle = \xi(0) \langle \Delta_{,ij} \Delta_{,ij} \rangle + 2 \langle \Delta_{,ij} \delta_0 \rangle \langle \Delta_{,ij} \delta_0 \rangle \, .
\tag{6.34}
$$

Using these equations, we obtain

$$
\langle \Delta_{,ij} \Delta_{,ij} \rangle = \langle \Delta_{,ii} \Delta_{,jj} \rangle = 16\pi^2 \xi(0) \, ,
\tag{6.35}
$$

$$
\langle \delta_0 \Delta_{,ij} \rangle = \frac{1}{3} \delta^{ij} \langle \delta_0 \Delta_{,kk} \rangle = -\frac{4\pi}{3} \delta^{ij} \xi(0) \, ,
\tag{6.36}
$$

$$
\langle \delta_0^2 \Delta_{,ij} \Delta_{,ij} \rangle = \frac{80\pi^2}{3} \xi(0)^2 \, .
\tag{6.37}
$$

From these results, the skewness is given by

$$
\begin{aligned}
\langle \delta^3 \rangle &\simeq 3 \langle \delta_0^3 \epsilon \rangle \\
&= 3 \left\langle \frac{5}{7} \delta_0^4 - \frac{1}{4\pi} \delta_0^2 \delta_{0,i} \Delta_{,i} + \frac{1}{56\pi^2} \delta_0^2 \Delta_{,ij} \Delta_{,ij} \right\rangle \\
&= 3 \left(\frac{15}{7} \xi(0)^2 - \xi(0)^2 + \frac{10}{21} \xi(0)^2 \right) \\
&= \frac{34}{7} \xi(0)^2 \, .
\end{aligned}
\tag{6.38}
$$

In generic FRW Universes, we consider from Eq. (6.9),

$$
\ddot{\delta}^{(2)} + 2\frac{\dot{a}}{a} \dot{\delta}^{(2)} - 4\pi G\rho_b \delta^{(2)} = 4\pi G\rho_b \left(\delta^{(1)} \right)^2 + \frac{1}{a^2} \delta^{(1)}_{,i} \phi^{(1)}_{,i} + \frac{1}{a^2} \left[p^{(1)}_i p^{(1)}_j \right]_{,ij} \, ,
\tag{6.39}
$$

where the superscript means the order of perturbation. The last term of the right-handed side can be rewritten by

$$
\begin{aligned}
\left[p^{(1)}_i p^{(1)}_j \right]_{,ij} &= p^{(1)}_{i,i} p^{(1)}_{j,j} + 2 p^{(1)}_{i,ij} p^{(1)}_j + p^{(1)}_{i,j} p^{(1)}_{j,i} \\
&= a^2 \left(\frac{\dot{g}_1}{g_1} \right)^2 \left(\delta^{(1)} \right)^2 - \frac{a^2}{2\pi} \left(\frac{\dot{g}_1}{g_1} \right)^2 \delta^{(1)}_{,i} \Delta_{,i} + \frac{a^2}{16\pi^2} \left(\frac{\dot{g}_1}{g_1} \right)^2 \Delta_{,ij} \Delta_{,ij}
\end{aligned}
\tag{6.40}
$$

Using this equation, we obtain

$$
\begin{aligned}
\ddot{\delta}^{(2)} + 2\frac{\dot{a}}{a} \dot{\delta}^{(2)} - 4\pi G\rho_b \delta^{(2)} &= \left[4\pi G\rho_b + \left(\frac{\dot{g}_1}{g_1} \right)^2 \right] \left(\delta^{(1)} \right)^2 - \left[G\rho_b + \frac{1}{2\pi} \left(\frac{\dot{g}_1}{g_1} \right)^2 \right] \delta_{,i} \Delta_{,i} \\
&\quad + \frac{a^2}{16\pi^2} \left(\frac{\dot{g}_1}{g_1} \right)^2 \Delta_{,ij} \Delta_{,ij} \, .
\end{aligned}
\tag{6.41}
$$

Making use of Friedmann equation to eliminate ρ_b from the above equation, we obtain

$$\ddot{\delta}^{(2)} + 2\frac{\dot{a}}{a}\dot{\delta}^{(2)} - \frac{3}{2}\Omega_{M0}H_0^2\left(\frac{a_0}{a}\right)^3\delta^{(2)}$$

$$= \left[\frac{3}{2}\Omega_{M0}H_0^2\left(\frac{a_0}{a}\right)^3 + \left(\frac{\dot{g}_1}{g_1}\right)^2\right]\left(\delta^{(1)}\right)^2 - \left[\frac{3}{8\pi}\Omega_{M0}H_0^2\left(\frac{a_0}{a}\right)^3 + \frac{1}{2\pi}\left(\frac{\dot{g}_1}{g_1}\right)^2\right]\delta_{,i}\Delta_{,i}$$

$$+ \frac{a^2}{16\pi^2}\left(\frac{\dot{g}_1}{g_1}\right)^2\Delta_{,ij}\Delta_{,ij}. \tag{6.42}$$

Here we assume that the second-order perturbation is divided to two components,

$$\delta^{(2)} = \delta_a^{(2)} + \delta_b^{(2)}, \tag{6.43}$$

where $\delta_a^{(2)}$ and $\delta_b^{(2)}$ satisfies

$$\ddot{\delta}_a^{(2)} + 2\frac{\dot{a}}{a}\dot{\delta}_a^{(2)} - \frac{3}{2}\Omega_{M0}H_0^2\left(\frac{a_0}{a}\right)^3\delta_a^{(2)} = \frac{3}{2}\Omega_{M0}H_0^2\left(\frac{a_0}{a}\right)^3\left[\left(\delta^{(1)}\right)^2 - \frac{1}{4\pi}\delta_{,i}\Delta_{,i}\right], \tag{6.44}$$

$$\ddot{\delta}_b^{(2)} + 2\frac{\dot{a}}{a}\dot{\delta}_b^{(2)} - \frac{3}{2}\Omega_{M0}H_0^2\delta_b^{(2)} = \left(\frac{\dot{g}_1}{g_1}\right)^2\left[\left(\delta^{(1)}\right)^2 - \frac{1}{2\pi}\delta_{,i}\Delta_{,i} + \frac{1}{16\pi^2}\Delta_{,ij}\Delta_{,ij}\right] \tag{6.45}$$

respectively. We divide the second-order perturbations into the time and spacial components.

$$\delta_a^{(2)} = \widetilde{g}_{2a}\delta_a^{(2)}(t_0), \tag{6.46}$$

$$\delta_b^{(2)} = \widetilde{g}_{2b}\delta_b^{(2)}(t_0). \tag{6.47}$$

In general, the time components of these second-order perturbations do not coincide with those of Lagrangian second-order perturbations given by Eq. (4.22).

Then we notice that the time components. Eqs. (6.44) and (6.45) are not independent each other. If we set

$$\widetilde{g}_{2b} = \frac{g_1^2 - \widetilde{g}_{2a}}{2}, \tag{6.48}$$

Then, substituting Eqs. (6.45) and (6.48) into Eq. (6.44), we obtain

$$\ddot{\widetilde{g}}_{2b} + 2\frac{\dot{a}}{a}\dot{\widetilde{g}}_{2b} - \frac{3}{2}\Omega_{M0}H_0^2\widetilde{g}_{2b} - \dot{g}_1^2$$

$$= g_1\ddot{g}_1 + \dot{g}_1^2 - \frac{\ddot{\widetilde{g}}_{2a}}{2} + 2\frac{\dot{a}}{a}\left(g_1\dot{g}_1 - \frac{\dot{\widetilde{g}}_{2a}}{2}\right) - \frac{3}{4}\Omega_{M0}H_0\left(g_1^2 - \widetilde{g}_{2a}\right) - \dot{g}_1^2$$

$$= -\frac{1}{2}\left(\ddot{\widetilde{g}}_{2a} + 2\frac{\dot{a}}{a}\dot{\widetilde{g}}_{2a} - \frac{3}{2}\Omega_{M0}H_0^2\widetilde{g}_{2a}\right) + g_1\left(\ddot{g}_1 + 2\frac{\dot{a}}{a}\dot{g}_1 - \frac{3}{2}\Omega_{M0}H_0^2g_1\right)$$

$$= -\frac{1}{2}\left(\ddot{\widetilde{g}}_{2a} + 2\frac{\dot{a}}{a}\dot{\widetilde{g}}_{2a} - \frac{3}{2}\Omega_{M0}H_0^2\widetilde{g}_{2a}\right) + \frac{3}{4}\Omega_{M0}H_0g_1 = 0. \tag{6.49}$$

Therefore, when we analyze the evolution for the second-order perturbation, we have to solve Eqs. (3.4) and (6.44) simultaneously. In the E-dS Universe, the solutions are obtained as

$$\widetilde{g}_{2a} = \frac{3}{7}t^{4/3}, \ \widetilde{g}_{2b} = \frac{2}{7}t^{4/3}. \tag{6.50}$$

Now we derive the skewness,

$$\langle\delta^3\rangle = \left\langle\left(\delta^{(1)}\right)^3\right\rangle + 3\left\langle\left(\delta^{(1)}\right)^2\delta^{(2)}\right\rangle + O(\epsilon^5). \tag{6.51}$$

If we assume that the primordial fluctuation obeys Gaussian distribution,

$$\langle \delta^3 \rangle \simeq 3 \left\langle \left(\delta^{(1)} \right)^2 \delta^{(2)} \right\rangle . \tag{6.52}$$

We multiply $\left(\delta^{(1)} \right)^2$ to

$$
\begin{aligned}
\delta^{(2)} &= \widetilde{g}_{2a} \left[\left(\delta^{(1)} \right)^2 - \frac{1}{4\pi} \delta_{,i} \Delta_{,i} \right] \\
&\quad + \widetilde{g}_{2b} \left[\left(\delta^{(1)} \right)^2 - \frac{1}{2\pi} \delta_{,i} \Delta_{,i} + \frac{1}{16\pi^2} \Delta_{,ij} \Delta_{,ij} \right] \\
&= \frac{g_1^2 + \widetilde{g}_{2a}}{2} \left(\delta^{(1)} \right)^2 + \frac{g_1^2}{8\pi} \delta_{,i} \Delta_{,i} + \frac{g_1^2 - \widetilde{g}_{2a}}{32\pi^2} \Delta_{,ij} \Delta_{,ij} ,
\end{aligned}
\tag{6.53}
$$

and average over whole space. The relations between the average and the dispersion of the density in generic FRW Universes are not changed from that in the E-dS Universe. Therefore we can apply Eqs. (6.32), (6.33), (6.37).

$$
\begin{aligned}
\left\langle \left(\delta^{(1)} \right)^2 \delta^{(2)} \right\rangle &= \frac{g_1^2 + \widetilde{g}_{2a}}{2} \cdot 3\xi(0)^2 - g_1^2 \xi(0)^2 + \frac{g_1^2 - \widetilde{g}_{2a}}{2} \cdot \frac{5}{3} \xi(0)^2 \\
&= \frac{4g_1^2 + 2\widetilde{g}_{2a}}{3} \xi(0)^2 .
\end{aligned}
\tag{6.54}
$$

Therefore the skewness in generic FRW Universes are obtained as

$$S_3 = 3 \left\langle \left(\delta^{(1)} \right)^2 \delta^{(2)} \right\rangle = 4 + 2\mu = \frac{34}{7} + \frac{6}{7} \left(\frac{7}{3} \mu - 1 \right) , \tag{6.55}$$

$$\mu \equiv \frac{\widetilde{g}_{2a}}{g_1^2} . \tag{6.56}$$

In the E-dS Universe, because the ratio between the first- and second-order growth rate is $\mu = 3/7$, the skewness becomes $S_3 = 34/7$. In generic FRW Universes, we have to solve Eqs. (3.4) and (6.44) with numerical method simultaneously. Then we substitute the result into Eq. (6.55).

Fry [38] computed the fourth order cumulant of the density fluctuation. The fourth order cumulant of the density fluctuation is given by

$$
\begin{aligned}
\langle \delta^4 \rangle_c &\equiv \langle \delta^4 \rangle - 3 \langle \delta^2 \rangle^2 \\
&= 12 \left\langle \left(\delta^{(1)} \right)^2 \left(\delta^{(2)} \right)^2 \right\rangle_c + 4 \left\langle \left(\delta^{(1)} \right)^3 \delta^{(3)} \right\rangle_c .
\end{aligned}
\tag{6.57}
$$

In the E-dS Universe, the kurtosis is given by

$$\frac{\langle \delta^4 \rangle_c}{\sigma^6} = \frac{60712}{1323} . \tag{6.58}$$

For the analysis of non-Gaussianity in the density field, in order to avoid the divergence of the density fluctuation in the limit of small scale, it is necessary to consider the density field smoothed over some scale R.

$$
\begin{aligned}
\rho(\boldsymbol{x}; R) &= \int \mathrm{d}^3 \boldsymbol{y} W \left(|\boldsymbol{x} - \boldsymbol{y}| ; R \right) \\
&= \int \frac{\mathrm{d}^3 \boldsymbol{k}}{(2\pi)^3} \widetilde{W}(kR) \widetilde{\rho}(\boldsymbol{k}) e^{-i\boldsymbol{k}\cdot\boldsymbol{x}} ,
\end{aligned}
\tag{6.59}
$$

where W denotes a window function and $\widetilde{W}$ and $\tilde{\rho}$ represent the Fourier transforms of the corresponding quantities. Here we consider that the smoothing function is given by the top hat function, i.e.,

$$\widetilde{W} = \frac{3(\sin x - x \cos x)}{x^3}. \tag{6.60}$$

The filtered non-Gaussianity is derived by Bernardeau [5]. In the E-dS Universe, the filtered non-Gaissianity is given by

$$\gamma = \frac{34}{7} + \frac{d \log \sigma^2(R)}{d \log R} + \mathcal{O}\left(\sigma^2\right), \tag{6.61}$$

$$\eta = \frac{60712}{1323} + \frac{62}{3} \frac{d \log \left[\sigma^2(R)\right]}{d \log R} + \frac{7}{3} \left(\frac{d \log \left[\sigma^2(R)\right]}{d \log R}\right)^2 + \frac{2}{3} \frac{d \log \left[\sigma^2(R)\right]}{d (\log R)^2} + \mathcal{O}\left(\sigma^2\right). \tag{6.62}$$

In this subsection, we consider the case that the growth rate is independent of the scale of the fluctuation. When the linear growth rate depends on the scale, the derivation of the skewness becomes complicated. In this case, we have to consider mode-coupling for the second-order perturbation.

6.2. Baryon Acoustic Oscillations

In this section, we mention baryon acoustic oscillations. In the early universe, the acoustic peaks arise because the cosmological perturbations of the baryonic matter behave as the relativistic plasma and excite sound waves. These oscillations are the remnants of acoustic oscillations in the photon-baryon plasma. The recombination of the baryonic plasma to a neutral gas occurred at $z \simeq 1000$. After the recombination, the perturbations of the baryonic matter are promoted to grow by those of dark matter. On the other hand, the traces of the primordial acoustic oscillations are engraved on the present structure. These oscillations are called as baryon acoustic oscillations (BAO). The predictions of BAO are given by solutions of Boltzmann's equations for photon-baryon plasma and CDM. In 1990s, BAO is considered that it will be observed as excess power in the large-scale structure around 100 h^{-1} Mpc [35, 65].

Because the baryonic matter couples to photon closely until recombination, the oscillations in the distribution of the photon reflect BAO. The first peak of the oscillations was detected in the anisotropy of the power spectrum of the cosmic microwave background radiation [45, 66, 108]. Then BAO was found in the spacial two-point correlation function measured from a spectroscopic sample of luminous red galaxies from the Sloan Digital Sky Survey [37]. A peak in the correlation function at the separation $100h^{-1}$ Mpc was detected. The peak of BAO in the two-point correlation function or the power spectrum is given by the linear perturbation. During the nonlinear evolutions, BAO deviate from the primordial one. For the accurate description of the evolution of the oscillations, we should apply nonlinear perturbation theories. or cosmological N-body simulations.

Using perturbation theories, BAO has been predicted. In the early works, the low-redshift clustering with BAO was analyzed [36]. Because perturbation theories are valid until quasi-nonlinear region, the analyses of BAO in high-z clustering have been noticed. In the Eulerian description, the third-order perturbation has been used [50, 93] by Smith *et al.* [93] which were based on halo models and analyzed the scale dependence of halo and galaxy bias with the power spectrum. Using the third-order perturbation theory with non-linear and stochastic galasy bias, Jeong and Komatsu [50] showed that the power spectrum estimated from the perturbative calculations agreed with that from the Milennium Simulation [91] in the weakly nonlinear regime at high redshifts ($1 \leq z \leq 6$). Furthermore, the nonlinear power spectra in the real and redshift space are calculated [72].

BAO has also been analyzed by the applications of the renormalization approaches [17, 27, 62, 63, 74, 96]. Crocce and Scoccimarro [27] calculated both the power spectrum and the two-point correlation function with the Renormalised Perturbation Theory (RPT). They calculated these functions by both the RPT and N-body simulations and showed that the RPT successfully predicts the damping of the acoustic oscillations.

In general, for the detection of the oscillations, the power spectrum with the oscillations $P(k)$ is divided into a smoothed, no-wiggle linear power spectrum $P_{nw}(k)$ and the remaining part. At $k \simeq 0.1 h [\mathrm{Mpc}^{-1}]$, the ratio $P(k)/P_{nw}(k)$ becomes from 0.9 to 1.1. The scales of the peaks and the amplitude of the ratio depend on the cosmological models. Because these are changed by nonlinear evolution, accurate description for the nonlinear evolution is required.

According to the comparison of the ratio, the one-loop results are comparable to the one-loop results of the RPT [62]. Both the ratio by the RPT and the one-loop correction deviates from that by the 3LPT. In addition to the one-loop correction, biasing and redshift-space distortions have been considered [63]. Using the closure approximation, BAO is analyzed in the real and redshift-space, too [96].

When we apply perturbation theories for large-scale structure, we should notice the validity of the theories. At $z = 0$, the estimated validation scale in the power spectrum is larger than $\sim 0.07 h [\mathrm{Mpc}^{-1}]$ [62]. In this range, the second peak is narrowly included. In high-z region, the varidation scale expands. For example, the varidation scale becomes larger than $\sim 0.25 h [\mathrm{Mpc}^{-1}]$ at $z = 3$.

According to a recent paper, the two-loop correction can improve the power spectrum at $z = 1$ [17].

6.3.   Initial Condition Problem for N-body Simulations

For the strongly nonlinear evolution, cosmological N-body simulations are used. Although the growth of the density fluctuations starts at recombination era ($z \simeq 1000$), the initial condition for N-body simulations are set up around redshift $z \sim 30$. Because of the difficulty of the simulation, in the standard scheme, we must use the perturbative approach until the fluctuation come into the quasi-nonlinear regime ($z \simeq 30$) [58].

For the perturbative approach, ZA has been used for long time. As we mentioned in the previous section, even though the ZA describes the growth of the density fluctuations much better than the Eulerian linear perturbation theory, it does not reproduce the higher order statistics because the acceleration is always parallel to the peculiar velocity.

In the quesi-nonlinear region, the matter clusters and the direction of acceleration of the matter changes. Therefore, it is necessary to consider the nonlinear effect in the quasi-nonlinear region. In other words, if we set up the initial condition for cosmological N-body simulation at quasi-nonlinear stage, ZA seems insufficient for the description of the initial condition.

For this problem, recently, Crocce, Pueblas, and Scoccimarro [25] proposed the improvement by adopting different initial conditions. Basically, their initial conditions are based on the approximations valid up to the second-order Lagrangian perturbation theory (2LPT) which reproduce exact value of the skewness in the weakly nonlinear region. With these initial conditions, they calculate the statistical quantities and show the effects of transients related with 2LPT initial conditions decrease much faster than the ones related with ZA initial conditions, that is, the transients with 2LPT initial conditions are less harmful than ones with ZA initial conditions. By comparing with analytic solutions, it is confirmed that the initial condition based on the 2LPT is more accurate than the one based on ZA [49].

However, there still exist transients related with the initial conditions based on the 2LPT which prevent to reproduce the exact value of higher order statistical quantities like the kurtosis, and there is no guarantee that 2LPT initial conditions are accurate enough for these quantities.

We examined the impact of transients from initial conditions based on the higher-order Lagrangian perturbation theories [106]. In addition to the 2LPT, we also calculated non-Gaussianity with the initial conditions based on the third-order Lagrangian perturbation theory (3LPT). The cosmological parameters are given by WMAP 3-year result [92]. For the analysis of non-Gaussianity, we set the smoothing scale $R = 1$[Mpc].

For setting up the initial conditions, we use COSMICS code [58] which generates primordial Gaussian density field usually based on the ZA. Then we modified the initial conditions with 2LPT and 3LPT using Eqs. (4.26), (4.27), (4.29)-(4.32). By the code which generates the initial condition, it is more convenient to use Eqs. (4.45)-(4.47) instead of Eqs. (4.26), (4.29), (4.30). For the usage of Eqs. (4.26), (4.29), (4.30), the ambiguity about the divergence-free vectors remains. In generic cases, we ignore the divergence-free vectors.

In our analysis, when we set the initial condition of the maximal density contrast reaching unity ($\delta_{max} \sim 1$) at $z \simeq 20$, differences in the predicted non-Gaussianity between the three different initial conditions (ZA, 2LPT and 3LPT) are readily apparent. Since the accurate value of the skewness is reproduced by the 2LPT for the weakly nonlinear regime, this disagreement is because of the nonlinearity at this initial time. Therefore, 2LPT initial conditions are not sufficient to set up the initial condition for N-body simulations. There is also no guarantee that 3LPT initial conditions are accurate enough for precise determination of cosmological parameters.

Next, to avoid the problem with the initial nonlinearity, we set the initial conditions at $z \simeq 30$, corresponding to the maximal density contrast is about half, ($\delta_{max} \sim 0.5$). In this case, as is expected from the predictions of 2LPT in the weakly nonlinear region, we find no difference between the 2LPT and 3LPT results, confirming that the 2LPT initial conditions produce accurate skewness. For the kurtosis, the difference between the initial conditions with 2LPT and 3LPT is also very small (2%). On the other hand, the difference of non-Gaussianity with initial conditions based on the ZA and 3LPT is large (10% for the skewness and 25% for the kurtosis) until $z \sim 1$. This shows that transients from initial conditions with 2LPT have less impact than the ones with the ZA initial conditions.

This tendency is also obtained for the initial conditions set at $z \simeq 80$ when the maximal density contrast is about two-tenths, ($\delta_{max} \sim 0.2$). From the numerical calculations for the kurtosis, we can see that the effects of transients from 2LPT initial condition have completely disappeared until $z \sim 5$. However, there is still significant differences in the predicted non-Gaussianity with initial conditions based on the ZA and 3LPT (4% for the skewness and 10% for the kurtosis) until $z \sim 1$.

Therefore, for sufficiently early $z > 30$ initial conditions, the effects of transients on kurtosis from 2LPT initial conditions become negligible until roughly $z \sim 3$, while those from the ZA initial conditions survive until $z \sim 1$. As long as one considers typical N-body simulations, which start at $z \sim 50$, the predicted statistics are accurate enough up to the fourth-order (kurtosis) using 2LPT initial conditions.

We have considered the effect only the longitudinal mode in 3LPT. If we include the transverse mode in 3LPT for the initial conditions, the criterion when we can start cosmological N-body simulations would be changed. We will analyze the suitable initial conditions for cosmological N-body simulations with 3LPT (both the longitudinal and the transverse modes).

Joyce and Marcos [53] noticed the effect of discreteness in cosmological N-body simulations. They apply a standard method used to analyze phonons in a crystal for the displacement of the particles. The homogeneous distribution corresponds to particle distribution on a lattice. Therefore the inhomogeneity of distribution of particles is understood as the displacement from the grid points.

The equation of motion for particles in a dissipationless cosmological N-body system is rewritten with lattice vector and the displacement. Using Fourier transform for the equation of motion, they analyze growth factor for each modes. For the comparison to fluid limit, ZA is applied.

According to their analyses, the discreteness affects growth rate. In large scale, the effect is almost negligible. On the other hand, if the scale is comparable to mean particle interval, the effect is remarkable. This result affects the reliability of the spectrum which is obtained from cosmological N-body simulations.

In the algorithm for N-body simulations such as P^3M [47], the gravitational potential is smoothed by the characteristic scale. If the smoothing scale is smaller than the mean particle interval, the potential includes an artificial effects. When we compute the density field from the particle distribution, we apply a smoothing at small-scale such as Eq.(6.59). The smoothing scale R is chosen to be the characteristic scale of N-body simulations. According to the results above, if the smoothing scale is too small, the smoothed density field includes artificial effects greatly. It seems to be well to choose the smoothing scale longer than the mean particles interval.

7.   Summary

In this review, we summarised various analytic approaches to the evolution of the density fluctuations in Newtonian cosmology and show they can be helpful to distinguish the dark energy models when applied to the quasi-nonlinear region.

Analytic approaches are roughly divided into two category: the Eulerian description and the Lagrangian description. In the Eulerian description, the density fluctuation is regarded as the perturbative quantity. Although the statistical quantities such as the power spectrum is derived easily, the validation range of the approximation is narrow. On the other hand, the Lagrangian description is valid until quasi-nonlinear regime. As a shortcoming, because the Lagrangian description requires the coordinate transformation between the Eulerian comoving coordinates and the Lagrangian coordinates, in general cases, the transformation from the Lagrangian coordinates to the Eulerian comoving coordinates is difficult to obtain. Therefore, in the Lagrangian description, the streamlines of the material point on the coordinate grid are often computed. As we showed in Sec. 5.4., if the primordial fluctuations are Gaussian, we can calculate from the Lagrangian perturbation to the power spectrum in the Eulerian comoving coordinates directly.

It was shown that ZA describes the evolution of density fluctuations better than the Eulerian nonlinear perturbations [71, 83, 120]. Why is the Lagrangian description so accurate? According to recent analyses of several symmetric models [121], the evolution of the fluctuations are accurately described by the Lagrangian perturbation theory in the spherically symmetric models. As we mentioned in Sec. 4.2., ZA gives exact solutions in plane-symmetric case. For the comparison between the two-dimensional cylindrical and three-dimensional spherical collapse cases, the difference between exact solution and perturbative solution in three-dimensional case becomes larger. They concluded that ZA is more accurate in a lower dimensional collapse.

Because analytic approaches are valid until quasi-nonlinear regime, the predictions based on them can be compared to high-z observations. If the dark energy does not couple with dark matter, the dark energy model affect the structure formation mainly through the modification of the cosmic expansion. When we try to ascertain whether new dark energy model agrees with the observation in high-z region, we apply the analytic approaches and derive several statistical quantities following the previous methods.

Although the perturbative solutions are described with analytic form, we need to compute several quantities with numerical method. For example, when we take an ensemble average for statistical quantities, we need to compute averaged quantities over many samples, which are generated by numerical calculations. Huge numerical computations require supercomputers. Recently, another approach has been proposed. Applications of General purpose for Graphics Processing Unit (GPGPU) for scientific computing has been considered [73]. For example, an optimized algorithm for N-body simulations is proposed [44, 85]. Because Graphics Processing Unit (GPU) has many cores which execute numerical code in parallel pipelines, total performance is much higher than workstations. Furthermore, GPUs are inexpensive in comparison with supercomputers. Although the programming for GPUs is not easy, GPUs has the possibility that anyone learns to handle huge numerical computation.

In future plan of galaxy survey, high-z galaxies ($z \leq 3$) will be observed [19, 40, 46]. For comparisons between future observations and theoretical predictions, the analytic methods would be useful for constraining dark energy models. Even if we do not wait for the future plan, we can know the evolution of the large-scale structure. We notice quasi stellar objects (QSOs) or gamma-ray bursts (GRBs) instead of galaxies. For example, over 20,000 QSOs were observed in the 2dF QSO Redshift Survey [28]. Using the observational results, the evolution of two-point correlation functions for QSOs were measured [29]. Because there are a few samples, the error of the correlation function of each era is large. Such analyses are still important in the sense that they unveil the structure at $z = 3$. For short-lived GRBs, angular correlation function has been measured [59]. In future, not only galaxy surveys but also QSO surveys and GRBs mapping will be carried out. For comparisons between these observations and theoretical predictions, analytic approaches are expected to play inportant roles.

Acknowledgments

We thank to our collaborators Hiroki Anzai, Kei-ichi Maeda, Masaaki Morita, Hajime Sotani, and Momoko Suda. S. M. is supported by JSPS.

A. Beyond Shell-Crossing — One-Dimensional Sheet Model

The analytic approaches in this paper are based on fluid dynamics. Because streamlines do not cross each other, the description does not have physical meaning after shell-crossing. For a realistic matter fluid, pressure prevents diverging the mass density (cf. Sec. 5.6.).

For the plane-symmetric model, ZA gives exact solutions. In this model, if we have a collisionless particle such as cold dark matter, we can go beyond shell-crossing. In the plane-symmetric model, the particles are described by plane-parallel sheets. These sheets will pass through each other without collisions. Then, after this crossing, we regard that ZA gives exact solutions again. We call this the one-dimensional sheet model.

When a collision by two sheets has occurred, those two sheets exchange their numbering, i.e., Lagrangian coordinates as follows. We consider two sheets whose Lagrangian coordinates are q_1 and q_2 ($q_1 < q_2$), respectively. Their Eulerian coordinates and velocities are $x(q_1, t), x(q_2, t)$ ($x(q_1, t) < x(q_2, t)$) and $v(q_1, t), v(q_2, t)$, respectively. Assume that these sheets cross over at t_{cross}, i.e.,

$$x(q_1, t_{\mathrm{cross}}) = x(q_2, t_{\mathrm{cross}}), \tag{A.1}$$

$$v(q_1, t_{\mathrm{cross}}) > v(q_2, t_{\mathrm{cross}}). \tag{A.2}$$

For the evolution after shell-crossing ($t > t_{\mathrm{cross}}$), their Lagrangian coordinates are exchanged as

$$x(q_1, t_{\mathrm{cross}}) \to x(q_2, t_{\mathrm{cross}}) \quad , \quad v(q_1, t_{\mathrm{cross}}) \to v(q_2, t_{\mathrm{cross}}), \tag{A.3}$$

$$x(q_2, t_{\mathrm{cross}}) \to x(q_1, t_{\mathrm{cross}}) \quad , \quad v(q_2, t_{\mathrm{cross}}) \to v(q_1, t_{\mathrm{cross}}). \tag{A.4}$$

We again find a natural ordering between the Lagrangian and Eulerian coordinates, i.e., $x(q_1, t) < x(q_2, t)$ for $q_1 < q_2$. By this exchange, we obtain a new distribution of sheets just after shell-crossing.

In this procedure, we cannot ignore decaying modes in ZA. From Eq. (4.16), we find that the growing and decaying modes of the Lagrangian displacement are

$$\partial_q \psi_+^{(1)} = -\frac{1}{g_+ g'_- - g'_+ g_-} \left(g_- v - g'_-(x - q) \right), \tag{A.5}$$

$$\partial_q \psi_-^{(1)} = \frac{1}{g_+ g'_- - g'_+ g_-} \left(g_+ v - g'_+(x - q) \right). \tag{A.6}$$

For example, in E-dS Universe, these solutions become

$$\partial_q \psi_+^{(1)} = \frac{3}{5t^{2/3}} \left((x-q) + vt \right) , \qquad (A.7)$$

$$\partial_q \psi_-^{(1)} = \frac{t}{5} \left(2(x-q) - 3vt \right) . \qquad (A.8)$$

These solutions are valid until next shell-crossing.

This method has been applied for several statistical analyses: evolutions for the power spectrum [42, 118], fractal structure [103], the distribution of caustics [119].

The property of the one-dimensional sheet model is different from three-dimensional gravitating systems. For example, the gravitational force is independent of relative distance. Therefore, the force is not asymptotically flat. Although the one-dimensional model is different from three-dimensional model in several points, since the one-dimensional model has analytic exact solutions, it serves as important toy models for gravitating system.

B.   Derivation of the Basic Equations from Vlasov Equation

In the Newtonian description, the basic equations for self-gravitating fluid are Continuous equation (2.1), Euler's equation (2.2), and Poisson's equation (2.3).

Both continuous equation and Euler's equation are derived from collisionless Boltzmann equation (Vlasov equation) [7, 14, 79]. Vlasov equation is described as,

$$\frac{\partial f}{\partial t} + u_i \frac{\partial f}{\partial x_i} + g_i \frac{\partial f}{\partial u_i} = 0 , \qquad (B.1)$$

where f denotes one-point distribution function. g_i is given by Poisson's equation. Here we define averaged quantity over the momentum space as

$$\langle A \rangle \equiv \int \frac{mN}{\rho(\boldsymbol{x}, t)} \int \mathrm{d}\boldsymbol{u} A f(\boldsymbol{x}, \boldsymbol{u}, t) , \qquad (B.2)$$

where m, N, $\rho = mN \int \mathrm{d}\boldsymbol{u} f$ means mass of a particle, number of particles, and mass density, respectively. Buchert and Domínguez [14] defined averaged velocity $\bar{v}$ and stress tensor Π as

$$\bar{u}_i \equiv \langle u_i \rangle , \qquad (B.3)$$

$$\Pi_{ij} = \rho(\langle u_i u_j \rangle - \bar{u}_i \bar{u}_j) . \qquad (B.4)$$

When we integrate Vlasov equation (B.1) over the momentum space, we obtain the continuous equation,

$$\partial_t \rho + (\rho \bar{u}_i)_{,i} = 0 . \qquad (B.5)$$

When we multiply u_j to Vlasov equation (B.1) and integrate over the momentum space, we obtain Euler's equation,

$$\partial_t \bar{u}_i + \bar{u}_j \bar{u}_{i,j} = g_i - \frac{1}{\rho} \Pi_{ij,j} . \qquad (B.6)$$

If we suppose that the stress tensor is isotropic,

$$\Pi_{ij} = P\delta_{ij} , \qquad (B.7)$$

this tensor corresponds to the pressure.

Moreover, Buchert and Domínguez considered that they multiply $u_j u_k$ to Vlasov equation (B.1) and integrate over the momentum space,

$$\partial_t \Pi_{ij} + \bar{u}_k \Pi_{ij,k} + \bar{u}_{k,k} \Pi_{ij} = -\bar{u}_{j,k}\Pi_{ik} - \bar{u}_{i,k}\Pi_{jk} - L_{ijk,k} , \qquad (B.8)$$
$$L_{ijk} = \rho \langle (u_i - \bar{u}_i)(u_j - \bar{u}_j)(u_k - \bar{u}_k) \rangle . \qquad (B.9)$$

They assumed that the third-order cumulant of the peculiar velocity and shear terms are 0, i.e.,

$$L_{ijk} = 0 , \qquad (B.10)$$
$$\sigma_{ij} = 0 . \qquad (B.11)$$

Under this assumption, the evolution equation for the stress tensor (B.8) is rewritten as

$$\dot{p} = \frac{5}{3}\frac{p}{\rho}\dot{\rho} . \qquad (B.12)$$

When we integrate this equation over the Lagrangian space, we obtain equation of state.

$$p(\boldsymbol{q}, t) = \kappa(\boldsymbol{q})\rho(\boldsymbol{q}, t)^{5/3} . \qquad (B.13)$$

When we take into account the third order momentum, how the equation of state is modified? In general, L_{ijk} is written as follows:

$$\begin{aligned}
L_{ijk} &\equiv \rho \langle (u_i - \bar{u}_i)(u_j - \bar{u}_j)(u_k - \bar{u}_k) \rangle \\
&= \rho (\langle u_i u_j u_k \rangle + 2\bar{u}_i\bar{u}_j\bar{u}_k - \langle u_i u_j \rangle u_k - \langle u_j u_k \rangle u_i - \langle u_k u_i \rangle u_j) \\
&= \rho (\langle u_i u_j u_k \rangle - \bar{u}_i\bar{u}_j\bar{u}_k) - (\Pi_{ij}\bar{u}_k + \Pi_{jk}\bar{u}_i + \Pi_{ki}\bar{u}_j) . \qquad (B.14)
\end{aligned}$$

However, unfortunately, the equations do not close, because of infinite hierarchy of equations for moments of the velocity.

For simple cases, we assume the moment takes to be the following form:

$$L_{ijk} = Q(\Pi_{ij}\bar{u}_k + \Pi_{jk}\bar{u}_i + \Pi_{ki}\bar{u}_j) , \qquad (B.15)$$

where Q is a constant. The spatial derivative of the moment is written by

$$L_{ijk,k} = Q(\Pi_{ij,k}\bar{u}_k + \Pi_{ij}\bar{u}_{k,k} + \Pi_{jk,k}\bar{u}_i + \Pi_{jk}\bar{u}_{i,k} + \Pi_{ki,k}\bar{u}_j + \Pi_{ki}\bar{u}_{j,k}) . \qquad (B.16)$$

We substitute it into Eq. (B.8). Because the stress tensor is symmetric, we derive the following equation:

$$\begin{aligned}
&\partial_t \Pi_{ij} + (1 + Q)(\bar{u}_k\Pi_{ij,k} + \bar{u}_{k,k}\Pi_{ij}) \\
&= -(1 + Q)(\Pi_{ik}\bar{u}_{j,k} + \Pi_{jk}\bar{u}_{i,k}) - Q(\Pi_{jk,k}\bar{u}_i + \Pi_{ik,k}\bar{u}_j) . \qquad (B.17)
\end{aligned}$$

If we set $Q = 0$, the equation becomes quite simple. From the off-diagonal component of Eq. (B.17), we obtain

$$p(\bar{u}_{i,j} + \bar{u}_{j,i}) = -Q\left\{(p\bar{u}_j)_{,i} + (p\bar{u}_i)_{,j}\right\}. \tag{B.18}$$

On the other hand, when we take the contraction with respect to $i = j$, we derive the diagonal component of Eq. (B.17),

$$3\dot{p} + 5(1 + Q)\bar{u}_{k,k}p + (3 + 5Q)\bar{u}_k p_{,k} = 0. \tag{B.19}$$

Using continuous equation (B.5) and introducing the Lagrangian derivative,

$$\dot{p} = \partial_t p + \bar{u}_k p_{,k}, \tag{B.20}$$

Eq. (B.19) reduces to

$$\dot{p} + \frac{5}{3}Q\bar{u}_k p_{,k} = \frac{5}{3}(1 + Q)\frac{\dot{\rho}}{\rho}p. \tag{B.21}$$

It is also necessary to take into account the anisotropic velocity dispersion. In general, because the stress tensor is symmetric, there are six independent terms. For simplicity, we assume that the stress tensor consists of diagonal term and off-diagonal term:

$$\Pi_{ij} = p_D\delta_{ij} + p_{OD}(1 - \delta_{ij}), \quad p_D > 0. \tag{B.22}$$

First, we ignore the third order term L_{ijk} in Eq. (B.8). According this procedure, for example, $(i, j) = (1, 1)$ component is reduced to

$$\begin{aligned}&\partial_t p_D + \bar{u}_k p_{D,k} + \bar{u}_{k,k} p_D \\&= -\left\{2p_D\bar{u}_{1,1} + p_{OD}(\bar{u}_{(1,2)} + \bar{u}_{(1,3)})\right\},\end{aligned} \tag{B.23}$$

where $\bar{u}_{(1,2)}$ means

$$\bar{u}_{(1,2)} \equiv \bar{u}_{1,2} + \bar{u}_{2,1}. \tag{B.24}$$

From such a procedure, three equations corresponding to the diagonal components are derived. For the off-diagonal components $(i \neq j)$, six equations are derived. For example, $(i, j) = (1, 2)$ component is reduced to

$$\begin{aligned}&\partial_t p_{OD} + \bar{u}_k p_{OD,k} + \bar{u}_{k,k} p_{OD} \\&= -\left\{p_D\bar{u}_{(1,2)} + p_{OD}(\bar{u}_{1,1} + \bar{u}_{2,2} + \bar{u}_{1,3} + \bar{u}_{2,3})\right\}.\end{aligned} \tag{B.25}$$

When we add up diagonal components and off-diagonal components respectively, we obtain two equations finally:

$$\partial_t p_D + \bar{u}_k p_{D,k} + \frac{5}{3}\bar{u}_{k,k} p_D = -\frac{2}{3}p_{OD}\left(\bar{u}_{(1,2)} + \bar{u}_{(2,3)} + \bar{u}_{(3,1)}\right), \tag{B.26}$$

$$\partial_t p_{OD} + \bar{u}_k p_{OD,k} + \frac{5}{3}\bar{u}_{k,k} p_{OD} = -\frac{1}{3}(p_D + p_{OD})\left(\bar{u}_{(1,2)} + \bar{u}_{(2,3)} + \bar{u}_{(3,1)}\right) \tag{B.27}$$

Eqs. (B.26) and (B.27) include both p_D and p_{OD}. Now we separate the pressure components to new two components. First, from the difference between Eq. (B.26) and Eq. (B.27), we obtain equation obeyed by p_D and p_{OD}:

$$\partial_t \tilde{P} + \bar{u}_k \tilde{P}_{,k} + \frac{5}{3}\bar{u}_{k,k}\tilde{P} = \frac{1}{3}\tilde{P}\left(\bar{u}_{(1,2)} + \bar{u}_{(2,3)} + \bar{u}_{(3,1)}\right), \tag{B.28}$$

where $\tilde{P}$ is defined by

$$\tilde{P} \equiv p_D - p_{OD} \, . \tag{B.29}$$

Next, from (B.26) $+2\times$ (B.27), we obtain another equation:

$$\partial_t \bar{P} + \bar{u}_k \bar{P}_{,k} + \frac{5}{3}\bar{u}_{k,k}\bar{P} = -\frac{2}{3}\bar{P}\left(\bar{u}_{(1,2)} + \bar{u}_{(2,3)} + \bar{u}_{(3,1)}\right) \, , \tag{B.30}$$

where $\bar{P}$ is defined by

$$\bar{P} \equiv p_D + 2p_{OD} \, . \tag{B.31}$$

There are so serious problem. How do we solve these two equations (B.28) and (B.30)? When we consider only isotropic velocity dispersion [14], we could solve the equation of the velocity dispersion, because the equation has only pressure and mass density terms. However when we consider also anisotropic velocity dispersion, the equation has the average velocity terms. Even with only one additional term, the equations become so complicated.

C.   Quantities Used in This Paper

In this paper, we use the following variables for the physical quantities, coordinates, and so on.

Table 2. Coordinate and velocity variables.

r	physical coordinates	u	velocity in physical coordinates
x	comoving Eulerian coordinates	v	peculiar velocity
q	Lagrangian coordinates		
k	comoving Eulerian wavenumber vector		
K	Lagrangian wavenumber vector		

Table 3. Time variables.

t	standard cosmological time	a	scale factor
z	red shift	η	logarithm of a (Eq. (5.42))

Table 4. Cosmological parameters.

$H = \dot{a}/a$	Hubble parameter	$\mathcal{K}$	curvature constant
Λ	cosmological constant	Ω_M	density parameter

Table 5. Physical quantities of matter.

ρ	matter density
ρ_b	background (averaged) matter density
P	pressure of matter
G	gravitational constant
Φ	gravitational potential
Δ	velocity potential (Eq. (6.15))
g	gravitational force
$\tilde{g}$	gravitational force in comoving coordinates
δ	density fluctuation (Eq. (2.12))
$\delta^{(n)}$	density fluctuation in n-th order perturbation
δ_0	primordial density fluctuation (Eq. (6.10))
ϵ	second-order perturbation in the density fluctuation (Eq. 6.10))
ξ	two-point spacial correlation function
ω	vorticity (Eq. (4.9))
Π_{ij}	velocity dispersion (stress tensor) (Eq. (B.4))
σ_{ij}	shear of peculiar velocity
L_{ijk}	third-order cumulant of the peculiar velocity (Eq. (B.9))
p_D	diagonal term of the stress tensor (Eq. (B.22))
p_{OD}	off-diagonal term of the stress tensor (Eq. (B.22))

Table 6. Eulerian perturbation.

θ	expansion (Eq. (3.21))
$\mathcal{F}_n$	kernel in Eulerian perturbation (Eq. (3.28))
$\mathcal{G}_n$	kernel in Eulerian perturbation (Eq. (3.29))
$\tilde{g}_2$	growing factor in second order Eulerian perturbation (Eqs. (6.46) and (6.47))

Table 7. Lagrangian perturbation.

s	Lagrangian displacement (perturbation) vector
S	Lagrangian perturbation potential (longitudinal mode)
s^T	Lagrangian perturbation (transverse mode)
J	Jacobian of the coordinate transformation from x to q
ψ	spacial part of Lagrangian perturbation potential
g_{1+}	growing factor in linear perturbation (Eqs. (3.6) and (4.16))
g_{1-}	decaying factor in linear perturbation (Eqs. (3.6) and (4.16))
g_2	growing factor in 2LPT (Eq. (4.24))
g_3	growing factor in 3LPT (Eq. (4.25))
g_{3T}	growing factor in the transverse mode of 3LPT (Eq. (4.57))
f_n	logarithmic derivative of the growth factor g_n (Eq. (4.39))
$\mathcal{S}$	matrix of $s_{i,j}$ (Eq. (4.48))
$X_{\alpha\beta}$	deformation tensor (Eq. (4.19))
λ_i^0	eigenvalue of $\partial\psi_{,\alpha}/\partial q_\beta$
w_i	eigenvalue of $X_{\alpha\beta}$ (Eq. (4.20))

Table 8. Some quantities appearing in Renormalization group approaches.

$g_{ij}(\eta)$	time part solution (Eq. (5.53))
γ_{ijK}	symmetric vertex matrix (Eqs. (5.45)-(5.47))
s_z	displacement vector in redshift space (Eq. (5.70))
R_{ij}	redshift space distortion tensor (Eq. (5.73))

Table 9. Some quantities appearing in wave mechanical approac.

φ	wave function (Eq. (5.87))
Θ	gravitational potential (Eq. (5.82))
$\tilde{\phi}$	velocity potential (Eq. (5.84))
$\mathcal{V}$	effective potential (Eq. (5.86))
U	re-defined peculiar velocity (Eq. (5.79))
ν	artificial parameter
$\mathcal{P}$	*quantum pressure* (Eq. (5.92))

Table 10. Some quantities appearing in EJN model.

γ	polytropic index in equation of state
κ	proportional constant in equation of state
K_J	Jeans wavenumber in Lagrangian coordinates (Eq. (5.110))
k_J	Jeans wavenumber in Eulerian coordinates

References

[1] Arnol'd, V. I. ; Shandarin, S. F. ; Zel'dovich, Ya. B. *Geophys. Astrophys. Fluid Dynamics* 1982, **20**, 111.

[2] Audit, E. ; Alimi, J.-M. *Astron. Astrophys.* 1996, **315**, 11.

[3] Barrow, J. D. ; Saich, P. *Class. Quantum Grav.* 1993, **10**, 79.

[4] Betancort-Rijo, J. ; Lopez-Corredoira, M. *Astrophys. J.* 2000, **534**, L117.

[5] Bernardeau, F. *Astrophys. J.* 1994, **433**, 1.

[6] Bernardeau, F. ; Colombi, S. ; Gaztanaga, E. ; Scoccimarro, R. *Phys. Rep.* 2002, **367**, 1.

[7] Binney, J. ; Tremaine, S. *Galactic Dynamics*; Princeton University Press; Princeton, NJ, 1987.

[8] Borgani, S. ; Plionis, M. ; Coles, P. ; Moscardini, L. *Mon. Not. R. Astron. Soc.* 1995, **277**, 1191.

[9] Bouchet, F. R. ; Juszkiewicz, R. ; Colombi, S. ; Pellat, R. *Astrophys. J.* 1992, **394**, L5.

[10] Bouchet, F. R. ; Colombi, S. ; Hivon, E. ; Juszkiewicz, R. *Astron. Astrophys.* 1995, **296**, 575.

[11] Buchert, T. *Astron. Astrophys.* 1989, **223**, 9.

[12] Buchert, T. *Mon. Not. R. Astron. Soc.* 1992, **254**, 729.

[13] Buchert, T. *Mon. Not. R. Astron. Soc.* 1994, **267**, 811.

[14] Buchert, T. ; Domínguez, A. *Astron. Astrophys.* 1998, **335**, 395.

[15] Buchert, T. ; Domínguez, A. *Astron. Astrophys.* 2005, **438**, 443.

[16] Buchert, T. ; Ehlers, J. *Mon. Not. R. Astron. Soc.* 1993, **264**, 375.

[17] Carlson, J. ; White, M. ; Padmanabhan, N. *Preprint: arXiv:0905.0479.*

[18] Catelan, P. *Mon. Not. R. Astron. Soc.* 1995, **276**, 115.

[19] Cosmic Inflation Probe
http://www.cfa.harvard.edu/cip/

[20] Coles, P. *Mon. Not. R. Astron. Soc.* 2003, **330**, 421.

[21] Coles, P. ; Lucchin, F. *Cosmology: The Origin and Evolution of Cosmic Structure*; John Wiley & Sons: Chichester, 1995.

[22] Coles, P. ; Melott, A. L. ; Shandarin, S. F. *Mon. Not. R. Astron. Soc.* 1993, **260**, 765.

[23] Coles, P. ; Spencer, K. *Mon. Not. R. Astron. Soc.* 2003 **342**, 176.

[24] Copeland, E. J. ; Sami, M. ; Tsujikawa, S. *Int. J. Mod. Phys. D* 2006 **15**, 1753.

[25] Crocce, M. ; Pueblas, S. ; Scoccimarro, R. *Mon. Not. R. Astron. Soc.* 2006, **373**, 369.

[26] Crocce, M. ; Scoccimarro, R. *Phys. Rev. D* 2006, **73**, 063519.

[27] Crocce, M. ; Scoccimarro, R. *Phys. Rev. D* 2008, **77**, 023533.

[28] Croom, S. M. *et al. Mon. Not. R. Astron. Soc.* 2004, **349**, 1397.

[29] Croom, S. M. *et al. Mon. Not. R. Astron. Soc.* 2005, **356**, 415.

[30] Davis, M. ; Efstathiou, G. ; Frenk, C. S. ; White, S. D. M. *Astrophys. J.* 1985, **292**, 371.

[31] Davis, M. ; Peebles, P. J. E. *Astrophys. J. Supp.* 1977, **34**, 425.

[32] Domínguez, A. *Phys. Rev. D* 2000, **62**, 103501.

[33] Domínguez, A. *Mon. Not. R. Astron. Soc.* 2002, **334**, 435.

[34] Doroshkevich, A. G. ; Ryaben'kin, V. S. ; Shandarin, S. F. *Astrofizica*, 1973, **9**, 257.

[35] Eisenstein, D. J. ; Hu, W. ; Silk, J. ; Szalay, A. S. *Astrophys J.* 1998, **494**, L1.

[36] Eisenstein, D. J. ; Seo, H.-J. ; White, M. *Astrophys J.* 2007 **664**, 660.

[37] Eisenstein, D. J. *et al.*, *Astrophys J.* 2005, **633**, 560.

[38] Fry, J. N. *Astrophys. J.* 1984, **279**, 499.

[39] Gaztanaga, E. ; Lobo, A. *Astrophys J.* 2001, **548**, 47.

[40] Glazebrook. K. *et al. Preprint: astro-ph/0507457.*

[41] Goroff, M. H. ; Grinstein, B. ; Rey, S.-J. ; Wise, M. B. *Astrophys J.* 1986, **311**, 6.

[42] Gouda, N. ; Nakamura, T. *Prog. Theor. Phys.* 1989, **81**, 633.

[43] Gurbatov, S. N. ; Saichev, A. I. ; Shandarin, S. F. *Mon. Not. R. Astron. Soc.* 1989, **236**, 385.

[44] Hamada, T. ; Iitaka, T. *Preprint: astro-ph/0703100.*

[45] Hanany, S. *et al. Astrophys. J* 2000, **545**, L5.

[46] Hill, G. J. *et al. ASP Conf. Ser.* 2008, **399**, 115.

[47] Hockney, R. W. ; Eastwood, W. *Comupter Simulation Using Particles*; McGraw-Hill: New York, NY, 1981.

[48] Hunter, C. *Astrophys. J.* 1964, **139**, 570.

[49] Jenkins, A. *Preprint: arXiv:0910.0258.*

[50] Jeong, D. ; Komatsu, E. *Astrophys.J.* 2009, **691**, 569.

[51] Jones, B. J. T. *Mon. Not. R. Astron. Soc.* 1999, **307**, 376.

[52] Jones, B. J. ; Martínez, V. J. ; Saar, E. ; Trimble, V. *Rev. Mod. Phys.* 2004, **76**, 1211.

[53] Joyce, M. ; Marcos, B. *Phys. Rev. D* 2007, **76**, 103505.

[54] Kayo, I. ; Taruya, A. ; Suto, Y. *Astrophys. J.* 2001, **561**, 22.

[55] Kofman, L. ; Bertschinger, E. ; Gelb, J. M. ; Nusser, A. ; Dekel, A. *Astrophys. J.* 1994, **420**, 44.

[56] Kofman, L. ; Pogosyan, D. ; Shandarin, S. F. ; Melott, A. L. *Astrophys. J.* 1992, **393**, 437.

[57] Liddle, A. R. ; Lyth, D. H. *Cosmological Inflation and Large-Scale Structure*; Cambridge University Press: Cambridge, 2000.

[58] Ma, C.-P. ; Bertschinger, E. *Astrophys. J.* 1995, **455**, 7; Bertschinger, E. *Preprint astro-ph/9506070*.

[59] Magliocchetti, M. ; Ghirlanda, G. ; Celotti, A. *Mon. Not. R. Astron. Soc.* 2003, **343**, 255.

[60] Makler, M. ; Kodama, T. ; Calvão, M. O. *Astrophys. J.* 2001, **557**, 88.

[61] Matarrese S. ; Mohayaee, R. *Mon. Not. R. Astron. Soc.* 2002, **329**, 37.

[62] Matsubara, T. *Phys. Rev. D* 2008, **77**, 063530.

[63] Matsubara, T. *Phys. Rev. D* 2008, **78**, 083519.

[64] Matsubara, T. ; Yoshisato, A. ; Morikawa, M. *Astrophys. J.* 1998, **504**, 7.

[65] Meiksin, A. ; White, M. ; Peacock, J. A. *Mon. Not. R. Astron. Soc.*, 1999, **304**, 851.

[66] Melchiorri, A. *et al. Astrophys J.*, 2000, **536**, L63.

[67] Melott, A. L. ; Shandarin, S. F. ; Weinberg, D. H. *Astrophys. J.* 1994, **428**, 28.

[68] Melott, A. L. ; Pellman, T. F. ; Shandarin, S. F. *Mon. Not. R. Astron. Soc.* 1994, **269**, 626.

[69] Miyoshi, K. ; Kihara, T. *Pub. Astron. Soc. J.* 1975, **27**, 333.

[70] Morita, M. ; Tatekawa, T. *Mon. Not. R. Astron. Soc.* 2001, **328**, 815.

[71] Munshi, D. ; Sahni, V. ; Starobinsky, A. A. *Astrophys J.* 1994, **436**, 517.

[72] Nishimichi, T. *et al. Pub. Astron. Soc. Japan* 2007, **59**, 1049.

[73] NVIDIA CUDA Zone,
http://www.nvidia.com/object/cuda_home.html

[74] Padmanabhan, N. ; White, M. ; Kohn, J. D. *Phys. Rev. D* 2009, **79**, 063523.

[75] Padmanabhan, T. *Mon. Not. R. Astron. Soc.* 1996, **278**, L29.

[76] Padmanabhan, T. *Structure formation in the universe*; Cambridge University Press: Cambridge, 1993.

[77] Peacock, J. A. *Cosmological Physics*; Cambridge University Press: Cambridge, 1999.

[78] Peebles, P. J. E. *Astron. Astrophys.* 1974, **32**, 197.

[79] Peebles, P. J. E. *The Large-Scale Structure of the Universe*; Princeton University Press: Princeton, 1981.

[80] Plionis, M. ; Borgani, S. ; Moscardini, L. ; Coles, P. *Astrophys J.* 1995, **441**, L57.

[81] Reisenegeer, A. ; Miralda-Escudé, J. *Astrophys. J.* 1995, **449**, 476.

[82] Sahni, V. ; Coles, P. *Phys. Rep.* 1995, **262**, 1.

[83] Sahni V. ; Shandarin, S. F. *Mon. Not. R. Astron. Soc.* 1996, **282**, 641.

[84] Sasaki, M. ; Kasai, M. *Prog. Theor. Phys.* 1998, **99**, 585.

[85] Schive, H.-Y. ; Tsai, Y.-C. ; Chiueh, T. *Preprint: arXiv:0907.3390*.

[86] Scoccimarro, R. *Mon. Not. R. Astron. Soc.* 1998, **299**, 1097.

[87] Shandarin, S. F. ; Zel'dovich, Ya. B. *Rev. Mod. Phys.* 1989, **61**, 185.

[88] Short, C. J. ; Coles, P. *J. Cosmol. Astropart.* 2006, **12**, 012.

[89] Short, C. J. ; Coles, P. *J. Cosmol. Astropart.* 2006, **12**, 016.

[90] Spanier, E. H. ; Springer, G. *Advanced Mathematical Methods for Scientists and Engineers*; McGraw-Hill: New York, NY, 1978; Sec. 8.

[91] Springel, V. *et al.*, 2005, *Nature*, **435**, 629.

[92] Spergel. D. N. *et al.*, *Astrophys. J. Suppl.* 2007, **170**, 377.

[93] Smith, R. E. ; Scoccimarro, R. ; Sheth, R. K. *Phys. Rev. D.* 2007, **75**, 063512.

[94] Sotani, H. ; Tatekawa, T. *Phys. Rev. D* 2006, **73**, 024024.

[95] Taruya, A. ; Hiramatsu, T. *Astrophys. J.* 2008, **674**, 617.

[96] Taruya, A. ; Nishimichi, T. ; Saito, S. ; Hiramatsu, T. *Preprint: arXiv:0906.0507*.

[97] Tatekawa, T. *Phys. Rev. D* 2004, **70**, 064010.

[98] Tatekawa, T. *Phys. Rev. D* 2005, **71**, 044024.

[99] Tatekawa, T. *Phys. Rev. D* 2005, **72**, 024005.

[100] Tatekawa, T. *J. Cosmol. Astropart.* 2005, **04**, 018.

[101] Tatekawa, T. *Recent Research Development in Astrophysics* 2005, Vol. 2, 1-26 (Preprint: astro-ph/0412025).

[102] Tatekawa, T. *Phys. Rev. D* 2007, **75**, 044028.

[103] Tatekawa, T. ; Maeda, K. *Astrophys. J.* 2001, **547**, 531.

[104] Tatekawa, T.; Mizuno, S. *J. Cosmol. Astropart.* 2006, **02**, 006.

[105] Tatekawa, T.; Mizuno, S. *J. Cosmol. Astropart.* 2007, **02**, 015.

[106] Tatekawa, T.; Mizuno, S. *J. Cosmol. Astropart.* 2007, **12**, 014.

[107] Tatekawa, T. ; Suda, M. ; Maeda, K. ; Morita, M. ; Anzai, H. *Phys. Rev. D* 2002, **66**, 064014.

[108] Torbet, E. *et al. Astrophys J.* 1999, **521**, L79.

[109] Tomita, K. *Prog. Theor. Phys.* 1964, **37**, 831.

[110] Tomita, K. *Prog. Theor. Phys.* 1972, **47**, 416.

[111] Totsuji, H.; Kihara, T. *Pub. Astron. Soc. Japan* 1969, **21**, 221.

[112] Ueda, H. ; Yokoyama, J. *Mon. Not. R. Astron. Soc.* 1996, **280**, 754.

[113] Weinberg, S. *Rev. Mod. Phys.* 1989, **61**, 1.

[114] Weinberg, S. *Cosmology*; Oxford University Press: Oxford, 2008.

[115] Widrow, L. M. ; Kaiser, N. *Astrophys. J.* 1993, **416**, L71.

[116] Yano, T. ; Gouda, N. *Astrophys. J.* 1997, **487**, 473.

[117] Yano, T. ; Gouda, N. *Astrophys. J.* 1997, **495**, 533.

[118] Yano, T. ; Gouda, N. *Astrophys. J. Supp.* 1998, **118**, 267.

[119] Yano, T. ; Koyama, H. ; Buchert, T. ; Gouda, N. *Astrophys. J. Supp.* 2004, **151**, 185.

[120] Yoshisato, A. ; Matsubara, T. ; Morikawa, M. *Astrophys. J.* 1998, **498**, 48.

[121] Yoshisato, A. ; Morikawa, M. ; Gouda, N. ; Mouri, H. *Astrophys. J.* 2006, **637**, 555.

[122] Zel'dovich, Ya. B. *Astron. Astrophys.* 1970, **5**, 84.

In: Dark Energy: Theories, Developments and Implications   ISBN 978-1-61668-271-2
Editors: K. Lefebvre and R. Garcia, pp. 295-333   © 2010 Nova Science Publishers, Inc.

Chapter 12

SUPERNOVAE AND THE DARK SECTOR OF THE UNIVERSE

Nikolaos E. Mavromatos[1] and Vasiliki A. Mitsou[2]
[1] King's College London, Department of Physics, Theoretical Physics,
Strand, London WC2R 2LS, UK
[2] Instituto de Física Corpuscular (IFIC), CSIC – Universitat de València,
Edificio Institutos de Paterna, P.O. Box 22085,
E-46071 Valencia, Spain

Abstract

Type Ia supernovae, acting as standard candles, play a leading rôle in the exploration of the Universe evolution. Initiated by similar stellar explosions whose physics is known in detail, they provide simultaneous measurements of the (luminosity) distance versus the redshift. Observations of this type of supernovae at high redshifts, being sensitive to the Hubble expansion rate, provide the most direct evidence for the accelerating expansion of the Universe and they are consistent with cosmological models proposing a dark energy component dominating the Universe energy budget. These findings have been corroborated by several independent sources, such as measurements of the cosmic microwave background, gravitational lensing, and the large scale structure of the Cosmos.

This chapter focuses on the importance of supernova observations for exploring the nature of dark energy. It briefly outlines the procedure followed in order to extract information relevant to cosmology from measurements of supernova luminosity and spectra and addresses the statistical and systematic errors involved. A complete review is given on supernova observational evidence starting from the first observations presented in 1995 by the Supernova Cosmology Project and the High-z Supernova Search Team and reaching the recent developments by the Hubble Space Telescope, the Supernovae Legacy Survey and the ESSENCE project.

The cosmological implications of supernova observations in conjunction with evidence collected from other astrophysical probes are discussed. A survey of the theoretical approaches devised to address the dark energy problem and their relation to observational questions is given.

The prospects for future improved measurements by facilities such as the Supernova Acceleration Probe are also discussed, together with the possibility of exploiting observations of other types of supernovae to construct a Hubble diagram and determine the cosmological parameters.

1. Introduction

The announcement that distant type Ia supernovae (SNe Ia) indicate an accelerated expansion of the Universe initiated an avalanche of theoretical scenarios and experimental proposals attempting to explain and confirm this finding. The accelerated rate stems from the observation that the distant SNe appear fainter than expected in a freely coasting Universe. These results were announced independently by two different research collaborations in 1998 and have since given rise to a burst of scientific investigations aiming at confirming, complementing and explaining them. These interpretations range from introducing new cosmological approaches, e.g. the existence of a "dark energy" component to the Cosmos energy budget, to alternative observational explanations, such as the absorption due to dust at large redshifts.

The central rôle of SNe Ia [1] in the exploration of the Universe history lies with the fact that they serve as *standard candles,* i.e. they share an identical peak luminosity and a characteristic light-curve shape, which may lead to the determination of the distance to their host galaxies. The significance of supernovae as cosmological probes has been extensively reviewed in the past [2].

Besides the supernovae measurements, there is a plethora of astrophysical evidence today, such as the spectrum of fluctuations in the Cosmic Microwave Background (CMB), galaxy cluster observations and other cosmological data, indicating that the expansion of the Universe is currently accelerating. The energy budget of the Universe seems to be dominated at the present epoch by a mysterious dark energy component, but the precise nature of this energy is still unknown. Nevertheless, current astrophysical data are capable of placing severe constraints on the nature of the dark energy, whose equation of state may be determined by means of an appropriate global fit.

Since this discovery was announced, astrophysicists have been attempting to thoroughly explore the systematics of the measurements, while cosmologists and particle physics theorists have been investigating possible alternative explanations. Among the most exciting theoretical proposals is the postulate of a new form of dark energy, i.e. energy with a negative pressure, like the cosmological constant [3], time-varying "quintessence" scenarios [4] —with a decaying particle field providing the acceleration—, or other exotic scenarios. Among the proposed observational explanations, on the other hand, the faintness of the distant SNe may be attributed to dust absorption [5] or to luminosity evolution of the SNe [6]. Further possibilities include changes in the properties of the observed SN ensemble from the nearby sample to the distant dataset.

The structure of this chapter is as follows. Section 2. describes the general features of SNe Ia, addresses the systematic errors involved and outlines the method applied to extract cosmological constraints from measurements of supernova luminosity and spectra. In Section 3., we give a detailed review on the supernova observational evidence supporting the accelerated expansion of the Universe, starting from the first observations presented in 1995 and reaching up to the recent developments. Further observations of dark energy by other cosmological probes are briefly presented in Section 4.. In Section 5., various theoretical approaches proposed to explain the existence of dark energy are discussed. The future facilities designed to further explore supernovae and their observational prospects are reviewed in Section 6.. Finally, the conclusions and an outlook are presented in Section 7..

2. From Type-Ia Supernovae to the Evolution of the Universe

As astronomical objects, supernovae are classified according to the presence or absence of specific absorption lines in their spectrum near maximum light [7]. Type-I supernovae are distinct from type II in the sense that they lack hydrogen absorption lines in their peak light spectrum. In addition, type-Ia SNe exhibit a strong absorption line near 6100 Å, which comes from a doublet of singly ionized silicon at wavelengths of 6347 Å and 6371 Å [8] (for further information on spectra of SNe Ia, see Section 2.2.). The image of a typical SN Ia is shown in Figure 1; brightness of the supernova alone rivals that of the entire host galaxy.

Figure 1. Type Ia supernova 1994D in galaxy NGC 4526 (credit: NASA/ESA, HST, HZT).

A supernova of type I may form through several ways, however they all share a common underlying mechanism. If a carbon-oxygen white dwarf accreted enough matter to reach the Chandrasekhar limit of about 1.4 solar masses (for a non-rotating star), it would no longer be able to support the bulk of its plasma through electron degeneracy pressure and would begin to collapse [9]. However, the current view is that this limit is not normally attained; increasing temperature and density inside the core ignite carbon fusion as the star approaches the limit (to within about 1%), before collapse is initiated. Within a few seconds, a substantial fraction of the matter in the white dwarf undergoes nuclear fusion, releasing enough energy ($\sim 10^{44}$ J) to unbind the star in a supernova explosion. An outwardly expanding shock wave is generated, with matter reaching velocities of roughly $0.03\ c$. There is also a significant increase in luminosity, reaching an absolute magnitude of around -20 with little variation, an interesting feature that renders type-I SNe reliable distance indicators, as we shall see below.

A supernova of this category may originate from a close binary star system. The more massive of the two stars is the first to evolve off the main sequence, and it expands to form a red giant. The two stars share a common envelope, causing their mutual orbit to shrink. The giant star then sheds most of its envelope, losing mass until it can no longer continue nuclear fusion. At this point it becomes a white dwarf star, composed primarily of carbon and oxygen. Eventually the secondary star also evolves off the main sequence to become a red giant. Matter from the giant is accreted by the white dwarf, causing the latter to increase in mass [9].

Another model for the formation of a type-Ia explosion involves the merging of two white dwarf stars, with the combined mass temporarily exceeding the Chandrasekhar limit. Alternatively, a white dwarf could accrete matter from other types of companions, including a main sequence star, if it orbits sufficiently close to it. The rise time and decay time of their light curve (magnitude as a function of time) are, respectively, $15-20$ days and ~ 2 months, in the SN rest frame.

2.1. Standardized Candles

Type-Ia supernovae follow a characteristic "light curve," as the graph of luminosity as a function of time after the explosion is called. This luminosity is generated by the radioactive decay of ^{56}Ni to ^{56}Co, which subsequently β-decays into ^{56}Fe. The peak luminosity of the light curve is considered to be universal across type-Ia supernovae (the vast majority of which are initiated with a uniform mass via the accretion mechanism), allowing them to be used as secondary standard candles to measure the distance to their host galaxies [10]. Nonetheless, the peak luminosity exhibits a narrow, albeit considerable variation among individual SNe.

This issue was tackled when it was discovered in 1993 by Mark Phillips that there is a correlation between the absolute magnitudes at maximum light of SNe Ia and the rate of decline of the light curves [11]. This correlation basically states that both the time scale and the overall energy of the supernovae explosion depend on the amount of Ni present in the progenitor. He introduced the decline rate parameter, $\Delta m_{15}(B)$, which is the number of magnitudes that a SN Ia declines in its B-band light curve in the first 15 days after maximum light.

There are four basic methods of analyzing and describing the decline rate relation. In addition to the $\Delta m_{15}(B)$ method [12], there is the Multi-color Light Curve Shape (MLCS) method [13], the "stretch method" [14], and the recently developed "C-magic" method [15]. For the B-band and V-band light curves, the SNe Ia which are intrinsically brighter at maximum light have wider light curves. Different methods for the light-curve shape corrections, however, do not compare well with each other; significant differences in the implementations of the corrections are found [1, 16].

It is therefore possible to normalize the peak flux and also "stretch" the time axis so that all type-Ia SNe fit a universal light curve as shown in Figure 2 for nearby SNe. This calibration yields an approximately universal intrinsic luminosity $\mathcal{L}$ equivalent to an absolute blue sensitive magnitude of -19.6. Thus, if we observe a SN Ia in a distant galaxy and measure the peak light output $\mathcal{F}$, we can use the inverse square law to infer its luminosity distance d_{L} (in Euclidean space):

$$\mathcal{F} = \frac{\mathcal{L}}{4\pi d_{\mathrm{L}}^2}. \tag{1}$$

Subsequently, since a difference of five magnitudes corresponds to a factor 100 in brightness, we have the following relation between d_{L}, the apparent magnitude m, and

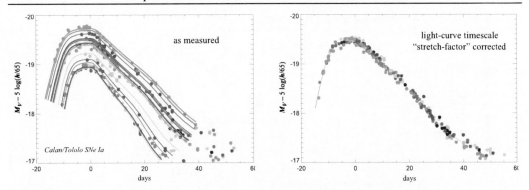

Figure 2. *Left:* Type-Ia SNe observed nearby showing a correlation between their peak absolute luminosity and the timescale of their light curve: the brighter SNe are slower and the fainter ones are faster. *Right:* the same SNe after fitting and removing the stretch factor and correcting the peak magnitude with a simple calibration relation. (Diagrams taken from Ref. [17].)

the absolute magnitude $\mathcal{M}$:

$$\mathcal{M} - m = 2.5 \log \frac{d_{\mathrm{L},0}^2}{d_{\mathrm{L}}^2} = 5 \log \frac{d_{\mathrm{L},0}}{d_{\mathrm{L}}}, \tag{2}$$

where $\mathcal{M} \simeq -19.6$ is the intrinsic magnitude of the standard candles at some nearby distance $d_{\mathrm{L},0}$ usually chosen equal to 10^{-5} Mpc. Thus one obtains for the "distance modulus" μ:

$$\boxed{\mu \equiv m - \mathcal{M} = 5 \log d_{\mathrm{L}} + 25.} \tag{3}$$

This way, the luminosity distance can be measured with an accuracy of $0.1 - 0.18$ mag [12].

In addition, utilizing the spectral information of the host galaxy, one can measure the redshift of the SN and hence draw a distance modulus – redshift diagram, so-called "Hubble diagram". Such a diagram is shown in Figure 3 for a sample compiled out of SNe observed by various ground-based telescopes. Predictions for various values of the cosmological constant component (to be discussed in Section 2.4.) are superimposed. These observations and their implications for cosmology will be thoroughly discussed in Section 3.1.. The measurement of the relation between distance and redshift permits the use of SNe as extragalactic distance indicators, thereby facilitating the exploration of the evolution of the Cosmos.

2.2. Systematic Uncertainties

The results of the cosmological parameters that will presented in the following section are based on the comparison of nearby and distant SNe, under the assumption that the brightness measurements correspond to the same standard candle. A variety of systematic effects, however, may introduce a bias in the brightness versus redshift relation, ranging from differences in the measurement techniques to differing SN sample selections reaching up to fundamental variations in the properties of the standard candles or their environment. Should these effects are large and not well-understood and corrected, they will hamper

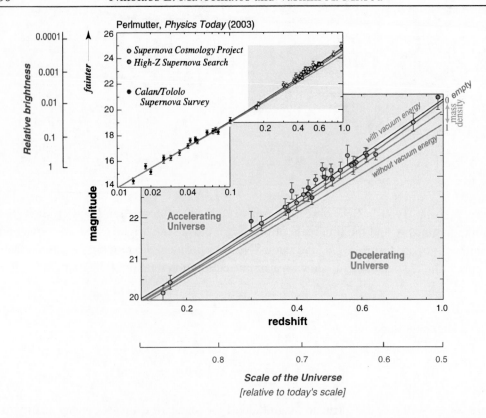

Figure 3. Observed magnitude versus redshift plotted for well-measured distant [18–20] and nearby [21] SN Ia. Curves represent cosmological model predictions (see text). (Diagrams reprinted with permission from [22]. ©2003 American Institute of Physics)

the accuracy of the cosmological parameters determination. A brief description of these potential sources of systematic uncertainties is given in the following.

Extinction by dust Intervening dust particles scatter and absorb light and can decrease the brightness of an object behind the dust screen. Extinction by dust particles along the line of sight may provide an alternative explanation to dark energy for the observed dimming of SNe Ia [23]. There are three possible contributions to extinction: dust in the host galaxy, intergalactic dust and dust in the Milky Way. Although the latter has been measured with sufficient accuracy, the estimation of the effect from the two former ones is more challenging.

Evolution Since the SN observations involve the observation of objects that exploded several billion years ago, it is possible that evolution has changed the explosions or their observable outcome [24]. There are two different processes that must be considered: the objects themselves may have changed, e.g. specific galaxies, or the average properties of the overall sample could have evolved. For the cosmological interpretation, a change of the SN Ia peak luminosity is the most important parameter to investigate.

Gravitational lensing The phenomenon of gravitational lensing, i.e. the deflection of SN light by massive objects while it propagates through the Universe towards the observer, has been studied thoroughly in the past [25]. Any considerably distant object experiences to some degree gravitational lensing. It appears that most distant sources are dimmed as light is scattered out of the line of sight, with only very few objects near deep potential wells being amplified [26]. In effect, the modal brightness is decreased, the dispersion is increased and the mean remains unchanged. For a limited sample of SNe, a bias may occur depending on the fraction of compact objects acting as lenses. Such a brightness bias may affect the cosmological interpretation of the SN measurements, especially those samples with high-redshift SNe.

Type contamination The measurements should be performed on type-Ia SNe, for which template light curves are available. Nevertheless, the light curve is not sufficient to distinguish the different types of SNe, especially between those of type Ia and type Ib/c. Spectroscopic data are needed to secure the classification, based on the most prominent spectral feature —the Si absorption line at 6100 Å—, as shown in Figure 4 for three different types of supernovae. Spectroscopy is less reliable for distant SNe due to higher noise or host-galaxy contamination in the spectra, as well as because the Si absorption line is red-shifted out of the optical regime for $z > 0.5$. In this case, the identification of SN Ia is based on other weaker spectral features on the blue part of the spectrum.

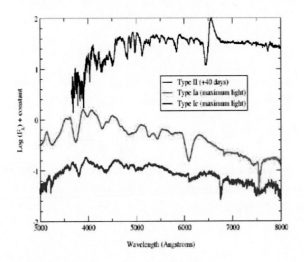

Figure 4. Measured spectra of three supernovae, from top to bottom: type II, type Ia, type Ic. The Si-II feature identifying a SNe Ia is clearly visible in the middle spectrum at about 600 nm. (From Ref. [27].)

Other possible sources of bias in the interpretation of SNe Ia observations include the Malmquist bias, i.e. an increase in average luminosity with distance due to the flux-limited nature of the distant SNe. Additionally, in the photometry and spectroscopy of an observed SN, other factors come into play such as the calculation of K-corrections [28].

2.3. The Redshift – Luminosity Distance Relation

A direct way to trace the evolution of the Universe observationally is to examine the redshift – luminosity distance relation. Redshift z, a reliably measured quantity by the spectroscopic study of a distant object, is directly related to the expansion of the Universe through the scale factor $a(t)$ by $1 + z = a^{-1}(t)$, if $a(0) \equiv 1$ is assumed for the present era.

The expanding Universe is described by the Friedmann-Lemaître-Robertson-Walker (FLRW) metric

$$ds^2 = -dt^2 + a^2(t) \left[\frac{dr^2}{1 - kr^2} + r^2 \left(d\theta^2 + \sin^2 \theta \, d\phi^2 \right) \right], \tag{4}$$

where k is a constant related to the spatial curvature: $k = 0$ for a flat Universe, $k > 0$ for a closed, and $k < 0$ for an open one. In this geometry, the measured flux $\mathcal{F}$ at the origin from the object of luminosity $\mathcal{L}$ located at a distance $r = r_1$ is given by

$$\mathcal{F} = \frac{\mathcal{L}}{4\pi r_1^2 (1 + z)^2}. \tag{5}$$

From the comparison of (1) and (5), the luminosity distance to the object is derived as $d_{\mathrm{L}} = r_1(1 + z)$.

Taking into account that photons propagate along a null geodesic ($ds^2 = 0$), the general expression for the luminosity distance can be derived:

$$d_L(z) = \frac{1+z}{\sqrt{|\Omega_k|}} \times \begin{cases} \sin \left(\sqrt{|\Omega_k|} \int_0^z \frac{dz'}{H(z')} \right), & \text{for a closed Universe;} \\ \sqrt{|\Omega_k|} \int_0^z \frac{dz'}{H(z')}, & \text{for a flat Universe;} \\ \sinh \left(\sqrt{|\Omega_k|} \int_0^z \frac{dz'}{H(z')} \right), & \text{for an open Universe,} \end{cases} \tag{6}$$

where $H \equiv \dot{a}/a$ is the Hubble parameter and $\Omega_k \equiv -k/H_0^2$ the density parameter of the curvature. If a spatially flat Universe is considered, (6) reduces to the simpler formula:

$$d_L(z) = (1 + z) \int_0^z \frac{dz'}{H(z')}. \tag{7}$$

Other observables relevant to the analysis of supernova data are the *angular diameter distance* and the *deceleration parameter*. The former, d_A, represents the co-moving distance perpendicular to the line-of-sight of the observer and is related to d_L by

$$d_A(z) = \frac{d_L}{(1 + z)^2}. \tag{8}$$

The deceleration parameter $q(t)$, on the other hand, reads

$$q(t) = -\frac{\ddot{a}}{aH^2}. \tag{9}$$

It has to be stressed that the relations (6)–(9) for the luminosity distance, the angular diameter distance and the deceleration parameter are derived by taking into account purely geometrical considerations, i.e. the FLRW metric (4), and hence they are independent of the dynamics of underlying cosmological model, which is concealed in the Hubble expansion rate as a function of the redshift.

2.4.   The Standard Cosmological Model

Within the standard framework of general relativity —according to which the dynamics of the gravitational field is described by the Einstein-Hilbert action— the gravitational (Einstein) equations in a Universe with cosmological constant Λ read

$$R_{\mu\nu} - \frac{1}{2}g_{\mu\nu}R + g_{\mu\nu}\Lambda = 8\pi G_{\mathrm{N}}T_{\mu\nu}, \tag{10}$$

where G_{N} is the gravitational constant, $T_{00} = \rho$ is the energy density of matter, and $T_{ii} = a^2(t)p$ with p the pressure. From the FLRW metric (4), we arrive at the Friedmann equation

$$\left(\frac{\dot{a}}{a}\right)^2 = \frac{8\pi G_{\mathrm{N}}}{3}\rho + \frac{\Lambda}{3} - \frac{k}{a^2}. \tag{11}$$

From this equation one obtains the expression for the critical density, defined as the total density required for a flat ($k = 0$) and cosmological-constant-free ($\Lambda = 0$) Universe

$$\rho_c = \frac{3H^2}{8\pi G_{\mathrm{N}}}. \tag{12}$$

If we assume that the Universe components behave like ideal fluids in a co-moving cosmological frame, where all cosmological measurements are assumed to take place, they obey the equations of state $p_i = w_i\rho_i$, with time-independent w_i: for radiation, $w_R = 1/3$; for matter $w_M = 0$; and for the cosmological constant $w_\Lambda = -1$. These parameters determine the evolution of the energy densities at various epochs, related to the Hubble parameter as follows:

$$H(z) = H_0\left(\Omega_k(1+z)^2 + \sum_i \Omega_i(1+z)^{3(1+w_i)}\right)^{1/2}, \tag{13}$$

with the notation $\Omega_i \equiv \rho_{i,0}/\rho_c$, $i = R, M, \Lambda$. In the present era, the densities Ω_i are linked by the relation $\Omega_k + \Omega_R + \Omega_M + \Omega_\Lambda = 1$.

As we shall see in the following, in some observational analyses, $w = w_\Lambda$ is treated as a free parameter —time-dependent in the general case—, with the intention to be determined by fitting the astrophysical data under study. The aforementioned cosmological scenario constitutes what is widely known as cosmological-constant model or, if the presence of a cold-dark-matter component is assumed, the ΛCDM concordance model.

In this context and in the case of a spatially flat Universe, the present-era value of the deceleration parameter, q_0, takes the form

$$q_0 = \frac{1}{2}\Omega_M - \Omega_\Lambda. \tag{14}$$

Thus, it becomes evident that Λ acts as "repulsive" gravity, tending to accelerate the Universe currently, and eventually dominates, leading to an eternally accelerating de-Sitter-type Universe, with a future cosmic horizon. This is schematically depicted in Figure 5. As we shall see in the following, the data favor a Universe that decelerated in the past, while it expands at an accelerated rate today!

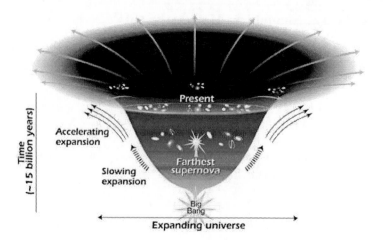

Figure 5. This diagram reveals changes in the rate of expansion since the Big Bang, which occurred $\sim$15 billion years ago. The more shallow the curve, the faster the rate of expansion. The curve changes noticeably about 7.5 billion years ago, when objects in the Universe began flying apart at a faster rate. Scientists theorize that the accelerated expansion rate is due to a mysterious repulsive force called dark energy. (Credit: NASA/STScI/Ann Feild.)

3.   Observations of Dark Energy by Supernovae

In the past decade, significant improvements in terrestrial and extraterrestrial instrumentation have led to spectacular progress in precision measurements in astrophysics. From the point of view of interest to particle physics, the most spectacular claims from astrophysics came in 1998 from the study of distant (redshift of $z \sim 1$) supernovae of type Ia by two independent groups [29]. These observations pointed towards a current-era acceleration of our Universe, something that could be explained either by a non-zero *cosmological constant* in a Friedmann-Robertson-Walker-Einstein Universe, or in general by a non-zero *dark energy* component, which could be even relaxing to zero (the data are consistent with this possibility). In the past five years, many more distant ($z > 1$) supernovae have been discovered, exhibiting similar features as the previous measurements, thereby supporting the geometric interpretation of the acceleration of the Universe today, and arguing against the nuclear physics or intergalactic-dust effects. The time-line of these findings is presented in this section.

3.1.   Accelerated Expansion of the Universe: First Evidence

The cosmological use of SNe Ia can be divided in two regimes. The first allows the determination of the Hubble constant at low redshifts ($z \lesssim 0.2$), where the effects of curvature are negligible. At these distances, SNe Ia can actually test the linearity of the expansion to a high degree [30], which in turn justifies their use as reliable distance indicators. In the second regime, at larger redshifts, the combination of the distinct cosmological models and evolution of SNe Ia peak luminosity can no longer be separated cleanly [16], and indirect evidence has to be deployed to tackle the lack of evolution of the SNe. A further difference between the determination of the current expansion rate, i.e. the Hubble constant H_0, and the measurement of the change of this parameter in the past, i.e. the deceleration parameter q_0, is the requirement to measure the absolute luminosity of the objects for H_0. In contrast to H_0, which requires an absolute measurement of the peak luminosity of SNe Ia, the determination of q_0 is independent of the absolute luminosity of SNe Ia, which, however, is assumed to be constant. In this chapter, emphasis is given on the high-redshift regime, where exciting discoveries for Cosmology arose.

Supernovae have a long history of employment initially in the measurement of the Hubble constant [30] and afterwards in the exploration of the expansion rate evolution. The first attempt to observe distant SNe was undertaken during the late 1980s in a ground-based effort by a Danish-British group, which discovered two distant SNe: SN 1988U, a SN Ia at $z = 0.31$ [31] and SN 1988T at $z = 0.28$, which according to the limited photometric information available was most probably a type-II SN [32]. This team employed modern image processing techniques to scale the brightness and resolution of images of distant clusters to match previous images and looked for supernovae in the difference frames.

Great progress was subsequently made by the Supernova Cosmology Project (SCP) [33], led by Saul Perlmutter, in the detection rate of high-redshift SNe Ia by employing large-format charged-coupled devices (CCDs), large-aperture telescopes, and more sophisticated image-analysis techniques [34]. These advances led to the detection of seven SNe Ia at $z \simeq 0.4$ between 1992 and 1994, yielding a confidence region that suggested a flat Universe without cosmological constant, but with a large range of uncertainty [35].

Brian Schmidt and Adam Riess, leading the High-z Supernova Search Team (HZT) [36], joined the hunt for high-redshift SNe Ia with their discovery of SN 1995K at $z = 0.48$ [37]. Both teams made rapid improvements in their ability to discover more type-Ia SNe at even larger redshifts. Before the samples of observed high-redshift SNe Ia become large enough to detect the acceleration signal, both teams found the data to be inconsistent with a Universe closed by matter [38, 39].

These first astrophysical findings were later followed by a thorough understanding of the SN data analysis, allowing thus the use of SN Ia observations to constrain the cosmological parameters. As mentioned in Section 2.1., empirical correlations between SN Ia light-curve shapes and peak luminosity improved the precision of distance estimates beyond the standard candle model. Degeneracies between Ω_M and Ω_Λ may be removed by means of SNe Ia measurements at different redshift bins [40]. Additional work on cross-filter K-corrections provided the ability to accurately transform the observations of high-redshift SNe Ia to the rest frame [28].

Nearby SNe Ia provided both the measure of the Hubble flow and the means to calibrate the relationship between light-curve shape and luminosity. The SCP used 20 nearby SNe Ia in the Hubble flow from the Calán/Tololo Survey [41], while the HZT adds to this set an equal number of SNe from the CfA sample [42]. The Hubble diagram compiled by these nearby SNe plus the distant ones observed by HZT [18, 19] and SCP [20] is shown in Figure 3 (measurements at the same redshift are combined) together with theoretical predictions (indicated by the curves). At redshifts beyond 0.1, the cosmological predictions begin to diverge, depending on the assumed cosmic densities of mass and dark ("vacuum") energy. The red (light grey) curves represent models with zero vacuum energy and mass densities ranging from the critical density ρ_c down to zero (an empty Universe). The best fit (blue/dark grey line) assumes a mass density of about $\rho_c/3$ plus a vacuum energy density twice that large —implying thus an accelerating cosmic expansion.

The determination of the cosmological parameters emerges from the fitting of these SN data to various models. For instance, a dark energy component Ω_X with an equation of state $w = p_X/\rho_X$ and a spatially-flat Universe ($\Omega_{tot} \equiv \Omega_M + \Omega_X = 1$) may be assumed. In this case, we obtain the confidence intervals in the (Ω_M, w) plane for the SN sample detected by the SCP team shown in Figure 6 (left). In such a diagram, Einstein's cosmological constant, Ω_Λ, corresponds to the equation of state $w = p_\Lambda/\rho_\Lambda = -1$. It is clear that the latter is (also) favored by the SN data at 68% confidence level, in conjunction with a matter density of $0.2 < \Omega_M < 0.4$.

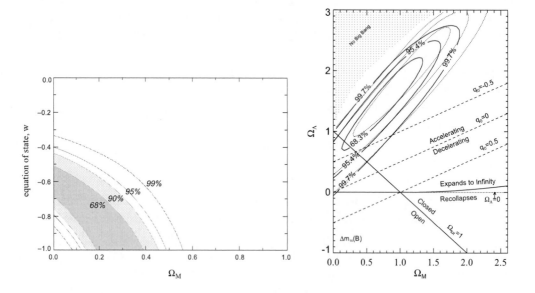

Figure 6. *Left:* Best-fit 68%, 90%, 95%, and 99% confidence regions in the (Ω_M, w) plane for an additional vacuum energy density component, characterized by an equation of state $w = p/\rho$, constrained to a flat cosmology (from SCP [20]). *Right:* Confidence intervals for $(\Omega_M, \Omega_\Lambda)$ from all SNe Ia of Ref. [18] (solid contours). The dotted contours are for the same objects excluding the unclassified SN 1997ck (from HZT [18]).

If we relax the requirement on flatness and we restrict the analysis to a cosmological constant component Ω_Λ, with $w = -1$, we acquire confidence contours for $(\Omega_M, \Omega_\Lambda)$, similar to those in Figure 6 (right), published by the HZT team. It is evident from this diagram, that the presence of a cosmological constant is favored by the SN indeed. Nevertheless, the possibility of a matter-only, open Universe is not entirely excluded. Regions representing specific cosmological scenarios are also illustrated, supporting the case for an accelerating expansion of the Universe today ($q_0 < 0$).

As we shall see in the following, these confidence intervals shrink considerably, when new SN data are added (c.f. Section 3.2.) and/or additional restrictions are posed by other astrophysical surveys (c.f. Section 4.). The reader should bear in mind that the statistical analysis followed to extract cosmological parameters from the absolute magnitude of observed SNe, does not require any assumption on priors, rendering it a direct observational method for probing the history of the Universe. Other astrophysical probes, on the other hand, do need such hypotheses for yielding estimations on observables, as discussed in Section 4..

3.2. Energy Budget of the Cosmos: Today's Picture

Following the traces of the first collaborations dedicated on the SNe Ia measurements, several teams are currently collect and analyze supernovae data to determine the cosmological parameters. During the Institute for Astronomy Deep Survey (Hawaii, US), which used the Canada-France-Hawaii 3.6 m Telescope (CFHT), 23 SNe were detected and monitored [43] leading to the confirmation of the findings of HZT and SCP made five years earlier. However, since only nine of these SNe were unambiguously classified as type Ia, the cosmological parameters were not constrained substantially.

The situation in the detection of SNe Ia drastically changed when the Hubble Space Telescope (HST) [44], as part of its scientific research program, was deployed to detect and monitor supernovae with very high redshift up to $z \sim 2$. The combination of its precision optics, location above the atmosphere, state-of-the-art instrumentation, and unprecedented pointing stability and control, allows HST to achieve the most detailed look at the farthest known objects in the Universe. This observational facility led, at a first stage, to the relatively precise measurement of the present value of the Hubble constant [45]: $H_0 = (72 \pm 8)$ km s^{-1}Mpc^{-1}. Although measurements of the cosmic microwave background provided a more accurate figure for H_0 (c.f. Section 4.1.), the SN-based determination does not strongly depend on the choice of assumptions and priors. The first results confirming the accelerated expansion of the Universe involving HST-detected SNe were released in 2003 by the SCP [46], which used 11 HST SNe of $z \lesssim 0.9$, but the cosmological parameters were severely constrained in 2004 when Riess *et al.* [47] analyzed a SN sample including 16 HST SNe, among which six were detected at very high redshift ($1.2 < z < 1.8$).

As discussed in Section 2.2., the dimming of the distant SNe Ia may be attributed to the astrophysics of this type of supernovae or in the propagation of their light to us, instead of purely cosmological explanations. An analysis of this kind is attempted in Ref. [47] for three such alternative models: (i) a gray dust scenario representing a smooth background of dust present at high-z ($z > 2$); (ii) a model of "replenishing" dust continually replenished at the same rate in which is diluted by the expanding Universe; and (iii) a simple evolution model scaling as z in percent dimming. A comparison between the predictions of these models and SN data is shown in Figure 7. Only the replenishing-dust model fits the data sufficiently well and appears indistinguishable from the Ω_Λ model. However, the fine tuning required in the dust model renders it unattractive as a rival to the dark energy [47].

The same SN sample is also confronted with various purely kinematic scenarios, as shown in Figure 8. The jerk parameter is defined proportional to the third derivative of the scale factor, as $j(t) \equiv +(\dddot{a}/a)(\dot{a}/a)^{-3}$. From Figure 8, it is evident that neither a constantly accelerating nor a constantly decelerating expansion are supported by the SN observations. A recently ($z < 0.46$) accelerating and previously decelerating expansion is favored instead.

This effort was joined by the Supernova Legacy Survey (SNLS) [48], an experi-

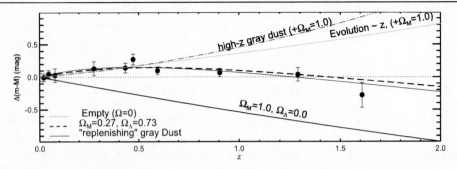

Figure 7. Residual Hubble diagram for weighted averages of SN Ia in fixed redshift bins. Comparison between cosmological models and astrophysical dimming predictions. Data and models are shown relative to a Milne Universe ($\Omega_{tot} = 0$, $\Omega_\Lambda = 0$). (From Ref. [47].)

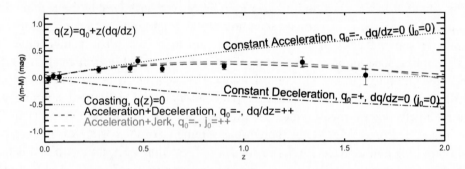

Figure 8. Residual Hubble diagram for weighted averages of SN Ia in fixed redshift bins. Comparison between specific kinematic models of the expansion history. Data and models are shown relative to an eternally coasting model, $q(z) = 0$. (From Ref. [47].)

ment/collaboration forming part of the CFHT Legacy Survey. A significant portion of its observation time is dedicated to the search and observation of distant supernovae, using the large field camera MegaPrime. The first year of operation yielded in 2005 71 SNe Ia of redshift up to $z \simeq 1$, which set further constraints on Ω_M, Ω_Λ and w [49] in agreement with earlier studies.

Early 2007 two new independent samples of distant supernovae became available. The first one released by Riess *et al,* [50] was enriched by 21 newly discovered SN Ia by HST, out of which 13 are of $z > 1$. This discovery, combined with nearby SNe and SNe detected by SNLS, narrowed even further the constraints on the early behavior of dark energy and is fully consistent with the existence of a cosmological constant.

In addition, the "Equation of State: SupErNovae trace Cosmic Expansion" (ESSENCE) [51], a NOAO (US) survey program, discovered and monitored 60 SNe Ia from 2002 through 2005 with redshifts up to $z = 0.78$ [52]. For comparison, two light-curve fitting methods were employed, namely the MLCS2k2 [53] and the SALT [54] fitters, to yield cosmological parameters constraints combined with HST [47] and SNLS [49] data.

A dataset of 192 SNe Ia has been compiled by all recently discovered type-Ia super-novae by HST [50], SNLS [49], ESSENCE [52] plus nearby SNe [42, 55] to yield the most restrictive cosmological parameters for the ΛCDM model, together with other astrophysical surveys [56,57]. Many studies make use of the SN sample compiled in Refs. [56,57] to test other possible theoretical proposals for the nature of the dark energy [58, 59] or to apply alternative statistical methods in the data analysis [60].

The cosmological constraints in the $(\Omega_M, \Omega_\Lambda)$ plane as imposed by the currently available SN sample is shown in Figure 9 (left). The enrichment of the SN sample with more and of higher redshift SNe Ia leads to a substantial narrowing of the confidence intervals as compared to the corresponding obtained after the first SN observations, shown in Figure 6 (right). The case for a spatially flat Universe continues to be favored at 68% confidence level, also supported by the measurements of baryon acoustic oscillations, which will be discussed in Section 4.2.. For a generally non-flat Universe, the favored matter and dark energy densities are $(\Omega_M, \Omega_\Lambda) \simeq (0.33, 0.85)$, while is spatial flatness is assumed, $\Omega_M = 0.259 \pm 0.019$ [59].

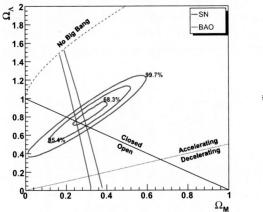 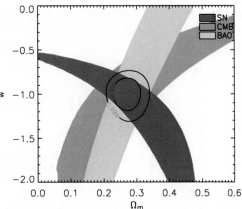

Figure 9. Cosmological constant model confidence intervals for the sample of 192 supernovae [56] from HST [50], SNLS [49], ESSENCE [52] plus nearby SNe [42, 55]. *Left:* 68.3%, 95.4% and 99.7% contours assuming $w = -1$. The result from BAO (1σ) is also superimposed. (From Ref. [59].) *Right:* Flat dark-energy model with constant w: 95% regions from each of the observational probes as shown in the legend. The combined contours (95% and 99.9% confidence) are overlayed in black. (From Ref. [56].)

Furthermore, if a flat dark-energy model with constant w is assumed, the confidence intervals are also more restricted as shown in Figure 9 (right). The SN constraints are depicted together with baryonic oscillations and CMB constraints (c.f. Section 4.1.). The complementarity of the different observational probes is clearly demonstrated in the differing angles of the overlapping contours. The combined data form a clear preference around the cosmological constant model ($w = -1$), however other scenarios for the equation of state of the dark energy may be assumed and tested.

In the case, for instance, of a time-dependent equation of state for the dark energy, the expansion history and geometry of the Universe can be expressed by the formula

$$\dot{a} = H_0\sqrt{\Omega_M/a + \Omega_R/a^2 + \Omega_k + \Omega_X[\rho_X(z)/\rho_X(0)]a^2}, \qquad (15)$$

where $\rho_X(z)$ is the density of dark energy. If the dark-energy equation of state as a function of redshift is computed using the Chevallier-Polarski-Linder parametrization [61]

$$w = w_0 + 2w'(1 - a),$$ (16)

the dark energy density scales as

$$\frac{\rho_X(z)}{\rho_X(0)} = (1 + z)^{3+3w_0+6w'} \exp\left(\frac{-6w'z}{1+z}\right).$$ (17)

This expansion history model has been tested with a combination of various astrophysical probes in many studies [62]. If a global fit is applied to SNe Ia data combined with observations of gamma-ray bursts, acoustic oscillations, nucleosynthesis, and large-scale structure [57], the contours in the (w, w') plane shown in Figure 10 are obtained. The bounds set by Big Bang nucleosynthesis, expressed by the stretch factor S^{BBNS}, severely constrain the allowed parameter space. In general, constraining a time-dependent equation of state with SN data only would be challenging, however a combination of various cosmological probes would make it feasible.

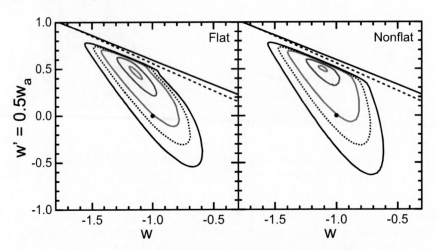

Figure 10. Confidence intervals in the (w, w') plane. The dotted contour shows the 95.4% C.L., the black dot shows the cosmological constant $w = -1$, and the dashed diagonal line shows where the dark energy is equal to the matter plus radiation density at last scattering. The solid diagonal line shows the 3σ limit on the stretch parameter S^{BBNS}. *Left:* The curvature Ω_k is fixed at zero and Ω_M is adjusted to minimize χ^2 at each point. *Right:* Both Ω_M and Ω_k are adjusted to minimize χ^2 at each point. (From Ref. [57].)

4. Complementary Constraints by Other Cosmological Probes

Although the supernova measurements provided the first hint for the accelerating expansion of the Universe, various other cosmological sources corroborated this finding. Such compelling evidence is discussed in the following.

4.1. CMB Anisotropy Measurements

After three years of running, the Wilkinson Microwave Anisotropy Probe (WMAP) [63] provided a much more detailed picture of the temperature fluctuations than its COBE predecessor [64], which can be analyzed to provide CMB-favored models for cosmology, leading to severe constraints on the energy content of various theoretical models, useful for particle physics, and in particular supersymmetric searches. Theoretically [65], the temperature fluctuations in the CMB radiation (shown in Figure 11, left) are attributed to: (i) our velocity with respect to the cosmic rest frame; (ii) gravitational potential fluctuations on the last scattering surface (Sachs-Wolf effect); (iii) radiation field fluctuations on the last scattering surface; (iv) velocity of the last scattering surface; and (v) dampening of anisotropies if Universe re-ionizes after decoupling.

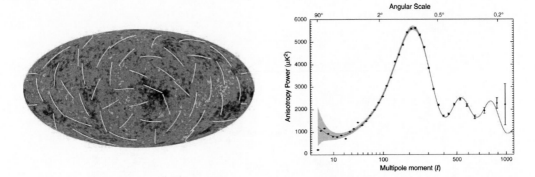

Figure 11. *Left:* CMB map of the Universe measured by WMAP, indicating the "warmer" (red) and "cooler" (blue) spots. The white bars show the "polarization" direction of the photons [66]. (Credit: NASA/WMAP Science Team.) *Right:* Angular power spectrum (black points), best-fit ΛCDM model (red curve), fit to WMAP data only [67], and 1σ cosmic variance error (grey band). (From Ref. [68].)

A Gaussian model of fluctuations [65], favored by inflation, is in very good agreement with the recent WMAP data [67], as shown in Figure 11 (right). The perfect fit of the first few peaks to the data allows a precise determination of the total density of the Universe, which implies its spatial flatness. The various peaks in the spectrum of Figure 11 (right) contain interesting physical signatures:

(i) The angular scale of the first acoustic peak (limited by the cosmic variance) determines the curvature (but not the topology) of the Universe.

(ii) The robust measurement of the second acoustic peak —truly the ratio of the odd peaks to the even peaks— determines the reduced baryon density.

(iii) The third acoustic peak can be used to extract information about the dark matter density. This is a model-dependent result, though; for instance, it requires the assumption of standard local Lorentz invariance.

The WMAP results constrain severely the equation of state $p = w\rho$ (with p the pressure), pointing towards $w = -1.08 \pm 0.12$, when the CMB data are combined with measurements of SNe Ia and large-scale structure. For comparison, we note that in the scenarios advocating the existence of a cosmological *constant* one has $w = -1$. The CMB results are in agreement with the type-Ia supernovae observations elaborated in Section 3.. The combination of WMAP three-year data plus the HST Key Project constraint on H_0 [45] implies $\Omega_k = -0.014 \pm 0.017$ and $\Omega_\Lambda = 0.716 \pm 0.055$, if $w = -1$ is assumed, due to high precision measurements of two secondary acoustic peaks as compared with previous CMB measurements (c.f. Figure 11, right). Essentially the value of Ω is determined by the position of the first acoustic peak in a Gaussian model, whose reliability increases significantly by the discovery of secondary peaks and their excellent fit with the Gaussian model [67].

The degeneracy of the cosmological parameter Ω_M and Ω_Λ from angular-size distances as measured by the cosmic microwave background is orthogonal to the one measured through luminosity distances, as shown in Figure 12.

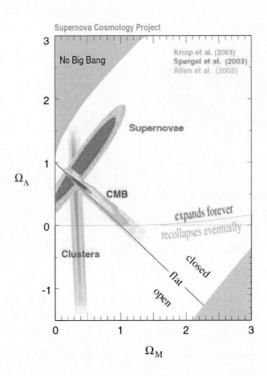

Figure 12. Confidence regions in the $(\Omega_M, \Omega_\Lambda)$ plane for supernovae [46], galaxy cluster [69] and CMB [70] data. The consistent overlap provides compelling evidence for a geometrically flat, dark-energy-dominated Universe (from [33]).

4.2. Baryon Acoustic Oscillations

Further evidence for the energy budget of the Universe is obtained by the detection of the baryon acoustic peak in the large-scale correlation function of luminous red galaxies (LRG) [71,72] measured by the Sloan Digital Sky Survey (SDSS) [73] and the 2dF Galaxy Redshift Survey (2dFGRS) [74]. The underlying physics of baryon acoustic oscillations (BAO), used to constrain the mass density of the Universe, Ω_M, can be understood as follows. Since the Universe has a significant fraction of baryons, cosmological theory predicts that the acoustic oscillations (CMB) in the plasma will also be imprinted onto the late-time power spectrum of the non-relativistic matter. From an initial point perturbation common to the dark matter and the baryons, the dark matter perturbation grows in place while the baryonic perturbation is carried outward in an expanding spherical wave. At recombination, this shell is roughly 150 Mpc in radius. Afterwards, the combined dark matter and baryon perturbation seeds the formation of large-scale structure. Since the central perturbation in the dark matter is dominant compared to the baryonic shell, the acoustic feature is manifested as a small single spike in the correlation function at 150 Mpc separation [71]. The resulting redshift-space correlation function is shown in Figure 13.

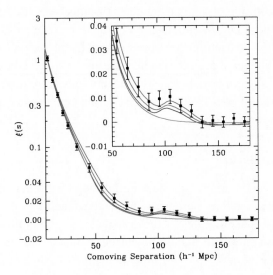

Figure 13. The large-scale redshift-space correlation function of the SDSS LRG sample. The inset shows an expanded view with a linear vertical axis. The models are $\Omega_M h^2 = 0.12$ (top, green), 0.13 (red), and 0.14 (blue). The magenta line (bottom) shows a pure CDM model ($\Omega_M h^2 = 0.105$), which lacks the acoustic peak (from [71]).

The acoustic signatures in the large-scale clustering of galaxies yield three more opportunities to test the cosmological paradigm with the Early-Universe acoustic phenomenon [75]:

1. They provide smoking-gun evidence for the theory of gravitational clustering, notably the idea that large-scale fluctuations grow by linear perturbation theory from $z \sim 1000$ to the present.

2. They give another confirmation of the existence of dark matter at $z \sim 1000$, since a fully baryonic model produces an effect much larger than observed.

3. They provide a characteristic and reasonably sharp length scale that can be measured at a wide range of redshifts, thereby determining purely by geometry the angular-diameter-distance-redshift relation and the evolution of the Hubble parameter.

In the current status of affairs of the BAO measurements, it seems that there is an underlying theoretical-model dependence of the interpretation of the results, as far as the predicted energy budget for the Universe is concerned. This stems from the fact that for small deviations from $\Omega_M = 0.3$, $\Omega_\Lambda = 0.7$, the change in the Hubble parameter at $z = 0.35$ is about half of that of the angular diameter distance. Eisenstein *et al.* [71] modeled this by treating the dilation scale as the cubic root of the product of the radial dilation times the square of the transverse dilation. In other words, they defined

$$D_V(z) = \left[D_M(z)^2 \frac{cz}{H(z)} \right]^{1/3}, \qquad (18)$$

where $D_M(z)$ is the co-moving angular diameter distance. As the typical redshift of the sample is $z = 0.35$, we quote the result for the dilation scale as $D_V(0.35) = 1370 \pm 64$ Mpc [71]. The BAO measurements from large galactic surveys and their results for the dark sector of the Universe are consistent with the WMAP data, as far as the energy budget of the Universe is concerned, but the reader should bear in mind that they based their parametrization on standard FLRW cosmologies, so the consistency should be interpreted within that theory framework.

4.3.   Large-Scale Structure

Accelerated expansion can be indirectly measured through its effect on the growth of large scale structures of matter, such as stars, quasars, galaxies and galaxy clusters. The recent accelerated expansion of space suppresses the growth due to gravitational attraction, forcing structure to have formed earlier to match the current pattern. This early growth and later slowdown should be measurable through the abundance of clusters of various masses at different epochs. This has been measured by two large galaxy surveys: the SDSS [76] and the 2dFGRS [77]. Both surveys converge to a value for matter density of $\Omega_M \simeq 0.3$ implying the existence of some form of dark energy. This constraint has been applied as a prior in SN analyses in the past [78] and has been combined with various astrophysical sources to probe dark energy models [57].

4.4.   Other Astrophysical Sources

By combining the precision measurements of the CMB by WMAP with radio, optical and X-ray probes of the large-scale distribution of matter, further evidence pointing toward acceleration of the expansion rate are provided [79]. It appears that the cluster gravitational potential wells in the Universe have been stretched and made shallower over time, as if under the influence of repulsive gravity. This phenomenon, known as the integrated Sachs-Wolfe (ISW) effect, leads to a correlation between the temperature anisotropies in the CMB and the large-scale structure of the universe [79].

X-ray images of multimillion degree Celsius gas in galaxy clusters taken by the Chandra [80] telescope may provide a powerful method to probe the mass and energy content of the Universe. A recent study [81] of 26 clusters of galaxies confirms that the expansion of the Universe stopped slowing down about 6 billion years ago, and began to accelerate. The value of matter density inferred from such studies, which indicate $\Omega_M = 0.28^{+0.05}_{-0.04}$ when the f_{gas} and CMB data are combined, are consistent with that inferred by combining CMB and supernova data.

Weak gravitational lensing is the statistically detectable distortion in the shapes of distant galaxies by the intervening dark matter. The shear correlations due to weak lensing are sensitive to the growth rate of clustering as well as to angular diameter distances. Recent analyses [82] on LSS measurements set an upper limit of $\Omega_M \lesssim 0.4$ on matter density in agreement with SN and CMB data.

Additional bounds to the cosmological model parameters, albeit weak, may be set by considering robust and model-independent measurements of the age of the Universe and the Hubble parameter. Limits to the former are imposed by measurements of the age of the elements, of the age of distant astronomical objects and the temperature of the coolest white dwarfs [83]. The latter can be determined with a limited accuracy by the differential ages of passively evolving galaxies [59, 84]. Furthermore, high-redshift ($z \lesssim 5$) gamma ray bursts (GRBs), being the most powerful astrophysical events in the Universe, hold great potential to bridge up the gap between the relatively "recent" SN Ia $z \lesssim 2$ and the much earlier CMB ($z \sim 1100$) [57, 85].

To recapitulate, the combination of the SN measurements with results from the CMB fluctuations and determinations of the mass density from galaxy clusters and flow fields has turned out to be a powerful tool to constrain Ω_M and Ω_Λ [57, 86]. These measurements, are completely independent, use different astrophysical objects, are applied at largely different scales, and constrain the cosmological parameters in different ways, as clearly depicted in Figure 12.

5. Survey of Theoretical Interpretations of Dark Energy

All these measurements point towards the fact that around 74% of the Universe vacuum energy consists of a dark (unknown) energy substance, as shown in Figure 14, in agreement with the supernovae observations. This claim, if true, could revolutionize our understanding of the basic physics governing fundamental interactions in Nature. Indeed, only a few years ago, particle theorists were trying to identify an exact symmetry of nature that could set the cosmological constant (or more generally the vacuum energy) to zero. Now, astrophysical observations point to the contrary.

Theoretically, there may be several possible explanations regarding the dark energy part of the energy budget of the Universe. A numerous, yet not exhaustive, list of such theoretical approaches is outlined here.

Cosmological constant The dark energy is a cosmological constant $\Lambda \sim 10^{-122} M_P^4$, which does not change through space and time [3]. This has been the working hypothesis of many of the best fits so far, but it should be stressed that it is *not* the only explanation consistent with the data.

Quintessence The cosmological constant is dynamical, mimicked by a slowly-varying field ϕ, whose time until it reaches its potential minimum is (much) longer than the age of the Universe. Simple quintessence models [4] assume exponential potentials

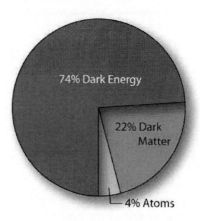

Figure 14. The energy content of our Universe as obtained by fitting data of WMAP satellite. The chart is in perfect agreement with earlier claims made by direct measurements of a current era acceleration of the Universe from distant supernovae of type Ia (credit: NASA).

$V(\phi) \sim e^{\phi}$, although more complicated combinations are most likely to characterize realistic quintessence cosmologies. In such a case the pertinent equation of state reads:

$$p_{\text{vac}} = w_{\text{vac}}\rho_{\text{vac}}, \qquad w = \frac{\frac{(\dot{\phi})^2}{2} - V(\phi)}{\frac{(\dot{\phi})^2}{2} + V(\phi)}, \tag{19}$$

and one has a relaxing-to-zero vacuum energy, with a relaxation rate such that $V(\phi)$ has the right order of magnitude today to be compatible with observations. Such a situation could be met in some models [87] of string theory, where the rôle of the quintessence field could be played by the dilaton [88, 89], i.e. the scalar field of the string gravitational multiplet. The basic theoretical problem in such string-inspired scenarios, where the dilaton plays the rôle of the quintessence field, is to explain the late-era rôle of the dilaton as driving the acceleration of the Universe, while in the past its rôle was subdominant, as compared with radiation and matter fields, so as not to disturb the delicate balance between the Universe expansion and particle production and annihilation rates characterizing the nucleosynthesis era.

Modified gravity The Einstein-Friedmann model is incorrect or insufficient to describe the Universe in its entirety, and one could have modifications in the gravitational law at galactic or supergalactic scales. Models of this kind have been proposed as alternatives to dark matter, for instance Modified Newtonian Dynamics (MOND) by Milgrom [90], and its field theory version by Bekenstein [91], known as Tensor-Vector-Scalar (TeVeS) theory, which is Lorentz violating, in the sense of involving a preferred frame. However, the simplest versions of such models seem to be incompatible with gravitational lensing [92] in certain samples of galaxies. Moreover, their compatibility with large scale structure in the Universe, such as CMB data (c.f. Section 4.1.), requires the presence of some form of Dark Matter, for instance massive neutrinos with masses higher than 0.31 eV [93]. Although WMAP imposes constraints on the upper limit of light neutrinos of order of 0.23 eV, the reader should bear in mind that such constraints are model dependent, and they were based on the standard ΛCDM Cosmology. Modified gravity models may become compatible with CMB spectrum under a different composition of Dark Matter, and indeed this is what is claimed in [93]. However, one should really constrain these models against the compilation of all available data, such as SN-type Ia, CMB, weak lensing and Baryon oscillations. Probably they will not survive a compilation of all such stringent tests, but this remains to be seen. Hence, the final word on such alternatives to dark matter has not been said yet and further astrophysical tests are required before definite conclusions are reached.

Brane cosmologies Other deviations from Einstein theory, which however maintain Lorentz invariance of the four-dimensional world, could be brane models for the Universe, which are characterized by a non-trivial —and in most cases time dependent— vacuum energy [94]. The above-mentioned stringy dilaton quintessence could be accommodated [88] in such a framework for the Universe. It should be noted that such alternative models may lead to completely different energy budget.

Gauss-Bonnet In string/brane-inspired cosmologies, in general, there are higher-curvature corrections to the Einstein term in the effective action, the most studied form of which is the so-called Gauss-Bonnet gravitational ghost-free form,

$$\int \sqrt{-g}R + f(\Phi)\left(R_{\mu\nu\rho\sigma}R^{\mu\nu\rho\sigma} - 4R_{\mu\nu}R^{\mu\nu} + R^2\right) + \dots, \tag{20}$$

where Φ indicates scalar fields, e.g. the dilaton, from the gravitational multiplet of the string/brane-inspired theory, and the other terms $(\dots)$ involve derivatives of this field [95]. Such higher-curvature corrections may give non trivial contributions to the dark energy sector of the model, and also can lead to singularity-free models [96].

$f(R)$ **models** In addition to the above modifications to standard General Relativity (GR), more *ad hoc* models, not necessarily derived from a microscopic string or brane theory, have been proposed in an attempt to account for the dark-energy sector of the Universe. As an example, one simple way to modify GR is to replace the Einstein-Hilbert Lagrangian density by a general function $f(R)$ of the Ricci scalar R. For appropriate choices of the function $f(R)$ it is then possible to obtain late-time cosmic acceleration without the need for dark energy [97]. However, evading bounds from precision solar-system tests of gravity turns out to be a much trickier matter, since such simple models are equivalent to a Brans-Dicke theory with $\omega = 0$ in the approximation in which one may neglect the potential, and are therefore inconsistent with experiment. To construct a realistic $f(R)$ model requires at the very least a rather complicated function, with more than one adjustable parameter in order to fit the cosmological data and satisfy solar system bounds.

It is natural to consider generalizing such an action to include other curvature invariants, such as the above-mentioned Gauss-Bonnet combination together with the $f(R)$ modifications [98], and it is straightforward to show these generically admit a maximally-symmetric solution: de Sitter space. Further, for a large number of such models (see e.g. [99]), solar system constraints, of the type I have described for $f(R)$

models, can be evaded. However, in these cases another problem arises, namely that the extra degrees of freedom that arise are generically ghost-like.

DGP braneworlds An alternative, and particularly successful approach, is that employed by Dvali and collaborators [100], in which an interesting modification to gravity arises from extra-dimensional models with both five and four dimensional Einstein-Hilbert terms. These Dvali-Gabadadze-Porrati (DGP) braneworlds allow one to obtain cosmic acceleration from the gravitational sector because gravity deviates from the usual four-dimensional form at large distances. One may also ask whether ghosts plague these models. However, it is our understanding that Dvali has claimed that this theory reaches the strong coupling regime before a propagating ghost appears. In fact, Dvali has shown that theories that modify gravity at cosmological distances must exhibit strong coupling phenomena, or else either possess ghosts or are ruled out by solar system constraints.

Liouville string We now remark that Cosmology, especially in the context of string and brane theory, may *not* necessarily be an *equilibrium* phenomenon, described entirely by means of on-shell fields, satisfying classical equations of motion. For instance, Early Universe cosmically catastrophic phenomena, such as the collision of two brane worlds, may result in significant departures from equilibrium, which from the point of view of string excitations on the brane world may be described [88] not by conformal theories on the world-sheet of the string, as is standard in the above-mentioned approaches to string cosmology, but by the so-called Liouville (non-critical) string [101]. The departure from equilibrium is described by specific terms, dictated by the requirement of restoration of world-sheet conformal invariance by means of the Liouville mode, which in such cosmologies plays the rôle of (an irreversible) cosmic time variable. For instance, the off-shell variations with respect to the graviton field $g^{\mu\nu}$, which replace the standard Friedmann equation for equilibrium string cosmologies, are of the form [59, 88, 102]:

$$0 = -\frac{\delta S^{G+\text{matter}}}{\delta g^{\mu\nu}} = \ddot{g}_{\mu\nu} + Q(t)\dot{g}_{\mu\nu} + \dots, \tag{21}$$

where $Q(t)$ denotes the world-sheet conformal anomaly (central charge deficit of the Liouville string [101]). $S^{G+\text{matter}}$ is the total gravitational plus matter string-inspired action, including higher-curvature modifications, if appropriate, to the standard Einstein gravitational term. The dot indicates derivative with respect to the Liouville mode, which is identified dynamically in these models with (a function of) the cosmic time [88]. The dots indicate corrections that may describe Early epochs of the Universe. Theoretically, the above form, as given in (21), is actually valid for late eras, corresponding to redshifts $z \lesssim \mathcal{O}(10)$.

One such case of a non-critical string inspired cosmology (termed Q-cosmology) has been studied in detail in the context of brane models in Ref. [59, 102], entailing a relaxation dark energy contribution to the Universe energy budget. The latter is due to the (time dependent) dilaton field, $\Phi(t)$, that is required to be non trivial in this kind of non-equilibrium string models [88]. As it will be discussed below, such models still fit the astrophysical data with, however, exotic forms of "dark matter," not scaling like dust with the redshift at late epochs, and different percentages of dark (dilaton quintessence) energy. Most importantly, one cannot disentangle dark energy from dark matter contributions in such models. For instance, there are negative-dust contributions from the dark energy sector of the theory, which are crucial for consistency with the data, and may be attributed either to higher-string-loop contributions [59, 102], which are important due to the non-trivial dilaton $\Phi(t)$ configuration in this kind of models, given that the string coupling is $g_s \sim e^{\Phi(t)}$, or to the existence of bulk Kaluza-Klein graviton modes exerting pressure on the brane world pushing it outwards, thereby appearing effectively as negative energy density on the brane world for an observer on the brane [103].

Back-reaction models It has been suggested [104] that the currently observed acceleration of the Universe is caused not be the existence of a dark energy component, but it may appear as a result of back-reaction effects on the geometry of the Cosmos due to cosmological perturbations. The energy budget of these models is therefore characterized by $\Omega_M = 1$, if flatness is assumed. Unfortunately such models seemed to be ruled out when one combines SN Ia observations [102] with measurements of the Hubble parameter by differential ages of galaxies [59].

Space-time foam models The structure of quantum space-time at microscopic scales (of the order of the Planck length, 10^{-35} m), may be characterized by topologically non trivial stochastic fluctuations, responsible for giving space-time a "foamy" structure [105]. In such stochastic models of quantum gravity, it is possible that Gravitons are in a squeezed vacuum state, and they introduce metric fluctuations. These metric fluctuations will introduce fluctuations of the lightcone itself [106], with the result that some photons propagate faster than the classical light speed, whereas others propagate slower. Thus there is a non-trivial refractive index in such vacua. A similar effect arises in string theory, as a result of interactions of photons with localized space-time defects of membrane type [107], which arise naturally in the modern version of Brane Theory. The difference of the string induced foam from light cone fluctuation models in general lies with the fact that the induced vacuum refractive index in this case does not depend on the photon polarization and is always subluminal, thereby not exhibiting birefringence. The strength of these fluctuations depends heavily on the underlying microscopic model, and whether there are extra space-time dimensions or not.

In some models, such stochastic structures also affect the dark sector of the Universe, because of non-trivial contributions of the metric fluctuations to the (dark) vacuum energy [107, 108]. The disentanglement (or constraining) of this type of dark energy contributions from the ones encountered in standard ΛCDM Cosmology can be made through complimentary astrophysical observations associated with the detection of the arrival times of high energy cosmic photons. Indeed, as a result of the associated refractive index, there are in general energy-dependent delays or advances in the arrival times of these photons, due to their propagation in the stochastic quantum gravity vacuum (for the string foam models there are only delays of the more energetic photons, as a result of the specific type of sub-luminal induced refractive index, which is larger for higher energy photons [107]). Such effects may be constrained in general by observations of energetic photons from distant sources, such as Gamma Ray Bursts or Active Galactic Nuclei, modulo however our understanding of the emission mechanisms, which is still pending [109].

In this respect, testing such models (wherever the fluctuations lead to dark sector contributions) via astrophysical measurements, such as CMB and supernova data, will provide complementary information on their parameters, such as the space-time density of foam defects, and ultimately lead to their falsification.

The aforementioned theoretical scenarios —among others— can provide alternative interpretations of the dark energy other than the cosmological constant. For instance, the behavior of a dust-dominated inhomogeneous Lemaître-Tolman-Bondi Universe model [110] may be directly confronted with supernova observations. It is found that such a model can easily explain the observed luminosity distance-redshift relation of supernovae without the need for dark energy, when the inhomogeneity is in the form of an under-dense bubble centered near the observer [111]. This is evident in the Hubble diagram of Figure 15, where the predictions of the aforementioned model, the standard ΛCDM model, and an Einstein-de Sitter model ($\Omega_M = 1$, $\Omega_\Lambda = 0$) are compared. It turns out that the statistics χ^2 is slightly better (lower) for the inhomogeneous-Universe model than the concordance one, rendering it as a possible interpretation of the observed accelerated expansion of the Universe [111]. Similar conclusions are extracted for other cosmological scenarios, such as a back-reaction and a Liouville string model, tested against SN data [59, 102].

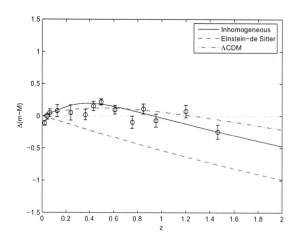

Figure 15. Distance modulus versus redshift for the ΛCDM model, the Einstein-de Sitter model and an inhomogeneous-Universe model (see text) together with type-Ia supernova observations. (Reprinted figure with permission from [111]. ©2006 by the American Physical Society)

We next remark that, since from most of the standard best fits for the Universe up to now, it follows that the energy budget of our Cosmos today is characterized by 73–74% vacuum energy, i.e. an energy density of order $\rho_{vac} \simeq (10^{-3} \text{ eV})^4 = 10^{-8} \text{ erg/cm}^3$, and about 27–26% matter (mostly dark), this implies the *Coincidence Problem*: "*The vacuum energy density today is approximately equal (in order of magnitude) to the current matter density.*" As the Universe expands, this relative balance is lost in models with a cosmological constant, such as the standard ΛCDM model, since the matter density scales with the scale factor as $\frac{\Omega_\Lambda}{\Omega_M} = \frac{\rho_\Lambda}{\rho_M} \propto a^3$. In this framework, at early times we have a vacuum energy much more suppressed as compared with that of matter and radiation, while at late times it dominates. There is only one brief epoch for which the transition from domination of one component to the other can be witnessed, and this epoch, according to the ΛCDM model, happened to be the present one! This calls for a microscopic explanation, an issue which is still lacking and which the theoretical models ought to address.

The smallness of the value of the dark energy today is another big mystery of particle physics. For several years the particle physics community thought that the vacuum energy was exactly zero, and in fact they were trying to devise microscopic explanations for such a vanishing by means of some symmetry. One of the most appealing, but eventually failed in this respect, symmetry justifications for the vanishing of the vacuum energy was that of *supersymmetry* (SUSY): if unbroken, supersymmetry implies strictly a vanishing vacuum energy, as a result of the cancelation among boson and fermion vacuum-energy contributions, due to opposite signs in the respective quantum loops. However, this cannot be the correct explanation, given that SUSY, if it is to describe Nature, must be broken below some energy scale M_{susy}, which should be higher than a few TeV, as partners have not been observed as yet. In broken SUSY theories, in four dimensional space times, there are contributions to vacuum energy $\rho_{\mathrm{vac}} \propto \hbar M_{\mathrm{susy}}^4 \sim (\text{few TeV})^4$, which is by far greater than the observed value today of the dark energy $\Lambda \sim 10^{-122} M_P^4$, with $M_P \sim 10^{19}$ GeV. Thus, SUSY does not solve the *Cosmological Constant Problem*, which at present remains one of the greatest mysteries in Physics.

In this respect, the smallness of the value of the "vacuum" energy density today might point towards a relaxation problem. Our world may have not yet reached equilibrium, from which it departed during an early-epoch cosmically catastrophic event, such as a Big Bang, or —in the modern version of string/brane theory —a collision between two brane worlds. This non-equilibrium situation might be expressed today by a quintessence-like exponential potential e^ϕ, where ϕ could be the dilaton field, which in some models [87, 88] behave at late cosmic times as $\phi \sim -2 \ln t$. This would predict a vacuum energy today of order $1/t^2$, which has the right order of magnitude, if t is of order of the age of the Universe, i.e. $t \sim 10^{60}$ Planck times. Supersymmetry in such a picture may indeed be a symmetry of the vacuum, reached asymptotically, hence the asymptotic vanishing of the dark energy. SUSY breaking may not be a spontaneous breaking but an *obstruction*, in the sense that only the excitation particle spectrum has mass differences between fermions and bosons. To achieve phenomenologically realistic situations, one may exploit [112] the string/brane framework, by compactifying the extra dimensions into manifolds with non-trivial "fluxes" (these are not gauge fields associated with electromagnetic interactions, but pertain to extra-dimensional unbroken gauge symmetries characterizing the string models). In such cases, fermions and bosons couple differently, due to their spin, to these flux gauge fields (a sort of generalized "Zeeman" effects). Thus, they exhibit mass splittings [113] proportional to the square of the "magnetic field," which could then be tuned to yield phenomenologically acceptable SUSY-splittings, while the relaxation dark energy has the cosmologically observed small value today.

In such a picture, SUSY is needed for stability of the vacuum, although today, in view of the landscape scenarios for string theory, one might not even have supersymmetric vacua at all. However, there may be another reason why SUSY could play an important physical rôle, that of providing candidates (e.g. neutralinos in Minimal Supersymmetric Standard Model extensions or gravitinos in other models) for (cold) dark matter. We shall not discuss this important issue here. Instead we only remark that stringent constraints on such models can be provided by combining cosmological and collider searches (Tevatron, Large Hadron Collider, etc.) of dark matter [114].

6. Future of Supernova Cosmology

The evidence for the existence of dark energy discovered so far call for a continuation in the exploration of supenovae with new, powerful apparatus. Such experiments under study are discussed in this section, as well as the use of type II supernovae.

6.1. The SNAP Satellite

A space observatory called SNAP (SuperNova Acceleration Probe) [115,116] is proposed, designed to probe further into the expansion of the Universe and the nature of the mysterious dark energy that is accelerating this expansion. SNAP is being proposed as part of the Joint Dark Energy Mission (JDEM) [117] and, if selected, it will be launched before 2020. Other projects proposed in the frame of JDEM are Destiny [118], an infrared survey telescope and ADEPT, which will provide measurements on supernovae and the distribution of galaxies.

The SNAP satellite and mission design [119] has been optimized for efficient supernova detection and high quality follow-up measurements. The combination of a three-mirror, two-meter telescope and a $\sim$600-million-pixel optical to near-infrared imaging camera with a 0.7-square-degree field of view will allow simultaneous discovery and recording of multiple supernovae. The imaging system comprises 36 large format (3512×3512 pixels) CCDs and the same number of 2048×2048 HgCdTe infrared sensors. Both the CCDs and the near-infrared (NIR) detectors are placed in four symmetric 3×3 arrangements. Both the imager and a low resolution ($R \sim 100$) high-throughput spectrograph cover the waveband from 350 to 1700 nm, allowing detailed characterization of supernovae up to $z = 1.7$. This deep reach in redshift is essential to the mission as it will allow to resolve degeneracies in cosmological parameters and to discriminate between models of dark energy. Nine special filters fixed above the imaging sensors will provide overlapping red-shifted B-band coverage in the range $350 - 1700$ nm. As SNAP repeatedly steps across its target fields in the north and south ecliptic poles, every supernova will be seen in every filter in both the visible and NIR. Because of their larger linear size, each NIR filter will be visited with twice the exposure time of the visible filters. This, combined with the time-dilated light curve, will ensure that type-Ia supernovae out to redshift 1.7 will be detected with a signal-to-noise ratio higher than six at least two magnitudes below peak brightness. A schematic view of the SNAP satellite components is shown in Figure 16.

SNAP will conduct two primary surveys, a $\sim$15-square-degree ultra-deep supernova survey, and a $\sim$300-square-degree-deep weak-lensing survey. With this wealth of detailed

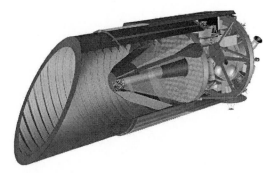

Figure 16. Cross-sectional view of the SNAP satellite. The principal assembly components are the telescope, optical bench, instruments, propulsion deck, bus, stray light baffles,
thermal shielding and entrance door (from [115]).

data, SNAP will construct a Hubble diagram with unprecedented control over systematic
uncertainties, addressing all known and proposed sources of error. The first goal is to provide precision measurements of the cosmological parameters: the matter density Ω_M, will
be measured to ± 0.02, while Ω_Λ, and the curvature parameter Ω_k, will both be determined
to an accuracy of ± 0.04. The SNAP measurements will be largely orthogonal to the CMB
measurements in the $(\Omega_M, \Omega_\Lambda)$ plane, and the curvature measurement at $z \sim 1$ will test
cosmological models by comparison with the CMB determination at $z \sim 1000$. The scientific reach of SNAP will then extend to an exploration of the nature of the dark energy,
measuring the present equation of state, w, with an uncertainty of 5%. Of even more interest is a determination of w as a function of redshift. SNAP will maintain a tight control
over systematics and the high statistics in each redshift bin will allow the determination of
the dynamical variation of w, as shown in Figure 17.

To complement its supernova cosmology observations, SNAP will conduct a wide-area
weak lensing survey. These weak lensing observations provide important independent measurements and complementary determinations of the dark matter and dark energy content
of the Universe. They will substantially enhance ability of SNAP to constrain the nature
of dark energy [120]. SNAP weak lensing observations benefit enormously from the high
spatial resolution, the accurate photometric redshifts, and the very high surface density of
resolved galaxies available in these deep observations.

6.2. Other Future Missions

Besides the JDEM/SNAP mission, the DUNE mission (Dark UNiverse Explorer) [121],
due to be launched around 2012, should provide measurements of $\sim 10\,000$ SNe up to a
redshift of $z \sim 1$, within a 18-month period. In addition, the Large Synoptic Survey Telescope (LSST) [122] is a proposed ground-based 8.4-meter, 10-square-degree-field telescope
that will provide digital imaging of faint astronomical objects across the entire sky. The
Dark Energy Survey (DES) [123] collaboration, finally, proposes to build an extremely red-
sensitive 500-Megapixel camera and a one-meter-diameter, 2.2-degree field-of-view prime
focus corrector, with a data acquisition system fast enough to take images in 17 seconds,
used to conduct a large scale sky survey. It aims at extracting information on the dark

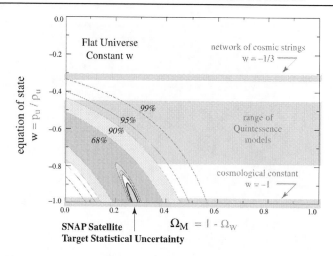

Figure 17. Best-fit 68%, 90%, 95%, and 99% confidence regions in the (Ω_M, w) plane for an additional energy density component w, characterized by an equation of state $w = p/\rho$ (from SCP [20]). For Einstein's cosmological constant Λ, $w = -1$. The fit is constrained to a flat cosmology ($\Omega_M + \Omega_W = 1$). Also shown is the expected confidence region allowed by SNAP assuming $w = -1$ and $\Omega_M = 0.28$ (from SNAP [115]).

energy from: (i) cluster counting and spatial distribution of clusters at $0.1 < z < 1.3$; (ii) the shifting of the galaxy spatial angular power spectra with redshift; (iii) weak lensing measurements on several redshift shells to $z \sim 1$; and (iv) about 2000 supernovae at $0.3 < z < 0.8$.

6.3. Type II Supernovae

Although the detailed study of type Ia SNe is the main focus of the future SN projects, the use of type II supernovae as cosmological probes is also feasible. In recent studies [124], the construction of a Hubble diagram for type II-P SNe at cosmologically significant redshifts has been demonstrated. The root-mean-square scatter of this method, 13% in distance, compares favorably to the 7–10% scatter typically seen in the SN Ia measurements. Further improvements may be sought with a view to reducing the scatter and increasing the cosmological power of the high-redshift data.

Exploring the utility of measuring distances to SNe II-P has potential benefits well beyond simply verifying, independently, the acceleration seen at redshifts $z < 1$. Several plausible models for the time evolution of the dark energy require distance measures to $z \gtrsim 2$. At such high redshifts, other cosmological probes may become less effective than at $z < 1$. However, current models for the cosmic star-formation history predict an abundant source of SNe II at these epochs and future facilities, such as the proposed JDEM telescope, SNAP, could potentially use SNe II-P to determine distances at these very high redshifts.

7. Conclusion

Type-Ia supernovae have provided so far "smoking gun" evidence for the accelerated expansion of the Universe and the existence of the dark energy. It constitutes a well-understood and relatively simple technique, allowing the direct probing of the history of the Universe. More precise measurements are expected to follow, should further improvements on the control of systematic errors, such as redshift evolution of SN properties and dust extinction corrections, are achieved. There is a vigorous current and future program of SN surveys, ranging from medium-z SNe from the ground, to high-z surveys from space.

On the interpretation of the SN findings front, a wide spectrum of theoretical models have been proposed from the existence of Einstein's cosmological constant, though models predicting a modified theory of gravity, to scenarios involving out-of-equilibrium strings. We should expect more insight on the nature of dark energy from current and future studies of type-Ia supernova samples.

Une affaire à suivre...

References

[1] For a comprehensive review, see Leibundgut, B. *Astron. Astrophys. Rev.* 2000, 10, 179–209.

[2] Gibson, B. K.; et al. *Astrophys. J.* 1999 529, 723–744.
Parodi, B. R.; Saha, A.; Sandage, A.; Tammann, G. A. *Astrophys. J.* 2000, 540, 634–651.
Leibundgut, B. *Ann. Rev. Astron. Astrophys.* 2001, 39, 67–98.

[3] Einstein, A. Sitzungsber. *Preuss. Akad. Wiss. Berlin (Math. Phys.)* 1917, 1917, 142–152.
Weinberg, S. *Rev. Mod. Phys.* 1989, 61, 1–23.
Carroll, S. M.; Press, W. H.; Turner, E. L. *Ann. Rev. Astron. Astrophys.* 1992, 30, 499–542.
Carroll, S. M. eConf C0307282 2003, *TTH09 & AIP Conf. Proc.* 2005, 743, 16–32, and references therein.

[4] Steinhardt, P. J. *Phil. Trans. Roy. Soc. Lond. A* 2003, 361, 2497–2513.

[5] Totani T.; Kobayashi C. *Astrophys. J. Lett.* 1999, 526, L65–L68.
Rowan-Robinson, M. *Mon. Not. Roy. Astron. Soc.* 2002, 332, 352–360.

[6] Umeda, H.; Nomoto, K.; Kobayashi, C.; Hachisu I.; Kato, M. *Astrophys. J. Lett.* 1999, 522, L43–L47.
Nomoto, K.; Umeda, H.; Kobayashi, C.; Hachisu, I.; Kato, M.; Tsujimoto, T. In *Cosmic Explosions;* Holt, S. S. and Zhang, W. W.; Ed.; American Institute of Physics: New York, NY, 2000; pp 35–52.

[7] Cappellaro, E.; Turatto, M. In *The influence of binaries on stellar population studies;* Vanbeveren, D.; Astrophysics and space science library (ASSL), xix, Kluwer Academic Publishers: Dordrecht, Netherlands, 2001; Vol. 264, pp 199–213.

[8] Minkowski, R. *Ann. Rev. Astron. Astrophys.* 1964, 2, 247–266.
Filippenko, A. V. *Ann. Rev. Astron. Astrophys.* 1997, 35, 309–355.

[9] Hillebrandt, W.; Niemeyer, J. C. *Ann. Rev. Astron. Astrophys.* 2000, 38, 191–230.

[10] Elias, J. H.; Matthews, K.; Neugebauer, G.; Persson, S. E. *Astrophys. J.* 1985, 296, 379–389.
Meikle, W. P. S. *Mon. Not. Roy. Astron. Soc.* 2000, 314, 782–792.

[11] Phillips, M. M. *Astrophys. J.* 1993, 413, L105–L108.

[12] Phillips, M. M.; Lira, P.; Suntzeff, N. B. a.; Schommer, R. A.; Hamuy, M.; Maza, J. *Astron. J.* 1999, 118, 1766–1776.

[13] Riess, A. G.; Press, W. H.; Kirshner, R. P. *Astrophys. J.* 1996, 473, 88–109.

[14] Perlmutter, S.; et al. *Astrophys. J.* 1997, 483, 565–581.
Goldhaber, G.; et al. *Astrophys. J.* 2001, 558, 359–368.

[15] Wang, X. F.; Wang, L. F.; Zhou, X.; Lou, Y. Q.; Li, Z. W. *Astrophys. J.* 2005, 620, L87–L90.

[16] Drell, P. S.; Loredo, T. J.; Wasserman, I. *Astrophys. J.* 2000, 530, 593–617.

[17] Perlmutter, S.; et al. *Bull. Am. Astron. Soc.* 1997, 29, 1351–1359.

[18] Riess, A. G.; et al. *Astron. J.* 1998, 116, 1009–1038.

[19] Garnavich, P. M.; et al. *Astrophys. J.* 1998, 509, 74–79.
Riess, A. G.; et al. *Astrophys. J.* 2001, 560, 49–71.

[20] Perlmutter, S.; et al. *Astrophys. J.* 1999, 517, 565–586.

[21] Hamuy, M.; et al. *Astron. J.* 1993, 106, 2392–2407.
Hamuy, M.; Phillips, M. M.; Maza, J.; Suntzeff, N. B.; Schommer, R. A.; Aviles, R. *Astron. J.* 1995, 109, 1–13.

[22] Perlmutter, S. *Physics Today* 2003, 56, 4, 53–60.

[23] Cardelli, J. A.; Clayton, G. C.; Mathis, J. S. *Astrophys. J.* 1989, 345, 245–246.

[24] Howell, D. A.; Sullivan, M.; Conley, A.; Carlberg, R. *Astrophys. J. Lett.* 2007 667, L37–L40.

[25] Blandford, R. D.; Narayan, R. *Ann. Rev. Astron. Astrophys.* 1992, 30, 311–358.
Mellier, Y. *Ann. Rev. Astron. Astrophys.* 1999 37, 127–189.

[26] Wambsganss, J.; Cen, R. y.; Xu, G. h.; Ostriker, J. P. *Astrophys. J.* 1997, 475, L81–L84.

[27] Miquel, R. *J. Phys. A* 2007, 40, 6743–6755.

[28] Kim, A.; Goobar, A.; Perlmutter, S. *Publ. Astron. Soc. Pac.* 1996, 108, 190–201.

[29] For a detailed review on the discovery, see: Riess, A. G. *Publ. Astron. Soc. Pac.* 2000, 112, 1284–1299.

[30] For a review, see: Branch, D. *Ann. Rev. Astron. Astrophys.* 1998, 36, 17–55.

[31] Norgaard-Nielsen, H. U.; Hansen, L.; Jorgensen, H. E.; Aragon Salamanca, A.; Ellis, R. S. *Nature* 1989, 339, 523–525.

[32] Hansen, L.; Jorgensen, H. E.; Norgaard-Nielsen, H. U.; Ellis, R. S.; Couch, W. J. *Astron. Astrophys.* 1989, 211, L9–L11.

[33] Supernova Cosmology Project (SCP): http://panisse.lbl.gov

[34] Perlmutter, S.; et al. *Astrophys. J.* 1995, 440, L41–L44.

[35] Perlmutter, S.; et al. *Astrophys. J.* 1997, 483, 565–581.

[36] High-Z Supernova Search Team (HZT): http://cfa-www.harvard.edu/supernova//HighZ.html

[37] Schmidt, B. P.; et al. *Astrophys. J.* 1998, 507, 46–63.

[38] Garnavich, P. M.; et al. *Astrophys. J.* 1998, 493, L53–L57.

[39] Perlmutter, S.; et al. *Nature* 1998, 391, 51–54.

[40] Goobar, A.; Perlmutter, S. *Astrophys. J.* 1995, 450, 14–18.

[41] Hamuy, M.; Phillips, M. M.; Suntzeff, N. B.; Schommer, R. A.; Maza, J. *Astron. J.* 1996, 112, 2408–2437.
Hamuy, M.; et al. *Astron. J.* 1996, 112, 2438–2447.

[42] Riess, A. G.; et al. *Astron. J.* 1999, 117, 707–724.

[43] Barris, B. J.; et al. *Astrophys. J.* 2004, 602, 571–594.

[44] Hubble Space Telescope (HST): http://hubble.nasa.gov

[45] Freedman, W. L.; et al. *Astrophys. J.* 2001, 553, 47–72.

[46] Knop, R. A.; et al. *Astrophys. J.* 2003, 598, 102–137.

[47] Riess, A. G.; et al. *Astrophys. J.* 2004, 607, 665–687.

[48] Supernova Legacy Survey (SNLS): http://cfht.hawaii.edu/SNLS

[49] Astier, P.; et al. *Astron. Astrophys.* 2006, 447, 31–48.

[50] Riess, A. G.; et al. *Astrophys. J.* 2007, 659, 98–121.

[51] Equation of State: SupErNovae trace Cosmic Expansion (ESSENCE): http://www.ctio.noao.edu/essence

[52] Wood-Vasey, W. M.; et al. *Astrophys. J.* 2007, 666, 694–715.

[53] Jha, S.; Riess, A. G.; Kirshner, R. P. *Astrophys. J.* 2007, 659, 122–148.

[54] Guy, J.; Astier, P.; Nobili, S.; Regnault, N.; Pain, R. *Astron. Astrophys.* 2006, 443, 781–792.

[55] Hamuy, M.; Phillips, M. M.; Schommer, R. A.; Suntzeff, N. B.; Maza, J.; Aviles, R. *Astron. J.* 1996, 112, 2391–2397.
Jha, S.; et al. *Astron. J.* 2006, 131, 527–554.

[56] Davis, T. M.; et al. *Astrophys. J.* 2007, 666, 716–725.

[57] Wright, E. L. *Astrophys. J.* 2007, 664, 633–639.

[58] Sullivan, S.; Cooray, A.; Holz, D. E. *JCAP* 2007, 0709, 004.
Chongchitnan, S.; Efstathiou, G. *Phys. Rev. D* 2007, 76, 043508.
Wu, Q.; Gong, Y.; Wang, A.; Alcaniz, J. S. *Phys. Lett. B* 2008, 659, 34–39.
Lazkoz, R.; Majerotto, E. *JCAP* 2007, 0707, 015.
Tartaglia, A.; Capone, M.; Cardone, V.; Radicella, N. *AIP Conf. Proc.* 2008, 1059, 39–47.

[59] Mavromatos, N. E.; Mitsou, V. A. *Astropart. Phys.* 2008, 29, 442–452.
Mitsou, V. A. *J. Phys. Conf. Ser.* 2010, 203, 012054.
Mavromatos, N. E.; Mitsou, V. A. In *The Identification of Dark Matter: Proceedings of the Sixth International Workshop;* Axenides, M.; Fanourakis, G.; and Vergados, J.; Ed.; World Scientific: Singapore, 2007; pp 623-634.

[60] Gong, Y.; Wu, Q.; Wang, A. *Astrophys. J.* 2008 681, 27–39.
Kurek, A.; Szydlowski, M. *Astrophys. J.* 2008 675, 1–7
Szydlowski, M.; Kurek, A. Preprint arXiv:0801.0638 [astro-ph] 2008.

[61] Chevallier, M.; Polarski, D. *Int. J. Mod. Phys. D* 2001, 10, 213–224.
Linder, E. V. *Phys. Rev. Lett.* 2003, 90, 091301.

[62] For some recent studies, see:
Johri, V. B.; Rath, P. K. *Phys. Rev. D* 2006, 74, 123516.
Wu, P.; Yu, H. *JCAP* 2007, 0710, 014.
Xu, L. X.; Zhang, C. W.; Liu, H. Y. *Chin. Phys. Lett.* 2007, 24, 2459–2462.
Nesseris, S.; Perivolaropoulos, L. *Phys. Rev. D* 2008, 77, 023504.

[63] Wilkinson Microwave Anisotropy Probe (WMAP):
http://map.gsfc.nasa.gov

[64] Bennett, C. L.; et al. *Astrophys. J.* 1996, 464, L1–L4.
Jaffe, A. H.; et al. *Phys. Rev. Lett.* 2001, 86, 3475–3479.

[65] Kolb, E. W.; Turner, M. S. *The Early Universe;* Frontiers in Physics, Addison-Wesley: Reading, MA, 1988; 719 pages.

[66] Page, L.; et al. *Astrophys. J. Suppl.* 2007, 170, 335–376.

[67] Spergel, D. N.; et al. *Astrophys. J. Suppl.* 2007, 170, 377–408.

[68] Hinshaw, G.; et al. *Astrophys. J. Suppl.* 2007, 170, 288–334.

[69] Bahcall, N. A.; Ostriker, J. P.; Perlmutter, S.; Steinhardt, P. J. *Science* 1999, 284, 1481–1488.

[70] Spergel, D. N.; et al. *Astrophys. J. Suppl.* 2003, 148, 175–194.

[71] Eisenstein, D. J.; et al. *Astrophys. J.* 2005, 633, 560–574.

[72] Percival, W. J.; Cole, S.; Eisenstein, D. J.; Nichol, R. C.; Peacock, J. A.; Pope, A. C.; Szalay, A. S. *Mon. Not. Roy. Astron. Soc.* 2007, 381, 1053–1066.

[73] Sloan Digital Sky Survey (SDSS): http://www.sdss.org

[74] 2dF Galaxy Redshift Survey (2dFGRS): http://www.mso.anu.edu.au/2dFGRS

[75] Tegmark, M.; et al. Phys. Rev. D 2006, 74, 123507.

[76] Tegmark, M.; et al. *Astrophys. J.* 2004 606, 702–740.

[77] Cole, S.; et al. *Mon. Not. Roy. Astron. Soc.* 2005, 362, 505–534.

[78] Perlmutter, S.; Turner, M. S.; White, M. J. *Phys. Rev. Lett.* 1999, 83, 670–673. Tonry, J. L.; et al. *Astrophys. J.* 2003, 594, 1–24.

[79] Boughn, S.; Crittenden, R. *Nature* 427, 2004, 45–47. Giannantonio, T.; et al. *Phys. Rev. D* 2006, 74, 063520.

[80] Chandra X-ray Observatory: http://chandra.harvard.edu

[81] Allen, S. W.; Schmidt, R. W.; Ebeling, H.; Fabian, A. C.; van Speybroeck, L. *Mon. Not. Roy. Astron. Soc.* 2004, 353, 457–467.

[82] Benjamin, J.; et al., *Mon. Not. Roy. Astron. Soc.* 2007, 381, 702–712.

[83] Chaboyer, B.; Demarque, P.; Kernan, P. J.; Krauss, L. M. *Astrophys. J.* 1998, 494, 96–110. Hansen, B. M. S.; et al. *Astrophys. J. Suppl.* 2004, 155, 551–576.

[84] Simon, J.; Verde, L.; Jimenez, R. *Phys. Rev. D* 2005, 71, 123001.

[85] Li, H.; Su, M.; Fan, Z.; Dai, Z.; Zhang, X. *Phys. Lett. B* 2008, 658, 95–100.

[86] The following works consist an indicative yet not exhaustive list:
Nesseris, S.; Perivolaropoulos, L. *JCAP* 2007, 0701, 018.
Nesseris, S.; Perivolaropoulos, L. *Phys. Rev. D* 2005, 72, 123519.
Melchiorri, A.; Paciello, B.; Serra, P.; Slosar, A. *New J. Phys.* 2006, 8, 325.

Huterer, D.; H. V.Peiris, H. V. *Phys. Rev. D* 2007, 75, 083503.

Wang, Y.; Mukherjee, P. *Astrophys. J.* 2006, 650, 1–6.

da Conceicao Bento, M.; Bertolami, O.; Santos, N. M. C.; Sen, A. A. *J. Phys. Conf. Ser.* 2006, 33, 197–202.

Capozziello, S.; Cardone, V. F.; Elizalde, E.; Nojiri, S.; Odintsov, S. D. *Phys. Rev. D* 2006, 73, 043512.

Jassal, H. K.; Bagla, J. S.; Padmanabhan, T. *Phys. Rev. D* 2005, 72, 103503.

Zhang, X.; Wu, F. Q. *Phys. Rev. D* 2005, 72, 043524.

Rapetti, D.; Allen, S. W.; Weller, J. *Mon. Not. Roy. Astron. Soc.* 2005, 360, 555–564.

[87] Antoniadis, I.; Bachas, C.; Ellis, J. R.; Nanopoulos, D. V. *Phys. Lett. B* 1988, 211, 393–399.

Antoniadis, I.; Bachas, C.; Ellis, J. R.; Nanopoulos, D. V. *Nucl. Phys. B* 1989, 328, 117–139.

Antoniadis, I.; Bachas, C.; Ellis, J. R.; Nanopoulos, D. V. *Phys. Lett. B* 1991, 257, 278–284.

Ellis, J. R.; Mavromatos, N. E.; Nanopoulos, D. V. *Phys. Lett. B* 2005, 619, 17–25.

Ellis, J. R.; Mavromatos, N. E.; Nanopoulos, D. V. Preprint arXiv:hep-th/0105206 2001.

[88] Diamandis, G. A.; Georgalas, B. C.; Mavromatos, N. E.; Papantonopoulos, E. *Int. J. Mod. Phys. A* 2002, 17, 4567–4589.

Diamandis, G. A.; Georgalas, B. C.; Mavromatos, N. E.; Papantonopoulos, E.; Pappa, I. *Int. J. Mod. Phys. A* 2002, 17, 2241–2266.

Diamandis, G. A.; Georgalas, B. C.; Lahanas, A. B.; Mavromatos, N. E.; Nanopoulos, D. V. *Phys. Lett. B* 2006, 642, 179–186.

Ellis, J. R.; Mavromatos, N. E.; Nanopoulos, D. V.; Westmuckett, M. *Int. J. Mod. Phys. A* 2006, 21, 1379–1444, and references therein.

[89] Gasperini, M. *Phys. Rev. D* 2001, 64, 043510.

Gasperini, M.; Piazza, F.; Veneziano, G. *Phys. Rev. D* 2002, 65, 023508.

Bean, R.; Magueijo, J. *Phys. Lett. B* 2001, 517, 177–183.

Gasperini, M.; Veneziano, G. *Phys. Rept.* 2003, 373, 1–212.

[90] Milgrom, M. *Astrophys. J.* 1983, 270, 365–370.

[91] Bekenstein, J. D. *Phys. Rev. D* 2004, 70, 083509 [Erratum-ibid. D 2005, 71, 069901].

[92] Ferreras, I.; Mavromatos, N. E.; Sakellariadou, M.; Yusaf, M. F. *Phys. Rev. D* 2009, 80, 103506.

Ferreras, I.; Mavromatos, N. E.; Sakellariadou, M.; Yusaf, M. F. *Phys. Rev. D* 2009, 79, 081301.

Ferreras, I.; Sakellariadou, M.; Yusaf, M. F. *Phys. Rev. Lett.* 2008, 100, 031302.

[93] Ferreira, P. G.; Skordis, C.; Zunckel, C. *Phys. Rev. D* 2008, 78 , 044043.

Skordis, C.; Mota, D. F.; Ferreira, P. G.; Boehm, C. *Phys. Rev. Lett.* 2006, 96,

011301.

Skordis, C. *Class. Quant. Grav.* 2009, 26, 143001.

[94] Maartens, R. *Living Rev. Rel.* 2004, 7, 7, and references therein.

[95] For a partial list of references, see:

Mavromatos, N. E.; Rizos, J. *Phys. Rev. D* 2000, 62, 124004.

Mavromatos, N. E.; Rizos, J. *Int. J. Mod. Phys. A* 2003, 18, 57–84.

Neupane, I. P. In *Dark Matter in Astroparticle and Particle Physics: Proceedings of the 6th International Heidelberg Conference;* Klapdor-Kleingrothaus, H. V.; and Lewis, G. F.; Ed.; World Scientific: Singapore, 2008; pp 228-242.

Leith, B. M.; Neupane, I. P. *JCAP* 2007, 0705, 019.

Nojiri, S.; Odintsov, S. D.; Sami, M. *Phys. Rev. D* 2006, 74, 046004.

Copeland, E. J.; Sami, M.; Tsujikawa, S. *Int. J. Mod. Phys. D* 2006, 15, 1753–1936.

Neupane, I. P. *Class. Quant. Grav.* 2006, 23, 7493–7520.

Nojiri, S.; Odintsov, S. D. *Int. J. Geom. Meth. Mod. Phys.* 2007, 4, 115–146.

Kofinas, G.; Maartens, R.; Papantonopoulos, E. *JHEP* 2003, 0310, 066.

Mavromatos, N. E.; Papantonopoulos, E. *Phys. Rev. D* 2006, 73, 026001.

Nojiri, S.; Odintsov, S. D.; Sasaki, M. *Phys. Rev. D* 2005, 71, 123509, and references therein.

[96] Antoniadis, I.; Rizos, J.; Tamvakis, K. *Nucl. Phys. B* 1994, 415, 497–514.

Binetruy, P.; Charmousis, C.; Davis, S. C.; Dufaux, J. F. *Phys. Lett. B* 2002, 544, 183–191.

Sami, M.; Singh, P.; Tsujikawa, S. *Phys. Rev. D* 2006, 74, 043514.

[97] Carroll, S. M.; Duvvuri, V.; Trodden, M.; Turner, M. S. *Phys. Rev. D* 2004, 70, 043528.

Capozziello, S.; Carloni, S.; Troisi, A. In *Recent Research Developments in Astronomy & Astrophysics;* Pandalai, S. G.; Ed.; Research Signpost: Kerala, India, 2003; Vol. 1, Part II, pp 625–670.

Bertolami, O.; Paramos, J. *Phys. Rev. D* 2008 77, 084018.

Bertolami, O.; Boehmer, C. G.; Harko, T.; Lobo, F. S. N. *Phys. Rev. D* 2007, 75, 104016.

[98] Carroll, S. M.; De Felice, A.; Duvvuri, V.; Easson, D. A.; Trodden, M.; Turner, M. S. *Phys. Rev. D* 2005, 71, 063513.

Nojiri, S.; Odintsov, S. D.; Tretyakov, P. V. *Phys. Lett. B* 2007, 651, 224–231.

[99] Navarro, I.; Van Acoleyen, K. *Phys. Lett. B* 2005, 622, 1–5.

[100] Dvali, G. R.; Gabadadze, G.; Porrati, M. *Phys. Lett. B* 2000, 485, 208–214.

Deffayet, C. *Phys. Lett. B* 2001, 502, 199–208.

Deffayet, C.; Dvali, G. R.; Gabadadze, G. *Phys. Rev. D* 2002, 65, 044023.

[101] David, F. *Mod. Phys. Lett. A* 1988, 3, 1651–1656.

Distler, J.; Kawai, H. *Nucl. Phys. B* 1989, 321, 509–527.

Mavromatos, N. E.; Miramontes, J. L. *Mod. Phys. Lett. A* 1989, 4, 1847–1853.

D'Hoker, E.; Kurzepa, P. S. *Mod. Phys. Lett. A* 1990, 5, 1411–1422.

[102] Ellis, J. R.; Mavromatos, N. E.; Mitsou, V. A.; Nanopoulos, D. V. *Astropart. Phys.* 2007, 27, 185–198.
Mitsou, V. A. In *Fundamental Interactions: Proceedings of the 22nd Lake Louise Winter Institute;* Astbury, A.; Khanna, A. F.; and Moore, R.; Ed.; World Scientific: Singapore, 2008; pp 363-367.

[103] Minamitsuji, M.; Sasaki, M.; Langlois, D. *Phys. Rev. D* 2005, 71, 084019.

[104] Kolb, E. W.; Matarrese, S.; Riotto, A. *New J. Phys.* 2006, 8, 322.
Kolb, E. W.; Matarrese, S.; Notari, A.; Riotto, A. Preprint arXiv:hep-th/0503117 2005.

[105] Wheeler, J. A.; Ford, K. *Geons, Black Holes and Quantum Foam: A Life in Physics;* Norton: New York, NY, 1998; 380 pages.

[106] Ford, L. H. *Phys. Rev. D* 1995, 51 , 1692–1700.
Yu, H. W.; Svaiter, N. F.; Ford, L. H. *Phys. Rev. D* 2009, 80 , 124019, and references therein.

[107] Ellis, J. R.; Mavromatos, N. E.; Nanopoulos, D. V.; *Gen. Rel. Grav.* 2000, 32, 127–144.
Ellis, J. R.; Mavromatos, N. E.; Westmuckett, M. *Phys. Rev. D* 2004, 70, 044036.
Ellis, J. R.; Mavromatos, N. E.; Westmuckett, M. *Phys. Rev. D* 2005, 71, 106006.
Ellis, J. R.; Mavromatos, N. E.; Nanopoulos, D. V.; *Phys. Lett. B* 2008, 665, 412–417.
Ellis, J. R.; Mavromatos, N. E.; Nanopoulos, D. V.; Preprint arXiv:0912.3428[astro-ph.CO] 2009.
Li, T.; Mavromatos, N. E.; Nanopoulos, D. V.; Xie, D. *Phys. Lett. B* 2009, 679, 407–413.

[108] Ng, Y. J. *Phys. Lett. B* 2007, 657, 10–14.
Kirillov, A. A.; Savelova, E. P.; Zolotarev, P. S. *Phys. Lett. B* 2008, 663, 372–376.

[109] Amelino-Camelia, G.; Ellis, J. R.; Mavromatos, N. E.; Nanopoulos, D. V.; Sarkar, S. *Nature* 1998, 393, 763–765.
Ellis, J. R.; Farakos, K.; Mavromatos, N. E.; Mitsou, V. A.; Nanopoulos, D. V. *Astrophys. J.* 2000, 535, 139–151.
Ellis, J. R.; Mavromatos, N. E.; Nanopoulos, D. V.; Sakharov, A. S.; Sarkisyan, E. K. G. *Astropart. Phys.* 2006, 25, 402–411 [Erratum-ibid. 2008, 29, 158–159].
For a review, see: Mavromatos, N. E. *J. Phys. Conf. Ser.* 2009, 174, 012016, and references therein.

[110] Lemaitre, G. *Gen. Rel. Grav.* 1997, 29, 641–680 & *Annales Soc. Sci. Brux. Ser. I Sci. Math. Astron. Phys. A* 1933, 53, 51–85.
Tolman, R. C. *Proc. Nat. Acad. Sci.* 1934 20, 169–176.
Bondi, H. *Mon. Not. Roy. Astron. Soc.* 1947, 107, 410–425.

[111] Alnes, H.; Amarzguioui, M.; Gron, O. *Phys. Rev. D* 2006, 73, 083519.

[112] Gravanis, E.; Mavromatos, N. E. *Phys. Lett. B* 2002, 547, 117–127.

[113] Bachas, C. Preprint arXiv:hep-th/9503030 1995.

[114] Lahanas, A. B.; Mavromatos, N. E.; Nanopoulos, D. V. *Int. J. Mod. Phys. D* 2003, 12, 1529–1591, and references therein.
Mavromatos, N. E. In *Fundamental Interactions: Proceedings of the 22nd Lake Louise Winter Institute;* Astbury, A.; Khanna, A. F.; and Moore, R.; Ed.; World Scientific: Singapore, 2008; pp 80-127, and references therein.
Munoz, C. *Int. J. Mod. Phys. A* 2004, 19, 3093–3170.

[115] SuperNova Acceleration Probe (SNAP): http://snap.lbl.gov

[116] Albert, J.; et al. Preprint arXiv:astro-ph/0507458 2005, white paper to Dark Energy Task Force.

[117] Joint Dark Energy Mission (JDEM):
http://universe.nasa.gov/program/probes/jdem.html

[118] Dark Energy Space Telescope (Destiny): http://destiny.asu.edu

[119] For a recent review, see: Levi, M. E. *Nucl. Instrum. Meth. A* 2007, 572, 521–525.
Lampton, M.; et al. *Proc. SPIE Int. Soc. Opt. Eng.* 2002, 4849, 215–226.
Lampton, M.; et al. *Proc. SPIE Int. Soc. Opt. Eng.* 2003, 4854, 632–639.

[120] Albert, J.; et al. *Astropart. Phys.* 2004, 20, 377–389.
Albert, J.; et al. Preprint arXiv:astro-ph/0507460 2005, white paper to Dark Energy Task Force.

[121] Dark Universe Explorer (DUNE): http://www.dune-mission.net

[122] Large Synoptic Survey Telescope (LSST):
http://www.lsst.org/lsst_home.shtml

[123] Dark Energy Survey (DES): https://www.darkenergysurvey.org

[124] Baron, E. A.; Nugent, P. E.; Branch, D.; Hauschildt, P. H. *Astrophys. J.* 2004, 616, L91–L94.
Nugent, P.; et al. *Astrophys. J.* 2006, 645, 841–850.

INDEX